TRENDS IN COGNITIVE PSYCHOLOGY

TRENDS IN COGNITIVE PSYCHOLOGY

SERGE P. SHOHOV
EDITOR

Nova Science Publishers, Inc.
New York

Senior Editors: Susan Boriotti and Donna Dennis
Coordinating Editor: Tatiana Shohov
Office Manager: Annette Hellinger
Graphics: Wanda Serrano
Editorial Production: Vladimir Klestov, Matthew Kozlowski, Tom Moceri,
Alexandra Columbus and Maya Columbus
Circulation: Ave Maria Gonzalez, Vera Popovic, Luis Aviles, Raymond Davis,
Melissa Diaz and Jeannie Pappas
Marketing: Cathy DeGregory

Library of Congress Cataloging-in-Publication Data
Available Upon Request

ISBN: 1-59033-665-8.

400 Oser Ave, Suite 1600
Hauppauge, New York 11788-3619
Tele. 631-231-7269 Fax 631-231-8175
e-mail: Novascience@earthlink.net
Web Site: http://www.novapublishers.com

Printed in the United States of America

CONTENTS

PREFACE

Cognitive psychology is concerned with several mental processes, including those involved in perception, attention, learning, memory, problem solving, decision making and the use of language. It is often said that cognitive psychology tries to understand how people represent their experience and then use these representations to operate effectively. Cognitive psychology holds that people are not passive organisms whose mental representations are simple or direct reflections of the outside world. Rather, they are active processors of environmental events, and as such they bring their past knowledge and their biases to bear on how they perceive and understand all current events. Thus perceiving, imagining, thinking, remembering, forming concepts, and solving problems, indeed all aspects of people's mental lives, define the domain of cognitive exploration.

In this book, we present important research which was carefully selected and screened for both current relevance and long-term advancement of the field.

Chapter 1

THE RELATIONAL PREMISES OF JUNG'S THEORY OF PSYCHOLOGICAL VALUE AND THE EMBODIED SYMBOL

R.A. Jones
School of Social Sciences
Cardiff University
Glamorgan Building
King Edward VII Avenue
Cardiff CF10 3WT, UK

ABSTRACT

Jung's radical departure from Freud's libido theory has been largely overlooked. This chapter aims to bring to light Jung's theoretical propositions and to indicate their relational foundations. Jung distinguished between "force" and "energy", arguing that energy denotes relations of intensity and polarity, and therefore cannot constitute a causal factor. He posited the psyche as an abstract system of psychological values, and contended that this constitutes the context in which the function of specific responses should be investigated. There are parallels between Jung's thesis and Gestalt field theories, though ultimately they diverge. However, Jung's retention and application of libidinal jargon to the issue of symbol forming have confounded his relational conception of the micro-genesis of embodied symbols. The "energy" metaphor aside, the issue identified by Jung in this context is highly relevant in the postmodern world, potentially bearing on current attempts to conceptualize how our bodily lived experiences become meaningful to us in symbolic productions.

INTRODUCTION

Carl Gustav Jung, the founder of analytical psychology, is best known for his theories of archetypes and the psychological types. Both pivot on a conception of "psychic energy" which constitutes a radical departure from Freud's notion of libido. Whereas Freud endowed the libido with the quality of sexual desire, Jung argued that as a scientific concept "energy" is to do with relations of affective intensity and direction among elements of experience. Jung maintained that this "quantitative" conception of energy explains how symbols are spontaneously formed, both at the level of individuals (e.g., dream images) and cultures (e.g., religious icons, even scientific ideas).

There are at least two reasons to consider Jung's concept nowadays. One is a matter of historical interest: Libidinal theories are an important chapter in the history of psychology. Jung's revision is interesting not only for its centrality in Jungian thought, but also because ultimately its dissemination has failed to make the "clean break" from the Freudian concept that Jung eschewed. Jung's concept of psychic energy was not subjected to rigorous scholarship on a significant scale. The history of the idea could be understood from a sociology-of-science perspective, taking into account the early institutionalization of analytical psychology as a "splinter" movement from the psychoanalytical establishment. The other reason for considering Jung's revision is the relevance of the issues that Jung grappled with—and purported to address in his libidinal theory—for current debates in psychology.

The aim of this chapter is to bring to light Jung's propositions and indicate their relational basis. Currently the term "relational" is used in psychology in two ways. Most commonly, it refers to the origin of the self in individuals' "embedding" in networks of social relationships, as in Gergen's (e.g., 1994) "relational self" or the "methodological relationalism" outlined by Ho and associates (e.g., Ho, Peng, Cheng Lai & Chan, 2001). Separately, some psychologists, such as Looren de Jong (1997), describe a relational approach to the mind, following Gibson's (1979) ecological perception theory in opposition to Fodor's computational approach. Looren de Jong's critique of Fodor begins with a defense of the functional explanation in psychology as opposed to causal explanations. Jung similarly had contested the reduction of the psyche to an array of discrete elements and their causal connections. "If indeed our concepts were exclusively those of moving bodies in space, there would be only a causal explanation; but we have also to deal conceptually with relations of movement, which require the energic standpoint" (Jung, 1928/1960, para. 4).

Jungian archetype theory continues to be researched, revised and debated afresh by Jungian and non-Jungian scholars (e.g., Pietikäinen, 1998, 1999, and commentaries by Hogenson, 1998, Mogenson, 1999, Solomon, 1998, and Stevens, 1998; Maloney, 1999, and commentary by Jones, 2000; McDowell, 2001; Saunders & Skar, 2001). Likewise, though to a lesser extent, Jung's typology continues to inspire original theoretical work, such as Beebe (2002), at the interface of analytical psychology and postmodern perspectives. Yet few such contributions mention Jung's concept of psychic energy. This might be in part due to the epistemological redundancy of the concept, in part due to the demise of the energy metaphor in psychology in general. However, in spite of the archaic metaphor, Jung's thesis remains potentially relevant for any attempt to conceptualize the psychological processes through which our bodily lived experiences become meaningful to us, and how such cognition is mediated by the symbolic systems of a culture.

JUNG'S REVISION BRIEFLY

Jung introduced his revision of the libido concept in the monograph marking his philosophical "break" from Freud, *Psychology of the Unconscious* (1912/1919; revised as

Symbols of Transformation, 1952/1967). He elaborated it in *Psychological Types* (1921/1993), which contains a Definitions chapter in which Jung provides a succinct statement of his revised concept:

> By libido I mean *psychic energy*. Psychic energy is the *intensity* of a psychic process, its *psychological value*. This does not imply an assignment of value, whether moral, aesthetic, or intellectual; the psychological value is already implicit in its *determining* power, which expresses itself in definite psychic effects. Neither do I understand libido as a psychic *force*, a misconception that has led many critics astray. (para. 778; author's italics.)

His concept is outlined most directly and extensively in the essay "On Psychic Energy" (Jung, 1928/1960). As identified in Jones (2001a), Jung's 1928 essay is structured around six propositions, out of which the first three expand upon the above definition: (1) energy is not a force; (2) the functional context of behavior is a closed system of psychological values; and (3) libido is not desire. The subsequent three propositions outline his application of the revised libido concept to the explanation of symbol formation: (4) symbolic activity indicates libidinal transformations; (5) symbols transform energy; and (6) transformations occasioned by symbols reflect libidinal excess.

To expand, Jung began with the distinction between *force*, associated with causal-mechanistic explanations of phenomena, and *energy*, which he argued was a quantitative concept central to a kind of explanation which he called "energic" or (interchangeably) "final." The energic approach concerns the description of structural transformations in a complex whole over time. In psychology, this approach would seek to understand why people behave, feel, talk, dream, emote, and so forth, as they do, in the wider context of their relatedness to the world—rather than describe the causal mechanisms that enable or culminate in the particular representations. Jung contended that Freud's libido confused the two kinds of explanation. Secondly, Jung submitted that the psyche is a dynamical complex whole on grounds that elements of experience (including inner experiences such as dream images, specific thoughts, memories, or emotions) always exist in some "quantitative" relation to each other. Irrespective of their contents, these elements are distinguished from each other in the person's consciousness by virtue of their comparative affective intensity (e.g., something is less or more pleasing than something else) and direction or valence (positive/negative). This is what Jung meant in the phrase *psychological value*. The implicated field of inter-relations among experiential elements (in a word, the psyche) organizes itself around a perceiver. It thus constitutes a systemic whole—a "closed system", as Jung put it—characterized by structural integrity both synchronically and diachronically. That is, some specific behaviors, thoughts, moods, and so forth, tend to co-occur or exclude each other (synchronic organization), whilst some behaviors tend to lead to other behaviors (diachronic organization). Thirdly, Jung stressed that the term libido (etymologically, from the Latin for desire) should not be designated to signify a motivational force that causes behavior. To him, libidinal flow indicated structural transformations in the psyche, in the context of which the function of the specific behavior for the person's adaptation (or not) could be understood.

However, despite the epistemological position stated in Propositions 1 – 3, at a glance Propositions 4 – 5 are readable like the postulation of a causal mechanism (see extended discussion in Jones, 2001a). Their expansion by Jung in the 1928 essay and elsewhere refers to a "flow" forward and backward of libido, and to "damming" or obstacles to the flow, thus inviting the visualization of a fluid substance. The description posits psychic transformations as contingent on libidinal flow and as effecting the production of symbolic representations. Although it might be possible to resolve the apparent inconsistency between Jung's thesis and its application by carefully rephrasing of Jung's theory of symbol forming, the existing dissonance has perpetuated misconceptions about what Jung really meant. Before turning to discuss his propositions in more depth, it would be instructive to identify the prevalent ambiguities about Jung's revision.

Current Status

Textbooks tend to cite Jung's concept of psychic energy perfunctorily, presenting it as a component of Jung's thought but one that might be deployed (if at all) as specialist "technical" jargon in clinical practice. In one of his introductory books, the prolific Jungian "populaizer" Anthony Stevens (1999) invites the reader "to enter into the imaginative conspiracy that all Jungians share with each other and with their patients" (p. 28). Stevens is aware of critics' portrayal of Jungianism as mystical twaddle (e.g., Noll, 1994; cf. Jones, 2001b), and his light-hearted assertion of an imaginative conspiracy rhetorically pre-empts such charge. At a deeper level, his assertion coincides with the postmodern view of scientific and other knowledge as socially constructed (see, e.g., Gergen, 1994). However, such nuances are likely to be overlooked by readers seeking to familiarize themselves with the basics of Jungian psychology. The point made here is that misconceptions are perpetuated by images conjured up in simplifications made for the sake of the lay or student reader.

For example, a reliable glossary compiled by Sharp (1998), available on the Internet, defines libido as "the psychic energy that *motivates* the psyche. ... The libido *invested* in a given item is indicated by the quantity of its 'value-charge,' either positive or negative" (italics added). Although this accurately uses Jung's terminology, the definition inadvertently reinforces the hypostasized concept of energy that Jung (1921/1993) rejected as "an example of the superstitious overvaluation of facts" (para. 699). Similarly, Jung's (1928/1960) contention that the "idea of energy is not that of substance moved in space" (para. 3) seems overlooked in Stevens' (1999) reference to "free psychic energy at the disposal of the ego" (see full quotation below). Jung's (1928/1960) spirited argument for the theoretical differentiation of *energy* and *force* has disappeared in the literature. In another introductory book, Stein (1998) defines libido as "having affinities with the philosophical concept of 'life force'" (p. 234)—despite Jung's (1921/1993) caveat, "Neither do I understand libido as a psychic *force*" (para. 778; author's italics). Elsewhere in his writings, however, Jung seems to endorse a contradictory impression. Jung's hostile critic, Noll (1994), commented that "vitalistic metaphors and ideas always remained a significant part of his work" despite "occasional efforts ... to make his ideas more palatable to a professional readership that had become increasingly skeptical of vitalism" (p. 24). Echoing Noll, the historian Pietikäinen (1999) has noted that "Jung wants to differentiate his term from the universal concept of energy, 'life force'" yet the specific term that Jung suggested (libido) "betrays his vitalist tendencies" (p. 80).

Jung's writings lend themselves to multiple interpretations, and some contemporary assertions are "distortions" reflecting the evolving nature of analytical psychology. Stevens (1999) asserts that, contra Freud's deterministic outlook, "Jung insisted on the freedom of the will, which he defined as free psychic energy at the disposal of the ego. Where Freud's orientation was causal, Jung's was teleological" (p. 263). In an academic paper, Horne (2002) compares Jung's teleology with Aristotle's. However, although Jung contested Freud's causal-mechanistic approach at length, and promoted what he called the "final" approach, he categorically stated that his "principle of finality" is *not* teleological, for it does not contain the idea of an anticipated goal (Jung, 1928/1960). He pointed out the fallacy of final causes, or means-ends directionality, which assumes that an end-state exerts an influence backward in time to cause the psychological transformation leading to it. Furthermore, although Jung often stated his indebtedness to Schopenhauer's concept of Will, this metaphysical concept does not have the semantic meaning of a mental power to control one's behavior that the word has in ordinary parlance, and which is implied in Stevens' assertion that Jung insisted on "freedom of will."

On the contrary, Schopenhauer's Will could be viewed as a precursor of Freud's concepts of the unconscious and id (Young & Brook, 1994). In any case, Schopenhaur (1844/1922) discussed Will as a property of the natural universe, not of human minds, regarding it the dynamic underlying the lawful organization of all types of objective phenomena. In citing Schopenhauer, Jung was trying to recover a more reified meaning of energy as "a concept abstracted from relations of movement" (1928/1960, para. 3). Hauke (2000) compares Jung's psychic energy formulation to Nietzsche's doctrine of the will to power. To Hauke, Nietzsche's doctrine "sounds like the precursor for Freud's concept of libidinal psychic energy except for one important difference" (p. 178). The difference is that "this energy for Nietzsche is not the satisfaction of some aim or desire as it is for Freud" but (as for Jung) it includes all forms of activity that serve to maintain or increase the power of the body (p. 178). The obvious scope for in-depth philosophical comparison of Jung's concept with those of Schopenhauer, Nietzsche, and Freud, seems to have been generally overlooked.

The ambivalence about Jung's concept could be attributed partly to misunderstanding that would not have risen had he coined a new term to express what he meant; e.g., abandoned the terms *psychic energy* and *libido* and adopted instead the term *psychological value*, which he regarded as synonymous. Taken in the historical context of his turbulent breakaway from Freud, however, it could have been in his professional interest to present his thesis as widening Freud's concept. His decision to retain libidinal terms could be seen as a strategic move to downplay his disagreement with Freud—he was merely amending Freud's theory—and thus to minimize his estrangement from the psychoanalytical community. In the long run, the rhetorical strategy has undermined the communication of his innovation. It is conventionally assumed that "Jung expanded the concept of libido *beyond* the sexual meaning it had for Freud to include psychic energy in general" (Homans, 1995, p. 68-9; italics added). Yet, as indicated in the above outline of Jung's thesis, it would be more accurate to say that Jung expanded the concept of libido *in opposition* to the sexual meaning it had for Freud and any qualitative meaning (including that of a "generic" force) that it could have.

Jung's assimilation of libidinal terminology has meant that those who read him from a Freudian perspective are likely to overlook his innovation, and to regard him as mistaken about the libido. Copjec (1994) accuses Jung of "evacuating the libido of all sexual content by associating it exclusively with cultural process," a move that "leads Jung to stress the essential plasticity of malleability of the libido" (p. 22). Kristeva (1980) dismisses "Jung's dead end with its archetypal configurations of libidinal substance taken out of the realm of sexuality" (p. 276). However, what Jung had in mind—libido as *psychological value*—cannot be "configured" or have properties of "malleability" and "plasticity":

> Value requires for its explanation a quantitative concept, and a qualitative concept like sexuality can never serve as a substitute. A qualitative concept is always the description of a thing, a substance; whereas a quantitative concept deals with relations of intensity and never with a substance or a thing. (Jung, 1928/1960, para. 51)

Lacan (1977) interpreted the Freudian concept of libido, not as substance, but as "the effective presence, as such, of desire" (p. 153). On this basis, he argued that "Jungianism ... is necessarily accompanied by a repudiation of the term *libido*, by the neutralization of this function by recourse to a notion of psychical energy, a much more generalized notion of interest" (p. 153; author's italics). Lacan's translator, Sheridan, explains that whereas the English word *desire* is limited to isolated acts of wishing, the French word *désir* "has the stronger implication of a continuous force," an implication that enters Lacan's theory as "a perceptual effect of symbolic articulation" (in Lacan, 1977, p. 278). Yet it could be argued that *désir* in this sense is quite similar to the "much more generalized notion of interest" of which Lacan accused Jung.

Psychoanalytical discourses aside, "mental energy" is seldom mentioned seriously anymore in psychological contexts. The usefulness of the concept has long been repudiated. B.F. Skinner (1974/1993) pointed out the explanatory redundancy: to say that mental processes "require and consume psychic energy ... is simply another way of representing the probability of behavior derived from contingencies of survival or of reinforcement" (p. 167-8). From the standpoint of post-behaviorist psychology, the libidinal theories of Freud and Jung might seem esoteric or mystical, and altogether alien to the project of psychology as the science of mind and behavior. It is therefore worth bearing in mind that in pre-World War II Europe, especially to German-speaking psychologists, "energy" did have scientific connotations. The metaphor was borrowed directly from physics. Jung did not question its heuristic appropriateness, only its application in psychoanalysis. Mental energy was widely discussed; e.g., in the early work of Kurt Lewin (1935, 1936). In an introductory textbook, Thouless (1951) proposed to regard human behavior "as if it were activated by a fund of energy which, like physical energy, can be directed into one channel or another", but cautioned, "It is an analogy that must not be pressed too far" (p. 30). Jung could be accused of pressing the analogy too far, especially in his attempt to apply the laws of thermodynamics to the dynamics of the psyche (Jones, 2001a).

Talk of energy as an explanation for human motivation persisted well into the 1960s with attempts to identify it with physiological arousal. According to Vernon (1969), the degree or level of activation depends fundamentally on the amount of energy that the body's metabolism can supply. The brain controls "how this energy is applied, and how activation develops in particular circumstances"—and this "direction of energy is associated with arousal or alerting to particular situations and activities" (p. 85). Yet it is not clear how brains direct the application of energy in ways that become meaningful to us through our use of language. Nor is it clear how, as human organisms, we become alerted to the symbolic meanings of particular situations and activities, meanings that powerfully motivate our actions and yet are not presented directly to our sensory modalities. Jung, on his part, had maintained a thoroughgoing separation of psychic and biological energy. Moreover, as indicated above and will be discussed in the next section, the term *energy* had a special meaning to Jung.

From Force to Field Dynamics

Jung (1928/1960) begins his argument for revising libido theory by drawing attention to two ways in which physical events are understood, namely the "causal-mechanistic" viewpoint and the "energic" or "final" viewpoint, and contends that the same distinction applies also to the description of psychic events. In this context he draws a strict distinction between *energy* and *force* as explanatory constructs. *Force* is a causal factor operating on a pre-existing inert structure, effecting structural transformations. It belongs in the causal-mechanistic view, which "conceives an event as the effect of a cause, in the sense that unchanging substances change their relations to one another according to fixed laws" (para. 2). Whereas a force is an external factor introduced into an inert structure, *energy* is an intrinsic property of a system in motion. The energic explanation is "in essence final; the event is traced back from effect to cause on the assumption that some kind of energy underlies the changes in phenomena" (para. 3). "Tracing" presupposes knowing the outcome, having the full ("final") picture at hand. Seeing the whole allows the identification of structural transformations over time, and thereby enables the discovery of how a pattern of interaction between specific parts functions within the temporally extended whole.

Jung's "final" approach closely resembles what is sometimes called the functional explanation in psychology. The term *functional* has two meanings that are incompatible with each other: In one sense, "function" means purpose, and the (teleological) explanation would identify some end-goal of the observed behavior. As mentioned earlier, Jung disassociated his "principle of finality" from teleological conceptions. Jung (1928/1960) explained in footnotes that the so-called final view does not "contain the idea of an anticipated end or goal" (n. 4), and that means-ends directionality is "inadmissible" because a future goal cannot initiate retroactively the psychological process culminating in it (n. 5). Using "function" in a mathematical or algebraic sense, instead, the explanation would posit the psychological process as reflecting the interaction between two or more variables—a sense in which "function" is used in Skinnerian functional analysis. It is also exemplified in Lewin's (1935) famous "$B = f(P, E)$" formula, where behavior (B) is a function of the relationship between personal (P) and environmental (E) variables. Jung's energic/final approach should be understood in the same way. If Jung's position were to be expressed in a similar formula, however, behavior would be placed along with environmental and personal variables, reserving the left side of the equation to the "gestalt" of their constellation. Elsewhere Jung (1916/1948, para. 456) equates "striving for a goal" with "sense of purpose" immanent in all psychological phenomena, but his point is that the purpose of something does not enter any sequence of cause and effect—a claim echoing Schopenhauer's concept of Will.

Jung's force/energy distinction thus touches upon a fundamental disagreement about what constitutes an explanation in psychology. Should the psychological explanation identify which stimuli cause what response (or cause inner states that in turn cause the response) or identify the function of the response in a relevant context—and, if we opt for the latter, what is the relevant context? Gestalt psychologists viewed the perceptual field as the context in which responses may be functionally explained. Lewin (1936) asserted that "the effect of a stimulus depends in part upon the nature of surrounding field" (p. 33). Köhler (1930) objected to what he called the "machine theory" professed by behaviorists and introspectionists, who assume that "what happens in the nervous system will depend either upon inherited machine arrangements or upon secondary *acquired* arrangements" (p. 89; author's italics). Köhler argued that "the machine theory excludes *organization of process* in the field" (p. 96; author's italics), and that "some specific and orderly function might occur without being controlled either by special arrangements pre-established *ad hoc* or by arrangements acquired in learning" (p. 89). Jung on his part did not reject the notion of forces. In the 1928 essay he acknowledges that there are forces of hunger, sexual desire, fear, and other instinctual needs; and that these cause behavior. His contention is merely that these should not be confused with energy. To him, both the causal-mechanistic and energic/final explanations are necessary and therefore complementary, though mutually exclusive as ways of explaining events. "The distinction between force and energy is a conceptual necessity ... The concept of quantity should never be qualitative at the same time, otherwise it would never enable us to expound the relations between forces" (Jung, 1928/1960, para. 26-7).

For example, the falling down "behavior" of an apple could be explained either by reference to an antecedent event, such as becoming detached from the tree—or, alternatively, by citing the law of gravity (the example mine). The latter alternative informs about a law of nature, not about the conditions that must exist in order for the apple to fall. Whilst we cannot claim that the apple fell *because* of the law of gravity, citing the natural law does not challenge the causal explanation (or vice versa). The same would apply for the behavior of people or animals. We may identify either causes or laws, as wished, but must be clear which type of explanation we are offering. For example (again mine), thirst impels an animal to seek water, but the energy "invested", as Jung would put it, in approaching a river reflects the extent to which the animal's interest in the water is greater than its fear of predators lurking there. This relation of affective intensity and valence between the two elements of the animal's experience of its approach to the water is what Jung called the *psychological value*. It explains the animal's behavior—not as caused by a mental process of weighing cost-benefit options—but as a function of the lawful organization of its experiential realm.

Motivation and Intentionality

In Freudian libidinal theory, the inner "dynamo" is fuelled by sexual desire. To Jung (1952/1967), libido was a concept that "expands into a conception of *intentionality* in general ... an energy-value which is able to communicate itself to any field of activity whatsoever, be it power, hunger, hatred, sexuality, or religion, without ever being itself a specific instinct" (para. 197; author's italics). Jung used "intentionality" not in the (causal) sense of intention, but in the phenomenological-philosophical sense, to do with one's relatedness in the world. Intentionality is to the intending subject as sight is to looking. The causal idea implies a sequence that terminates when the action fulfils the actor's intention, thus altering her relation to the world in some way. Yet however altered the personal situation might be, she remains in *some* relation to the world—and this self/world relation would be experienced as better or worse than other possibilities. Therefore, when Jung stressed that libido *qua* intentionality is present in all our psychological processes "without ever being itself a specific instinct," he must mean that this theoretical construct belongs to an altogether different category than instincts, intentions, and desires. According to Jung, it belongs with quantitative concepts such as energy: "the concept of libido in psychology has functionally the same significance as the concept of energy in physics" (1952/1967, para. 189).

Just as force and energy are not interchangeable conceptually, yet are not rival theories in Jung's thinking, so the postulation of desire does not necessarily contradict his concept of libido. There are clearly stronger and weaker desires, desires to avoid and desires to have, and so forth. It is not clear whether Jung viewed the "force" of (e.g.) fear as distinct from the actual experience of fear in particular situations. It is clear, however, that he distinguished between qualitatively different *states* such as feeling afraid, thirsty, sexually aroused, etc., on the one side, and *relations* (such as "stronger than...") that could be applied to any combination of states, on the other. The butt of his argument was that human motivation cannot be reduced to a single force, namely sexuality, as was claimed by Freud: "To substitute a qualitative concept for energy is inadmissible ... This would be in biology vitalism, in psychology sexualism (Freud)" (Jung, 1928/1960, para. 51). In Lacan's (1977) theory, the "effective presence of desire"—which, as mentioned, is how he defined libido—arises from an existential lack experienced by all humans. One way of understanding this would cast *désir* as a causal factor, a state that we must first enter, a mode of *desiring*, before we can desire this or that. By analogy, we must contract the illness before displaying the symptoms. Another way to understand it renders the assumption of causality nonsensical. We do not enter a "looking" mode before looking at this or that object. Looking, desiring, etc. are relational states, a property of subject–object interactions.

According to Jung, then, libido is the emergent property of *all* relational states—"power, hunger, hatred, sexuality, or religion," to borrow his list. Taxonomies of mental states or instinctual forces would be trivial in this respect (but not in the context of Jung's psychology on the whole, as discussed below). Whereas the causal-mechanistic explanation necessarily posits some states as mutually exclusive, the energic explanation identifies a relation between such states. The state of desiring something versus the state of not desiring it, for instance, cannot simultaneously motivate the same response; but these states can be represented in relation to each other. Although Jung did not pursue the quantification of psychological value, his reasoning here is quantitative: The presence/absence of a desire could be coded "1"/"0", respectively. The degree to which desiring something seems to influence behavior—that is, the degree to which the desire is effectively present, we may say with Lacan (1977)—can be represented with a numerical value.

The Psyche as a Dynamical System

Jung's revision might be controversial in the psychoanalytical context, but it converged with the Gestalt-psychological understanding of energy as field dynamics. Jung's concept of psychological value resembles Lewin's field theory, which maps the immediate environment in terms of approach/avoidance qualities, or "invitation qualities" (usually translated as valences), later revised by Gibson (1979) in his affordances theory. Like Lewin's topological psychology, Jung's thesis pivots on a quantitative idea of relations of intensity and direction of affect, whereby various elements of experience are distinguished from each other in the individual's experience:

> The idea of energy is not that of substance moved in space, it is a concept abstracted from relations of movement. The concept, therefore, is founded not on the substances themselves but on their relations, whereas the moving substance itself is the basis of the mechanistic view. (Jung, 1928/1960, para. 3.)

Neither Jung nor Lewin acknowledged the conceptual convergence, to my knowledge. The similarity is perhaps more noticeable from a historical distance than was the case at the time.

At the time, psychologists' preoccupation with the experiential realm appropriated directly (e.g., Lewin) or indirectly (Jung) the concept of *Umwelt*, attributed to the biologist von Uexküll. *Umwelt* is the ecological environment, or the world as organized in terms of food, shelter, mate, dangers, and so forth, as distinguished from the world of phenomena as such. Lewin (1935) likewise distinguished the "functional" world from the "phenomenal" world, citing von Uexküll. Nowadays *Umwelt* is criticized as implying a dualism of inner/outer realms, for it partitions the world of the animal into "the world as sensed and the world of actions" (von Uexküll, 1909, cited in Crook, 1980, p. 4). However, the original concept could be interpreted as positing a dialectical relationship between the world as sensed by an organism and the organism sensing the world: The dualism of the known object and knowing subject, which William James (1890) considered to be inevitable in psychology. The ensuing problematic is—not Cartesian dualism—but a version of the so-called Hermeneutic Circle; that is, the conundrum that in order to understand the whole, we must first know its parts, but to know the parts we must first "see" the whole. Such issues were "in the air" for German-speaking psychologists in the early decades of the twentieth century (Valsiner & van der Veer, 2000).

However, whereas the Gestaltists (and later Gibson, 1979) described objectively a *subject-centered* field—i.e., the perceivable world as organized in relation to a particular perceiver—Jung described a *subjective* field; that is, the world from the viewpoint of a self-aware subject. The conceptual switch from the subject-centered to the subjective field is crucial for understanding Jung's theory of symbol forming. Briefly, Jung assumed a tension between the world as unconsciously sensed (i.e., *Umwelt*) and the world as consciously apprehended by the human being. Symbolic activity is a function of the complex interactions between conscious and unconscious factors, interactions that constitute structural transformations of the psyche.

According to Jung (1928/1960), the "immediate experience of quantitative psychic relations" justifies "at least a provisional view of the psyche as a relatively closed system" (para. 11). This is not to say that the psyche is closed to the organism's environment; on the contrary, Jung (1935/1968) viewed consciousness as "very much the product of perception and orientation in the *external* world" (p. 8; author's italics). However, the idea that phenomena are differentiated in the perceiver's experience by virtue of their affective intensity and valence relative to each other "closes off" the psyche into a mathematical space constituted of relations among a finite set of variables. Defining the psyche as a closed system expresses the idea that transformations in the experiential realm could be quick or slow, vary in intensity, or involve a reversal of polarity—but cannot alter the total "energy quota" (in accordance with the law of energy conservation). These "relations of movement" are the abstract dynamics of emotional self-regulation, its quantitative aspect, and their abstraction is conceptually separate from the identification of *what* the person finds highly desirable, less desirable, or undesirable; i.e., qualitative aspects of experience. Jung (1928/1960) suggested that the explanation in psychology would benefit from the idea of a constant underlying all relations of movement, and proposed to call this constant *psychic energy*. Libidinal terminology aside, the idea of a constant introduces into the picture a third attribute in addition to affect and phenomenon—something that remains constant even when we systematically manipulate the presentation of phenomena and observe dependent emotional states. Arguably, the only thing that would remain constant is the fact that phenomena differ in terms of their affective implications—a trite comment as such, which becomes theoretically and practically meaningful only when we identify qualitative data; i.e., which phenomena has which affective implications and under what conditions. Jung (1928/1960) too acknowledged that the "*applied* concept of energy always deals with the behavior of forces, with substances in motion" (para. 52; author's italics).

A few decades later and independently of Jung, the founder of personal construct psychology George Kelly (1955/1991) would make similar contentions about Freudianism a few steps further. According to Kelly, if the personality is a dynamic system, there is no need to postulate drives that set it in motion; instead, we should discover how individuals construe inter-relations among elements of their world. Personal construct systems develop on the basis of lived experiences, and are "typical" (or not) insofar as people have (or do not have) similar experiences. Unlike Kelly, Jung believed that the psyche is pre-configured due to the species' evolution—a claim made in the context of his theory of archetypes as innate dispositions to form certain kinds of symbols (e.g., representations of the Mother, Hero, etc.). While archetype theory chiefly concerns qualitative aspects of human psychological functioning, it implies that some psychological values are hard-wired; e.g., the positive value that the mother's proximity has for the infant, the intense negative value of her disappearance. According to Jung, such contradictory experiences give rise to symbols of a specific kind—e.g., cultural and individual representations of the "loving and terrible Mother", in Jung's (1954/1959) phrase—which enable overcoming the existential contradiction (Jones, 2001a). Jung did not make the full conceptual conversion to systemic thinking. His theorizing about the micro-genesis of symbols—the topic of the next section—was channeled by the energy metaphor into a narrative that psychological science has since eschewed.

The entrapment of the energy metaphor is most conspicuous in Jung's attempts to find manifestations of the laws of thermodynamics in human "energized" processes. With hindsight, the universality of the laws of energy conservation and entropy could be called into question in view of relatively recent theories concerning the evolution of non-equilibrium systems in nature (e.g., Prigogine, 1976), but Jung's application is debatable even within its own referential context. According to Jung, the principle of energy conservation predicts that when the psychological value of something diminishes in the person's experience, there would be an equal increase in the psychological value of something else; energy is transferred to the new object, as it were. Jung calls this the principle of equivalence. To Jung, this principle was evident in the phenomena of transference in analyst–patient interactions: "when a conscious value ... decreases or actually disappears, you [the analyst] immediately look for the substitute formation, expecting to see an equivalent value spring up somewhere else" (1928/1960, para. 35). As a practitioner he was concerned with instances when "a considerable sum of libido disappears as though completely swallowed up by the unconscious, with no new value appearing in its stead" (para. 35), for this often means neurosis. Jung advised, "In such cases it is advisable to cling firmly to the principle of equivalence, for careful observation of the patient will soon reveal signs of unconscious activity" (para. 35). The advice is embedded in the two narratives, applied and theoretical, interwoven in the 1928 essay, and it could be interpreted differently in each context. In the applied context, Jung is providing a practical guideline: "Anyone who has had practical experience of this field knows that the equivalence principle is of great heuristic value in the treatment of neuroses" (para. 35). However, a heuristic does not prove a scientific theory.

Two recent papers by Jungians draw upon the idea of self-organizing systems to support Jung's position. Saunders and Skar (2001) emphasize the analogy between the psyche and physical systems, contending that because self-organization is a widely demonstrated property of physical systems, the psyche too must have this property. In contradistinction, McDowell (2001) proposes to conceptualize archetypes as principles governing the spontaneous organization of themes or motifs into archetypal images. Instead of the vexed question of what causes archetypes, evolution versus culture, it thus becomes a question of how the archetypal image functions within a dynamical whole. To McDowell, the dynamic system in question is the personality (rather than psyche), which he views as comprised of instinctual impulses, affects, feelings, sensory perceptions, images, thoughts, hopes, etc., and memories of these. He maintains that a mathematical principle of organization specifies a dynamic form that universally applies to all complex systems.

For instance, the principle of division is "a principle of organization and not just a unique or idiosyncratic event, we see it everywhere in the manifest world" (McDowell, personal communication, January 21, 2002). Therefore, "one of the transformations that I can subjectively experience within my own personality is that parts of myself can fall apart"; moreover,

> An image of falling apart will be charged, in part, because it represents a vivid internal experience. Such an image will be derived from the environment, usually through vision, though it might be an image not literally seen, but verbally described and then "seen" internally. (McDowell, personal communication, January 21, 2002; author's italics.)

In other words, the image of falling apart—the specific content of which would utilize composite fragments of things seen or heard—would have a significant psychological value when it coincides with an unconsciously experienced transformation.

The Micro-Genesis of Symbols

According to Jung (1921/1993), for an image (utterance, etc.) to be psychologically meaningful, there must be a *symbolic attitude* "which *assigns meaning* to events ... and attaches to this meaning a greater value than to bare facts" (para. 819; author's itlaics). The symbolic attitude is linked in Jung's thought to the idea of tensions between the world as unconsciously sensed and the world as consciously apprehended. He later came to call the resolution of this existential tension the *transcendent function* and to view it as the most fundamental of psychological processes. In his view it is "a *natural process*, a manifestation of the energy that springs from the tension of opposites and it consists in a series of fantasy-occurrences which appear spontaneously in dreams and visions" (Jung, 1916/1948, para. 121; author's italics). He believed that if unhindered, this process necessarily culminates in a state of *individuation,* a state of harmony between conscious and unconscious parts—an equilibrium towards which the psyche must move, like any dynamical system. Jung reasoned that the series of fantasy-occurrences could reveal the sequence of psychic transformations underlying the transcendent function. On this basis, Jung and his students carried out an extensive research agenda, investigating the dreams, fantasies, and art of individuals, and also the symbolic productions of cultures, past and present, such as myths, fairytales, religious icons and ceremonies, and so forth.

According to Jung, the study of collective productions allows psychologists to trace the evolution of what he called the *collective unconscious*; i.e., the biologically given psychic configuration, that which all humans have in common. His anthropological affectation was severely attacked for misunderstanding cultural processes (Pietikäinen, 1999), and the collective unconscious hypothesis rests uneasily with neo-Darwinism, but his research agenda coincides with the project of the postmodern movement in psychology. Jung was aware of the dilemma of "modern" psychology, namely the inaccessibility of its subject matter: "knowledge of the psychic substance is impossible for us ... our subtlest lucubrations can establish no more than is expressed in the statement: this is how the psyche behaves" (Jung, 1948/1959, para. 384). In other words, if we want to know human beings, we must (and can only) study their symbolic productions—we must turn to the interpretative methodologies of the human sciences (cf. Bruner, 1986; Polkinghorne, 1988). Jung did not overtly make this claim, although his researches are clearly a kind of text analysis (loosely defined) rather than the scientific method of hypothesis testing. Jung perhaps saw the distinction between the natural and human sciences as anyway tenuous or relatively trivial. He held that all scientific theories, along with religions and philosophies, are but symbolic productions of a culture—a viewpoint that today is associated with postmodernism.

Jung's theorizing about symbol forming should be regarded—not as a scientific discovery—but as seeking to find the best narrative in which to articulate the vicissitudes of individuation. When he presented his psychic energy revision, a libidinal-flow narrative had widespread influence in the German-speaking world, an influence that lingered well into the middle of the twentieth century. This can be seen in Lewin's work, whose strategy was to operationalize the notion of libidinal regression experimentally (Barker, Dembo & Lewin, 1941). Jung's strategy was to paraphrase the narrative, to retell it with the expectation that the retelling would accommodate his revised definition of libido—though in the long run it was his conceptual innovation that has been assimilated into the Freudian narrative, losing some of its distinctiveness.

Jung's biographer and friend, Bennet (1966/1983), attributed the parting of the ways between Freud and Jung to their different understanding of symbol. Freud gave symbol a "material significance, that is, treating it as though it referred to actual objects or persons" and using the term as a sign "of something quite well known, though not consciously recognized" (p. 40). In contradistinction, to Jung a symbol is "the best possible formulation of a relatively *unknown* thing ... standing for something that is only divined and not yet clearly conscious" (1921/1993, para. 815, author's italics). The primary psychological function of a symbol in the Jungian sense is to mediate the conscious awareness of one's unconsciously experienced situation. Turner's (1967) account of the notion of *symbol* among the Ndembu in Africa provides a convenient way of conceptualizing Jung's meaning (Jones, 1999). The Ndembu word that Turner translated as "symbol" is derived from a phrase meaning to blaze a trail through the forest by marking the route from the unknown bush to known paths. "A symbol, then, is a blaze or landmark, something that connects the unknown with the known" (Turner 1967, p. 48).

It is my thesis that Jung's understanding of the psychological function of symbols holds promise for a fusion of ideas with the premises of postmodern psychology. However, his theoretical narrative is derailed, so to speak, by the energy metaphor, and is no longer "hearable" as coherent in psychological science. Jung (1928/1960) discussed *symbol* as a "psychological mechanism that transforms energy" (para. 88). He conjectured that the capacity to form symbols evolved as means for channeling "surplus energy", an inexplicable excess of nature. Libidinal channeling takes the form of regression or progression. In regression, "the libido streams backwards, as it were, to its source" (1921/1993, para. 824). In progression,

> [The] damming up of libido is analogous to a specific obstruction in the direction of the flow, such as a dike, which transforms the kinetic energy of the flow into the potential energy of the reservoir. Thus dammed back, the water is forced into another channel. (Jung, 1928/1960, para. 72.)

Libidinal progression is purported to constitute "the daily advance" of psychological adaptation, "a continual satisfaction of the demands of environmental conditions … possible only by means of attitude, which is necessarily directed and therefore characterized by a certain one-sidedness", hence becomes maladaptive when environmental changes require a different attitude (para. 60-61).

The metaphor-talk obscures three important propositions. First, the presence of a symbolic activity is an evidence for a transformation happening in the system of psychological values. The transformation could emulate a previous configuration of subject–object relations (regression) or create a new relational configuration (progression). The second proposition is that apperceiving an image (utterance, etc.) as subjectively meaningful *ipso facto* transforms the closed system of psychological value, because the concrete image enters relations of intensity with other elements of experience. Third, transformations occasioned by symbols are "more" than required for an organism's adaptation to its immediate circumstances (libidinal excess). This claim could be elucidated with the example of fear of heights, a natural survival mechanism evolved in humans and other species. This aspect of our *Umwelt* is represented in dreams, fairytales of monsters lurking in deep places, linguistic metaphors such as falling from Grace or status, and so forth. The symbols do not directly buffer the physical danger inherent in standing at a precipice, merely allow us to *know* that we are afraid of falling. Rather, such embodied knowledge is used as a metaphoric "source domain" (Lakoff & Johnson, 1980) to describe personal and social contingencies that do not involve physical heights or threaten survival at the level of the organism.

To some degree, it is possible to undo Jung's web of libidinal-flow concepts. In his post-Freudian thought, the metaphor of libido *blockage* implies a person's preoccupation with some existential aspect that prevents consciousness of other possibilities—the "one-sidedness" of attitude that he referred to in his description of libidinal progression as adaptation. One obvious example comes from developmental research into infant–caregiver attachment: The caregiver's proximity matters highly to the infant. As widely documented in the Strange Situation Test, a mother's departure "blocks off" the child's interest in toys or other people present. The intensely positive value of the maternal presence and the intensely negative value of her absence are innately given in the biological system of attachment behavior and its physiological correlates. Symbolic productions do not necessarily issue forth out of the above state (attachment behavior is observed also in non-human mammals). However, Jung's theory predicts that the more intense and sharply polarized the experience is, the less likely its emotional tone and ideational contents are to dissipate in the individual's subsequent interactions with the external world. A third factor must be introduced. In humans, "there now emerges a new content, constellated by thesis and antithesis in equal measure and standing in *compensatory* relation to both. It thus forms the middle ground on which the opposites can be united" (Jung, 1921/1993, para. 825; author's italics). The new content "transforms energy" insofar as it alters the way that the contradictory experiences relate to each other in subjective experience. Young children who can symbolize—simply, verbally represent—the caregiver's return tend to be less anxious in the Strange Situation test (Sroufe, 1996).

As noted in Jones (2001a), an apt illustration of what Jung meant in his description of the symbol as a Hegelian synthesis—a new content arising from a thesis/antithesis tension—is ironically provided in Freud's famous observation of the so-called *Fort! Da!* (Gone! Here!) Game by his 18-month-old grandson. The child derived great delight from playing with a cotton reel, throwing it away from him and reeling it back. Freud linked this to the mother's regular absence during the day, and speculated that through the game the child gained a sense of control over his predicament: "the little boy has the mother on a string. He makes her go away, even throws her away, and then makes her come back at his pleasure" (Erikson, 1963, p. 195). The Freudian assumption is untenable in view of current knowledge about cognitive development, which suggests that using the cotton reel as a token of the mother is an object-substitution far beyond the ability of a toddler. Rather than understand the game as a symbol in the Freudian sense of a token for an actual person or known situation, the child's game seems a symbol in the Jungian sense. The delight he displayed was not necessarily due to a delusional mastery over his life predicament, but could be more simply the delight of discovering "how the world works", the formation of what Piagetians called the concept of object permanence. The "Jungian" point would be that, first, the here/gone principle is plainly "a new content, constellated by thesis and antithesis" between the *visual* absence and presence of things. Secondly, the child's fundamental discovery is not merely the intellectual appreciation of object permanence, but his *control* over the phenomena. This sense of control, in turn, could alleviate his separation anxiety insofar as his play activity reconfigured the subjective field. It created a new meaningful content, meaningful in the toddler's experience at the preverbal level of psychological value (evident in his delight); and this content has entered relations with other elements of the immediate experience.

CONCLUSION

Jung addressed an audience that did not require persuading about the appropriateness of the energy metaphor, but in his opinion needed weaning off the conflation of energy with force. He contested "those constant confusions of 'energy' with 'force'" but conceded that the "in practice, we speak of electrical energy and the like, as if energy were a definite force" (1928/1960, para. 52)—and continued this discursive practice in his application of libidinal explanations to symbolic activity. Unlike the Freudian idea, however, Jung's libidinal flow narrative does not posit energy consumption as a causal factor that directly effects behavioral change. Instead, the metaphor refers to transformational patterns in the subjective field, transformations that he regarded as the source of symbols.

To date, the scope for "postmodernizing" Jung has barely been tapped (but cf. Hauke, 2000). This chapter has aimed to demonstrate that although Jung understated the dialogical or social-relational context in which the symbolic attitude is actualized (Bovensiepen, 2002) and underestimated cultural processes (Pietikäinen, 1999), his thesis complements current dialogical and cultural theories with its focus on the embodied symbol.

REFERENCES

Barker, R., Dembo, T. & Lewin, K. (1941) Frustration and regression: an experiment with young children *University of Iowa Studies: Child Welfare* 18, xv-314

Bennet, E. A. (1983) *What Jung Really Said* New York: Schocken (originally published 1966)

Bovensiepen, G. (2002) Symbolic attitude and reverie: problems of symbolization in children and adolescents *Journal of Analytical Psychology* 47, 241-257

Bruner J. (1986) *Actual Minds, Possible Worlds.* Cambridge, Mass.: Harvard University Press

Copjec J. (1994) Sex and the euthanasia of reason. In Copjec J (ed.) *Supposing the Subject* (pp. 16-44) London: Verso

Crook, J. H. (1980) *The Evolution of Human Consciousness* Oxford: Clarendon Press

Erikson E. (1963) *Childhood and Society*. London: Paladin

Gibson J. J. (1979) *The Ecological Approach to Visual Perception* Boston: Houghton-Mifflin

Hauke, C. (2000) *Jung and the Postmodern* London: Routledge

Ho, D.Y. F., Peng, S-Q., Cheng Lai, A. & Chan, S-F. F. (2001) Indigenization and Beyond: Methodological Relationalism in the Study of Personality Across Cultural Traditions *Journal of Personality*, 69, 925-953

Hogenson, G. B. (1998) Response to Pietikainen and Stevens *Journal of Analytical Psychology* 43, 357-372

Homans, P. (1995) *Jung in Context* (Second edition) Chicago: University of Chicago Press

Horne, M. (2002) Aristotle's ontogenesis: a theory of individuation which integrates the classical and developmental perspectives *Journal of Analytical Psychology* 47, 613-628

James, W. (1890) *The Principles of Psychology*. London: Macmillan

Jones, R. A. (1999) Direct perception and symbol forming in positioning. *Journal for the Theory of Social Behavior* 29, 37-58

Jones, R. A. (2000) On the empirical proof of archetypes: commentary on Maloney (1999) *Journal of Analytical Psychology* 45, 599-605

Jones, R. A. (2001a) Psychological Value and Symbol Formation *Theory & Psychology* 11, 233-254

Jones, R. A. (2001b) Review essay: The relevance of C.G. Jung *Theory & Psychology* 11, 587-595

Jung, C. G. (1948) General aspects of dream psychology. *The Collected Works of C. G. Jung.* Vol. 8. London: Routledge and Kegan Paul (originally published 1916)

Jung, C. G. (1968) *Analytical Psychology* London: Ark (The Tavistock lectures, 1935)

Jung, C. G. (1993) Psychological Types. *The Collected Works of C. G. Jung*. Vol. 6. London: Routledge & Kegan Paul (originally published 1921)

Jung, C. G. (1919) *Psychology of the Unconscious* London: Kegan Paul (originally published 1912)

Jung, C. G. (1959) Psychological aspects of the mother archetypes *The Collected Works of C. G. Jung* Vol. 9i (pp. 75-112) London: Routledge (Originally published 1954)

Jung, C. G. (1959) The phenomenology of the spirits in fairy tales *The Collected Works of C. G. Jung* Vol. 9i (pp. 207-254) London: Routledge (originally published 1948)

Jung, C. G. (1960) On Psychic Energy *The Collected Works of C. G. Jung* Vol. 8 (pp. 3-66) Routledge & Kegan Paul (originally published 1928)

Jung, C. G. (1967) *Symbols of Transformation* (second edition) (The Collected Works of C. G. Jung Vol. 5) Princeton, NJ: Princeton University Press (originally published 1952)

Jung, C. G. (1992) *Psychological Types* (The Collected Works of C G Jung Vol. 6) London: Routledge (Original work published 1921)

Kelly, G. A. (1991) *The Psychology of Personal Constructs* (Vol.1 and 2) London: Routledge (originally published 1955)

Köhler, W. (1930) *Gestalt Psychology*. London: Bell and Sons

Kristeva, J. (1980) *Desire in Language*. Oxford: Blackwell

Lacan, J. (1977) *The Four Fundamental Concepts of Psychoanalysis*. Harmondsworth: Penguin

Lakoff, G. & Johnson, M. (1980) *Metaphors We Live By* Chicago: University of Chicago Press

Lewin, K. (1935) *A Dynamic Theory of Personality* Trans. D. K. Adams & K. E. Zener. New York: McGraw-Hill

Lewin, K. (1936) *Principles of Topological Psychology* Trans. F. Heider & G. M. Heider New-York: McGraw-Hill

Looren de Jong, H. (1997) Some remarks on a relational concept of mind. *Theory & Psychology* 7, 147-172

Maloney, A. (1999) Preference ratings of images representing archetypal themes: an empirical study of the concept of archetypes *Journal of Analytical Psychology* 44, 101-116

McDowell, M. J. (2001) Principle of organization: a dynamic-system view of the archetype-as-such *Journal of Analytical Psychology* 46, 637-654

Mogenson, G. (1999) Psyche's archetypes: a response to Pietikainen, Stevens, Hogenson and Solomon *Journal of Analytical Psychology* 44, 125-134

Noll, R. (1994) *The Jung Cult*. London: Fontana

Pietikainen, P. (1998) Archetypes as symbolic forms. *Journal of Analytical Psychology* 43, 325-344

Pietikäinen, P. (1999) *C.G. Jung and the Psychology of symbolic Forms* Helsinki: Academia Scientiarum Fennica

Polkinghorne, D. E. (1988) *Narrative Knowing and the Human Sciences* Albany: State University of New York Press

Prigorine, I. (1976) Order through fluctuation: self-organization and social system. In Jantsch E & Waddington C H (eds.) *Evolution and Consciousness: Human Systems in Transition*. (pp. 93-133) London: Addison-Wesley

Saunders, P. & Skar, P. (2001) Archetypes, complexes and self-organization *Journal of Analytical Psychology* 46, 3-20

Sharp, D. (1998) Glossary of Jungian terms *www.cgjung.com/glossary.html*

Shopenhauer, A. (1922) *The World as Will and Idea* (Vol. 1) London: Kegan Paul (Original work published 1844)

Skinner, B. F. (1993) *About Behaviorism* Harmondsworth: Penguin (originally published 1974)

Skinner B. F. (1984) Selection by consequences. *Behavioral & Brain Science* 7, 477-502

Solomon, H. M. (1998) Response to Petteri Pietikainen's "Archetypes as symbolic forms" *Journal of Analytical Psychology* 43, 373-378

Sroufe, L. A. (1996) *Emotional Development* Cambridge: Cambridge University Press

Stein, M. (1998) *Jung's Map of the Soul* Chicago: Open Court

Stevens, A. (1998) Response to P. Pietikainen *Journal of Analytical Psychology* 43, 345-356

Stevens, A. (1999) *On Jung* (Revised edition) Princeton, NJ: Princeton University Press

Thouless R H (1951) *General and Social Psychology* Third Edition. London: University Tutorial Press

Turner, V. W. (1967) *The Forest of Symbols* Ithaca: Cornell University Press

Valsiner, J. & van der Veer, R. (2000) *The Social Mind* Cambridge: Cambridge University Press

Vernon M D (1969) *Human Motivation.* Cambridge: Cambridge University Press

Young, C. & Brook, A. (1994) Schopenhauer and Freud *International Journal of Psychoanalysis*, 75:101-18

Chapter 2

PERCEPTION WITHOUT AWARENESS: THE QUALITATIVE DIFFERENCES APPROACH

Juan J. Ortells*[1], María T. Daza[1], Carmen Noguera[1], Encarna Carmona[1], Elaine Fox[2], and María J. F. Abad[3]

[1] University of Almería, Spain
[2] University of Essex, England
[3] University of Jaén, Spain

Is it possible to perceive stimulus information even when there is no awareness of perceiving? This question has been the focus of considerable research and discussion for more than 100 years (see Adams, 1957, for a review of earlier studies). However, despite the numerous experimental reports suggesting that our behavior and thoughts may be influenced by unconscious processes, research on perception without awareness has usually been plagued by continual controversy (Merikle, 1992). Much of this controversy stems directly from the logic underlying the task-*dissociation paradigm*, which has traditionally been the most widely followed approach to address that issue.

Although various versions of this paradigm have been used, the basic assumption underlying all versions is that unconscious perception can be demonstrated via a dissociation between two different indices or measures of perception. One *direct* measure (e. g., stimulus recognition) would reflect the effect of a perceived stimulus on the instructed responses to that stimulus (which is typically assessed by a measure of accuracy at the instructed task). Such a direct effect is assumed to indicate whether any relevant stimulus information is consciously perceived. A second *indirect* measure of stimulus perception would be an uninstructed effect of the task stimulus on behavior, which is assumed to be also sensitive to unconsciously perceived information. As an illustration, a well-known indirect effect is semantic priming (see also the Stroop's (1935) interference effect), which consists of a facilitation of responses (in terms of speed and/or accuracy) to a target stimulus (e.g., CAT)

* Please address correspondence to: Juan J. Ortells, Departamento de Neurociencia y Ciencias de la Salud, Facultad Humanidades y CC Educación, Universidad de Almería, 04120 Almería (Spain), jortells@ual.es

when it is preceded by a semantically related (prime) word (e.g., DOG), than when the preceding prime word is unrelated to the target (e.g., ARM). A dissociation is demonstrated whenever there is a reliable indirect effect by a stimulus that is unaccompanied by any direct effect of that stimulus. In other words, perception without awareness would be demonstrated whenever the indirect measure of perception (e.g., semantic priming) shows some sensitivity to stimuli under conditions on which the direct measure shows null sensitivity.

In spite of this relatively straightforward logic underlying the dissociation paradigm, it has proven difficult to design experiments that provide compelling, uncontroversial evidence for perception without awareness. On the one hand, there usually has been a lack of agreement about what constitutes an adequate direct measure of conscious perception. On the other hand, the means of using dissociation data to draw conclusions about unconscious perception have remained controversial. As noted by some authors (e.g., Holender, 1986; Reingold & Merikle, 1988), this is because the interpretation of a dissociation pattern mainly depends on assumptions being made about the relation of conscious and unconscious perception to performance on direct and indirect tasks. In an influential review, Holender (1986) suggested that in order to the occurrence of an indirect effect in the absence of a direct effect (i.e., an *indirect-without-direct-effect pattern*) provides compelling evidence for unconscious perception, the direct measure must be assumed to be sensitive to all relevant conscious effects of stimuli, and such a measure must exhibits null sensitivity. If this *exhaustiveness assumption* (as labelled by Reingold & Merikle, 1988) cannot be justified, then any dissociation between measures may simply indicate that the direct and indirect indices are sensitive to different aspects of consciously perceived stimuli. In addition, the direct measure should *exclusively* be influenced by conscious processes (i.e., an *exclusiveness assumption*). If a direct effect is sensitive to both consciously and unconsciously perceived information, then the attempt to establish null sensitivity for the direct measure could eliminate or underestimate any evidence for unconscious perception.

1. The Traditional Dissociation Paradigm: Subjective versus Objective Measures of Awareness

In efforts to find satisfactory exhaustive measures of conscious perception and to demonstrate that perception occurs even when these measures exhibit null sensitivity, two general classes of direct indices have been used, *subjective* and *objective*. With subjective measures, stimulus awareness is indexed by participants' self-reports of their conscious perceptual experiences. When these subjective reports show that participants do not believe any useful stimulus information was perceived, it is assumed that they did not have any relevant conscious experiences (i.e., a phenomenal lack of stimulus awareness). As an example of the general approach consider the classic experiment reported by Sidis (1898). He presented observers a series of cards containing single printed letters or digits. Participants saw the cards at a distance such that they informed all that they could see on each card was a dim, blurred spot or nothing at all. Yet, despite participants' reports indicated that they were unaware of perceiving either letter or digits, they were able to guess both the category (i.e., letter vs. digit) and the identity of the stimulus at a considerable better than chance level of performance (67% correct responses). On the basis of this dissociation between verbal self-reports and forced-choice guessing (i.e., direct vs. indirect measures of stimulus perception, respectively), Sidis concluded that he had found evidence for perception without awareness.

In general, the pattern of data obtained by these early studies indicate that relatively simple visual stimuli such as horizontal, vertical, and diagonal lines (e.g. Baker, 1937; Dunlap, 1900), circles, triangles, and squares (e.g. Miller, 1942; Williams, 1938), and letter and digits (e.g. Sidis, 1898; Stroh, Shaw & Washburn, 1908), can be perceived even when there is no awareness of perceiving. Likewise, latter studies using both behavioural and neuroimaging techniques (e.g., functional magnetic resonance imaging, positron emission tomography) have reported evidence that more complex information such as the lexical or semantic status or verbal stimuli (e.g. Cheesman & Merikle, 1986; Forster & Davis, 1984; Forster & Veres, 1998; Merikle, Joordens & Stolz, 1995), or the emotions expressed in faces (e.g. Esteves, Dimberg & Öhman, 1994; Esteves & Öhman, 1993; Whalen, Rauch, Etcoff, Mclnerney, Lee & Jenike, 1998), can be perceived even though participants' introspective reports indicated that they were unaware of critical stimuli.

If one accepts participants' verbal reports as valid indicators of awareness, then there is considerable evidence for unconscious perception. However, a number of investigators (e.g., Eriksen, 1960; Holender, 1986) have suggested that when awareness is assessed by subjective measures, there is always the concern that observers′ descriptions of their conscious experiences are influenced by many factors other than their awareness of the critical stimuli (cf. Eriksen, 1960). For example, statements or comments indicating an absence of relevant conscious experiences may simply reflect biases introduced by either the experimental setting or the participants' preconceived ideas concerning the value of particular types of conscious experiences for guiding decisions, rather a true absence of relevant conscious experiences. It this was actually the case, the studies using subjective measures may provide nothing more than additional demonstrations of the unreliability of introspective reports (cf. Merikle, 1992).

Because of these interpretative problems with introspective reports, a number of investigators have preferred to use objective behavioural measures of perceptual discrimination to distinguish between conscious and unconscious perception. Examples of objective measures are: forced-choice, presence-absence decisions and forced-choice decisions among a small set of stimulus alternatives. When an objective measure is used to index conscious awareness, a failure to discriminate between alternative stimulus states, such as the presence or absence of a stimulus (i.e., a forced-choice performance at chance), would indicate an absence of relevant, consciously perceived information.

At first sight, measures of discriminative capacities have both intuitive appeal and some methodological advantages relative to introspective self-reports. As noted by Merikle (1992), if a stimulus is presented under conditions such that observers are unable to distinguish between its presence or absence, then it seems reasonable to assume that presentation of this stimulus did not lead to conscious awareness. Additionally, unlike introspective reports, with an objective measure, perceptual sensitivity can be measured independent of the influence of preconceived biases (e.g., Eriksen, 1960; Holender, 1986).

Anthony Marcel (1980; 1983a; 1983b) was the first investigator to report that visually degraded stimuli can be perceived even when they are presented below an objectively defined threshold for awareness. For example, Marcel (1983a, Experiment 1) presented a display containing either a word or a blank field followed by a masking display, with perception of the masked words being assessed by both direct and indirect measures. The direct task consisted of a forced-choice presence vs. absence detection, such that on each trial the participants should decide whether the mask was preceded or not by a word. Participants were also required to select between two different words, what was more similar (either in semantic or in grapheme characteristics) to the masked word (the indirect task). The critical finding was that participants performed above chance in the similarity judgment (indirect) task, even at a word-mask stimulus onset asynchrony (SOA) such that it was impossible for them to discriminate better than chance when a word was present from when a word was absent. Further evidence for semantic encoding of words presented below an objective threshold for awareness was reported by Marcel using other indirect measures of word perception, such as semantic priming or the Stroop interference effect. In the variant of the Stroop (1935) color-word task used by Marcel (1983a, Experiment 3), participants named the color (red vs. green) of a central patch that was preceded by a masked prime word, which consisted of either a congruent color word, an incongruent color word or a control letter string. Previous research indicates that color-naming latencies in this task increase (i.e., interference effect) when the color patch is preceded by an incongruent color word, and decrease (i.e., facilitation effect) when it is preceded by a congruent color word (e.g., Dalrymple-Alford, 1972). The word-mask SOA was either of 400 ms, such that all participants were able to consciously perceive the word's identity, or a much shorter time. This latter SOA was for each participant 5 ms lesser than a previously established awareness threshold, which was defined by Marcel as the SOA interval at which presence/absence decisions were no better than chance. Marcel found reliable facilitation and interference effects at both SOA conditions, that is, even when participants were unable to discriminate between the presence or absence of the color words (see also Marcel, 1983b).

Whereas further studies have replicated and extended Marcel's findings (e.g. Balota, 1983; Bar & Biederman, 1998; Dagenbach, Carr & Wihelmsen, 1989; Dehaene, Naccache, Le Clec'H, Koechlin, Mueller, Dehaene-Lambertz, van de Moortele, & Le Bihan, 1998; Kemp-Wheeler & Hill, 1988; McCauley, Parmelee, Sperber & Carr, 1980), other investigators have been unable to successfully replicate the results reported y Marcel (e.g. Cheesman and Merikle, 1984). And more important, the results of studies using objective measures of awareness have been critized on the basis of several methodological concerns, which call into question whether the critical stimuli were actually presented below the participants' awareness thresholds (e.g., Cheesman & Merikle; 1984; 1985; Holender, 1986; Merikle, 1982; Nolan & Caramazza, 1982; Purcell, Stewart and Stanovich, 1983). For example, it has been argued that the participants' thresholds are usually based on such a small number of trials that it is not possible to obtain a reliable estimate of each participant's response distribution. Thus a non-discriminative responding (i.e., a 50% correct in a two-choice task) could not necessarily be attributed to a genuine inability to utilize the available perceptual information. It could, for example, be the case that participants performing a presence/absence detection on visually degraded stimuli showed a bias to say "no" on most trials, regardless of the stimulus was present or absent. On the other hand, the masking conditions used in some studies (e.g. McCauley et al., 1980) were less effective on the priming (indirect) trials than on the threshold (direct) trials, and hence, stimulus perception on the former trials could have occurred with, instead without, awareness. Furthermore, in some previous studies (e.g., Marcel, 1983a) the actual criterion for threshold performance was defined as "a performance less than 60% correct" (instead of 50%) in a two-choice task. Under these circumstances, some conscious detection of the masked stimuli could have occurred. In support of that argument, Cheesman & Merikle (1984; see also Nolan and

Caramazza, 1982; Purcell et al., 1983) used the Marcel's Stroop task methodology but introducing two modifications. First, participants' ability to identify the color words was measured using a forced-choice discrimination procedure involving four response alternatives (i.e., four possible color words). This method tried to avoid possible individual differences in the willingness to guess a response. Secondly, different word-mask SOAs were manipulated to estimate different levels of efficiency in word identification task: 25% (random performance because the existence of *four* response alternatives), 55%, 90% and 100% (in this latter case, the word was followed by no mask). Results showed reliable facilitation and interference effects at all the SOA conditions that allowed a word identification better than chance (i.e., 55%, 90%, and 100%), but not at the SOA interval inducing a non-discriminative responding (25% correct). Against Marcel's findings, this last result suggests that there is no semantic encoding for stimuli presented below an objective threshold for awareness. Interestingly, participants' self-reports indicated that they were unaware of word identity not only at the SOA condition associated with a random identification, but also at those SOA conditions producing identification scores reliably better than chance (i.e., 55% and 90% correct responses). Accordingly, Cheesman and Merikle suggest that there is compelling and consistent evidence for perception without awareness only when subjective, rather than objective, thresholds are used to define stimulus awareness.

2. Alternative Approaches to the Study of Unconscious Perception

Overall, traditional studies using objective direct measures of stimulus perception have not provided evidence, either for or against perception without awareness, that is any more definitive and convincing than has been provided by studies using subjective measures (Merikle, 1992). In fact, empirical approaches using the dissociation paradigm present a series of limitations, which make any data pattern difficult to interpret.

Firstly, it should be noted that studies of perception without awareness have traditionally used any kind of *facilitation paradigm*, whereby effects on unconscious perception of a stimulus produced the same data pattern as did conscious perception. According to Debner and Jacoby (1994), because both types of processes can contribute to performance in a similar vein (i.e., facilitating), it is difficult to determine whether the supposedly unconscious influences might partly or completely be attributed to any residual conscious perception. On the other hand, it has been argued that requiring direct measures that reflect both exclusively and exhaustively all relevant consciously perceived information (e.g. Holender, 1986) may be too stringent requirements for demonstrating unconscious influences. Reingold and Merikle (1988; 1990; see also Merikle & Reingold, 1998) point out that it appears unrealistic to expect that any particular direct measure of perception could actually be sensitive to all conscious stimulus effects. As stated by these authors, "...no matter what direct measure is selected, there is no way to demonstrate conclusively that this measure is a completely satisfactory measure of all relevant conscious experiences" (Merikle & Reingold, 1998, p. 305). Reingold and Merikle also considered implausible that unconscious stimulus effects should be excluded from contributing to performance on direct measures. It could rather be the case that any direct (like indirect) effect might include both conscious and unconscious contributions (see also Jacoby, Lindsay & Toth, 1992).

Finally, irrespective of either subjective or objective direct measures of conscious perception are used, traditional studies aimed to find an indirect effect in the complete absence of a direct effect (i.e., an *indirect-without-direct-effect* data pattern) would require accepting a null hypothesis (i.e., the hypothesis of no direct effect), and this is statistically problematic. As noted by Merikle and Joordens (1997a), it is impossible to demonstrate in a completely convincing fashion that null sensitivity necessarily indicates no relevant information was consciously perceived. There is the possibility that some amount of stimulus information was detected on at least some occasions that eluded detection by conventional direct tasks. Thus, claiming a null result on a direct measure is always susceptible to the familiar criticisms of inappropriately asserting the truth of a null hypothesis (e.g., the methodology may have been inadequate to reveal the true sensitivity of the measure).

Because of these limitations, an increasingly number of researchers abandoned the research strategy of establishing dissociations to demonstrate unconscious perception, and moved the methodology forward. On the assumption that it may just be impossible to design experiments that satisfy the logical requirements of the dissociation paradigm, a series of alternative approaches have been developed. These approaches provide different methods for studying perception without awareness that do not require a convincing demonstration of a null effect.

2.1. The Relative Sensitivity Approach

Reingold and Merikle (1988; 1990; Merikle & Reingold, 1991) proposed an alternative strategy in which, rather than searching an indirect-without-direct effect pattern, they advocate the search of an indirect effect that is statistically stronger (i.e., *more sensitive*) than a direct effect, when the two effects are comparable. This relative sensitivity approach is based on a relaxation of Holender's assumptions about direct measures. Firstly, Reingold and Merikle propose to replace the Holender's exclusiveness assumption with an "*inclusiveness*" assumption, according to which any direct (or indirect) measure of perception is potentially sensitive to both conscious and unconscious influences. In order to interpret any pattern of data from direct and indirect measures in terms of unconscious cognition, Reingold and Merikle introduce an additional *minimal* assumption, which replaced the exhaustiveness assumption. According to such a *relative sensitivity* assumption, the sensitivity of a direct measure to task-relevant information should be greater or at least equal than the sensitivity of the indirect measure to the same information. The rationale underlying this assumption is that consciously perceived information relevant to a particular discrimination (e.g., an old-new judgment), if exists, it should be used equally or more efficiently when participants are instructed explicitly to make the discrimination, than when they are instructed to judge the stimuli on some other basis (i.e., an indirect effect). Based on this relative sensitivity assumption, an unconscious influence is demonstrated whenever an indirect effect is statistically stronger than a comparable direct effect from the same stimulus information, irrespective of the direct measure exhibits a true null sensitivity or it does not. This is the case because the relative sensitivity assumption rules out the possibility that superior performance on the indirect task is attributable to conscious, task-relevant information.

It is important to note that the interpretation of an indirect-greater-than-direct pattern as evidence for unconscious cognition is only warranted if both direct and indirect measures are truly comparable. That is, both measures must be described on the same measurement scales and obtained in response to the same task stimuli. Otherwise, the greater sensitivity of an indirect measure might simply reflect a methodological artefact (e.g., differences in measurement scales, or in response bias across the two tasks), rather than an unconscious influence. As an example of this approach consider the now-classic study by Kunst-Wilson and Zajonc (1980; see also Mandler, Nakamura & Van Zandt, 1987). In a first phase of the experiment, they presented 10 meaningless, irregular geometric shapes, with each shape being presented five times for a 1-ms duration each time, such that no participants reported seeing any of the shapes. Following this initial presentation, the participants were shown 10 pairs of shapes, with each pair consisting of an "old" shape (which had been presented during the previous phase) and a "new" (not previously presented) shape. On each pair of shapes, participants were required to make two forced-choice tasks, namely, a recognition task and a preference task. For the recognition (direct) task, participants had to select the member of the pair that had been presented previously. For the preference (indirect) task, participants were told to choose the shape of each pair that they preferred. Whereas participants' performance in the recognition task was not better than chance (i.e., 50% correct), they performed significantly above chance in the preference task (i.e., 60%), thus suggesting that this latter (indirect) task was a more sensitive measure of the old-new discrimination that the recognition task. What makes the Kunst-Wilson and Zajonc's data a so compelling evidence for unconscious influence is that their direct (recognition) and indirect (preference) measures of memory were matched on all characteristics except the task's instructions. Thus, both measures were based on the same study and test conditions, the same pairs of stimuli and the same response metric. Another good example of the relative sensitivity approach is the study by Merikle and Reingold (1991), which examined the effects that allocation of visual attention may have on the relative sensitivity of direct and indirect measures of memory for words. Their experiments consisted on two phases. During a first study phase, participants were presented with pairs of words and were required to name one cued word in each pair. During the test phase, new words and either the previously cued words (Experiment 1) or the previously uncued words (Experiments 2a and 2b) were presented against a background mask. Participants were required to judge whether a word was old or new (a direct recognition task), or whether the contrast between a word and the background was high or low (an indirect task). Merikle and Reingold found that for the attended (cued) words, the direct task was much more sensitive than the indirect task. Yet, for the uncued (ignored) words the indirect task was initially more sensitive than the direct task, thus suggesting unconscious processing underlying memory for the ignored words.

An important advantage of using comparable direct and indirect measures to study unconscious processes, is that the methodological requirements are less stringent and *a priori* assumptions more minimal than those required by the traditional dissociation paradigm. As pointed by Greenwald, Klinger and Schuh (1995), by avoiding any need to justify an exhaustiveness assumption and to demonstrate a null direct effect, the relative sensitivity approach appeared to have it easier to develop more convincing evidence for unconscious cognition. However, it should be noted that the indirect-greater-than direct pattern has demonstrated to be replicable only in research using memory-based measures, in which both direct and indirect effects are measured at a substantial delay after stimulus presentation. But this is not the case in research aimed to study perception without awareness, mainly because it has proven difficult to find direct and indirect measures of perception that can be truly comparable (cf. Greenwald et al., 1995).

2.2. The Regression Analysis Method

Greenwald and colleagues (Draine & Greenwald, 1998; Greenwald & Draine, 1997; Greenwald, Draine & Abrams, 1996; Greenwald et al., 1995) have recently developed a new methodology for studying perception without awareness, which is based on a mixture of ideas taken from both the relative sensitivity approach and the classic dissociation paradigm. As already noted, one usual criticism to the traditional dissociation paradigm is that any convincing demonstration of an indirect effect in the absence of a direct effect requires the acceptance of a null hypothesis (i.e., a null result on the direct measure has been achieved). To bypass the statistical problems associated with asserting the truth of a null hypothesis, Greenwald et al. use a regression analysis strategy. The major innovation of this method is to analyse data using tests of the regression relation between direct and indirect measures of responses to near-threshold stimuli. In the regression analysis, the amount of semantic priming (an indirect effect) produced by near-threshold stimuli is plotted against a direct objective measure of the prime stimuli. In the priming task, participants perform a binary categorization judgment either for affectively polarized words (i.e., pleasant vs. unpleasant) or for common first names (to be categorized as male or female). The target word is preceded (by a variable time interval, depending on the experiment) by a briefly presented and masked prime word, which could be either related (i.e., belonging to the same category) or unrelated to the target. To increase the sensitivity to priming, Greenwald and colleagues (e.g., Draine and Greenwald, 1998; Greenwald et al., 1996) have recently used a "response-window procedure" in which participants are always required to respond to the target item within a moderately early response window (i.e., 250 ms ± 133 ms). This forced-responding method substantially controls response time and moves the priming effect into response accuracy (e.g., d' units of discriminability). Direct (objective) measures of prime perceptibility are obtained in separate (later) blocks of trials, in which participants make various discriminations of content for the masked primes, for example, deciding whether the prime stimulus was a word or a XGXG string. As occurred with the priming measures, the index of direct processing is also measured in terms of d' units (i.e., accuracy), such that both indirect and direct measures have rational zero points. Participants' performance on both tasks is then analysed as regression functions that relate priming scores to direct measures of perceptibility of the primes (where zero values on indirect and direct tasks indicate absence of priming and perceptibility, respectively). The height at which the regression function crosses the vertical axis (the *regression intercept*) provides a critical test for the hypothesis that priming has occurred unconsciously (Greenwald et al., 1996). This is because the intercept of the regression equation is an estimation of the level of performance on the indirect measure (i.e., the magnitude of semantic priming) associated with a zero value on the direct measure (i.e., d' = 0). Under these circumstances, the finding of a significant (i.e., a nonzero effect) regression intercept effect would indicate an unconscious contribution to the indirect measure. Note that with this regression method, evidence for unconscious perception (i.e., the indirect-without-direct effect pattern) occurs in the form of a null hypothesis rejection (i.e., an intercept effect significantly greater than zero), rather than the null hypothesis acceptance that characterized the traditional dissociation paradigm (Greenwald et al., 1995).

The approach developed by Greenwald and colleagues represents one of the most rigorous applications of the dissociation paradigm for studying perception without awareness, in that it reverses the usual difficulty associated with asserting the truth of a null hypothesis. However, their regression methodology relies on a series of assumptions, which must be satisfied to interpret a regression data pattern as providing compelling evidence for unconscious perception. For example, given that the exact value of the intercept parameter is a key issue in the Greenwald et al.'s application of the regression method, it is necessary to consider whether there is measurement variability (i.e., participant-to-participant response variation) in the direct (predictor) measure. Thus, the presence of a substantial variability in such a measure might lead to inflated intercepts even in the absence of any unconscious effect (Dosher, 1998). As already noted, in experiments by Greenwald et al. the predictor score is usually an objective direct measure of perceptual sensitivity (d') based on two-alternative forced-choice tasks, where a d' score = 0 is assumed to reflect stimulus unawareness. Yet, in some of their experiments (e.g., Draine & Greenwald, 1998) there is a noticeable d' variability for the direct task, including d' values below zero. As noted by Dosher (1998), such a values theoretically should arise because of measurement error, which could produce contaminated intercept values. It still remains unclear whether a d' score above zero in forced-choice tasks necessarily reflects a conscious perception of critical stimuli. In fact, neuropsychological research has consistently showed that brain-damaged (e.g. cortically blind) patients can perform significantly better than chance (e.g. d' > 0) when they are encouraged to make forced-choice discriminations on several attributes of stimuli (e.g., spatial location, movement, orientation, shape, or meaning) about which they report to be completely unaware (see Farah, 2001, for a recent review on neuropsychological research about consciousness). On the other hand, in a number of experiments examining the influence of different attentional instructions on semantic processing of parafoveally presented words (e.g., Ortells & Tudela, 1996; see also Ortells, Abad, Noguera & Lupiáñez, 2001), we found that the to-be-attended words produced facilitatory priming. Conversely, the words that participants were instructed to ignore produced an opposite pattern of negative priming, thus suggesting that these latter stimuli were truly ignored. But whereas participants' self-reports indicated no awareness about the ignored words' identity (unlike the attended words), their performance in a further recognition test about the unattended (and the attended) words yielded d' scores significantly greater than zero for those words.

Another implied assumption of the regression method by Greenwald et al. is that, in order that an *indirect-without-direct-effect pattern* provides evidence for unconscious processing, is also necessary that direct and indirect measures were comparable. That is, both measures should reflect some similar processing of the stimuli, and comparability requires at least a modest correlation between the direct and indirect measures. As argued by Dosher (1998; see also Merikle & Reingold, 1998), in the absence of a correlation between measures, the regression intercept may simply reveal the *overall mean* sensitivity of the indirect measure (with a number of the indirect data likely having associated direct scores above-zero). In order to justify the comparability of direct and indirect tasks, a number of criteria must be satisfied, such as the fact that both measures index the same perceptual discrimination, are based on the same response metric, and the experimental conditions for administering both tasks are completely matched. But this does not seem to be the case in some of Greenwald et al.'s experiments. For example, the semantic categorization (e.g., pleasant vs. unpleasant) indirect task used by Draine and Greenwald (1998) appears a quite different task that discriminating a word from a XGs string, with this latter task likely inducing no-semantic processing strategies. In addition, direct and indirect measures were assessed under considerably different conditions. Whereas the direct task did not impose time pressure on the participants' response (which could result in a substantial amount of variability across participants), the response-window procedure used in the indirect task required all the participants to respond within an experimenter-defined temporal interval. All of these differences in task quality, time testing and time pressure would call into question the comparability and appropriateness of the direct task, and contribute to the lack of a correlation between measures. To sum up, if the validity of the assumptions underlying an appropriate application of the regression method is not carefully considered (with each of the potential artefacts being rule out), then the regression strategy may be not any more informative than other previous approaches based on the dissociation paradigm (Dosher, 1998; Merikle & Reingold, 1998).

2.3. Qualitative Differences between Conscious and Unconscious Perception

It should be noted that the absence of a strong relationship between the direct and indirect measures, might reflect the fact that conscious and unconscious cognitive processes in certain domains can be independent. Based on that assumption and rather than to simply ask whether a stimulus can be unconsciously perceived, many investigators have addressed the question of how does unconscious perception differ from conscious perception (e.g. Dixon, 1981; Merikle & Daneman, 1998; Merikle, Smilek & Eastwood, 2001). Basically, the experimental logic is to contrast perception with and without awareness in order to determine whether a stimulus can have qualitatively different consequences on cognitive and affective reactions depending on whether it was either consciously or unconsciously perceived. Throughout the present section we describe a series of different approaches, which have consistently demonstrated that perception with and without awareness does lead to qualitatively different behavioural effects.[1]

On the one hand, a series of studies have showed that a word is coded differently depending on whether it is either consciously or unconsciously perceived. For example, Groeger (1984) presented visually (for a variable exposure time, depending on the group of participants) single target words and required participants to identify and select that stimulus from a subsequently presented matrix of 24 different words. The target word was never presented in the matrix, although this contained words that were either structurally (i.e., orthographically) or semantically similar to the target. When the target words were briefly presented such that participants experienced no awareness of them, they tended to select the semantically related words. But when the target words were presented for longer durations, thus allowing that they were perceived with a low level of awareness, participants tended to select the visually similar words. Groeger (1988) obtained a similar result pattern when words were presented auditorily. Namely, participants selected from the matrix either

[1] It is interesting to note that in most of approaches examining qualitative differences between conscious and unconscious perceptual processes, subjective measures of awareness are assumed to provide an adequate indication of the presence (or absence) of relevant conscious experiences.

phonologically similar or semantically related foils, depending on whether the target words had been perceived with or without awareness, respectively.

Other interesting series of studies have demonstrated that the influence of context in perception is limited to information that is consciously perceived. In contrast, unconscious perception would induce more automatic reactions, which would not be constrained by contextual information. To illustrate, Marcel (1980) presented sequences of three words (e.g. *hand – palm – wrist*) as follows: (a) A clearly visible context word (e.g., hand) aimed to bias interpretation of (b) an upcoming polysemus word with two different meanings (e.g., palm), and (c) a clearly visible target word related to one of the two possible meanings of the polysemous word (e.g., wrist). Unlike both the context and the target words, the polysemous word was presented under conditions so that they were either consciously or unconsciously perceived. In the former condition (conscious perception), time to respond to the target word was faster if it was preceded by a meaning-related context word (i.e., a semantic priming effect). Thus reaction times to the target "wrist" were faster for the sequence *hand-palm-wrist*, relative to the sequence *tree-palm-wrist*, as the target and the context were related to the same meaning of the polysemus word in the former, but not in the latter condition. However, when the polysemous word was not consciously perceived, time to respond to the target word was facilitated if it was preceded by a polysemous word with a related meaning (e.g., wrist preceded by palm, relative to wrist preceded by river), regardless of the meaning of the preceding biasing context word (e.g., palm vs. tree). As hypothesized by Marcel, it appears then that selective context-driven perception requires conscious awareness, as the context word biased interpretation of the polysemus word only when it was consciously perceived. But in the absence of conscious awareness the multiple meanings of a stimulus would automatically be activated.

On the other hand, there are several findings suggesting that affective reactions are more likely to be influenced by unconsciously perceived than by consciously perceived stimuli (e.g., Kunst-Wilson & Zajonc, 1980; Murphy & Zajonc, 1993). For example, Murphy and Zajonc (1993) required participants to indicate on a five-point scale whether each of a series of clearly visible Chinese ideographs represented a "good" or a "bad" concept. Each ideograph was preceded by a picture of a human face expressing either happiness (e.g., a smile) or anger (e.g., a scowl). The presentation duration of the faces was manipulated between-participants at two levels: 4ms and 1000 ms, such that participants reported either no awareness (4 ms) or a complete awareness (1000 ms) of the faces. Participants in the 1000-ms group were instructed to completely ignore the faces and to concentrate exclusively on rating the ideographs. Murphy and Zajonc observed that participants' ratings of the ideographs were reliably influenced by the emotion expressed by the preceding faces (i.e., an ideograph was more likely rated as representing a "good" concept if it was preceded by a smiling face). But such an influence only occurred when the faces were presented for 4 ms, such that participants reported no awareness of them, thus suggesting an important qualitative difference between conscious and unconscious perception.

In a series of recent experiments, Fox and colleagues have also used emotional stimuli to investigate possible attentional biases in subclinical anxiety (e.g., Correa, Fox, Carmona, Noguera, Lupiáñez & Tudela, 2002; Fox, 1996; Fox, Russo, Bowles & Dutton, 2001). To illustrate, Correa et al. (2002) have reported that high anxious people can show a faster shifting of spatial attention to threatening stimuli (e.g., angry faces), as these stimuli had advantage over positive or neutral stimuli in attracting attention to their own location.

Interestingly, such attention orienting biases occurred only when the faces were masked and participants were unaware of them. When the faces were consciously perceived, there was a delayed disengagement of attention in anxious people (see also Fox et al., 2001), but no evidence for enhanced cueing was found.

Several particularly interesting approaches to the study of unconscious perception have developed measures of perception that are influenced in opposite ways by consciously and unconsciously perceived information. The logic underlying to these studies is based on the assumption that conscious perception enables individuals to act intentionally on the world (e.g., they are able to follow instructions) and to produce effects on it. In contrast, information perceived without awareness leads to much more automatic reactions that cannot be controlled by individuals (e.g. Marcel, 1988; Searle, 1992). A good example is the stem completion task administered under *exclusion* instructions. The distinguishing characteristic of that measure, which was originally developed by Jacoby (1991; see also Debner & Jacoby, 1994; Jacoby, Toth & Yonelinas, 1993; Merikle & Joordens, 1997a; Merikle et al., 1995; Smith & Merikle, 1999), is that participants are instructed not to use particular responses when completing word stems. Consider, for example, the study by Debner and Jacoby (1994), in which, the likelihood that words were perceived with or without awareness was controlled by varying the stimulus duration. On each trial, a single word (e.g., *table*) was presented for either a short (i.e., 50 ms) or a somewhat longer (i.e., 150 ms) duration (depending on the experiment), and followed by a mask. Immediately following the mask offset, a three-letter stem consisting of the first three letters from the preceding word (e.g. *tab_ _*) was presented, and the participants were required to complete the word stem with any word that came to mind *except* the word that had just been presented (e.g. they could complete the word stem with *taboo*, but not *table*). Debner and Jacoby found that participants were able to follow these "*exclusion*" instructions when the words were presented for the longer 150-ms duration. Namely, target words were used to complete the word stems much less often than occurred in a baseline condition. Yet, when the words were presented for 50-ms, participants had difficulty following the exclusion instructions, as they used these words to complete the stems much more often than the baseline level, despite the explicit instructions not to use those words. These findings showed that the target words were always perceived, irrespective of they were presented for either 150 or 50 ms, but they influenced the participants' completions in opposite ways (relative to the baseline level), depending on its presentation duration. If, as already noted, an assumption is made that a stimulus must be consciously perceived in order that it can guide our actions in the world, then the results reported by Debner and Jacoby suggest that participants were generally aware of the words presented for 150 ms (i.e., a lower exclusion performance relative to the baseline level in the stem completion task), and generally unaware of those presented for the shorter 50-ms duration (i.e., higher than baseline exclusion performance). Debner and Jacoby (1994, Experiment 3) found a similar pattern of qualitative differences in the consequences of perception, when stimulus awareness was controlled instructing participants on how to focus or distribute their attention (see also Smith & Merikle, 1999). An important advantage of the exclusion procedure by Debner and Jacoby (1994; see also Merikle & Joordens, 1997a), as compared to traditional approaches using "facilitation" paradigms, is that conscious and unconscious influences are placed "in opposition". Under these circumstances, evidence of unconscious perception cannot be explained as truly resulting from conscious perception, as the latter would produce an opposite result pattern.

Converging evidence that unconscious perception leads to automatic reactions whereas conscious perception allows individuals to modify their reactions, comes from an elegant series of Stroop-priming studies by Merikle and colleagues, which showed that individuals can use predictive strategies based on stimulus redundancy only when the predictive stimuli are consciously perceived (e.g. Cheesman & Merikle, 1986; Merikle & Cheesman, 1987; Merikle and Joordens, 1997a; 1997b; Merikle et al., 1995). Their experiments are based on a variant of the Stroop (1935) color-word interference task, in which two color words (RED or GREEN) are used to prime responses to a probe target consisting of a color patch (or a series of ampersands) that was also *red* or *green*. The standard result found with this task is that responses to the probe target (e.g. *red*) are slower when it follows an incongruent color word (e.g. GREEN) that when it follows a congruent color word (e.g. RED). This interference effect presumably occurs because participants cannot avoid reading the word even if they are not required to read it, and reading a word that refers to a conflicting color concept interferes with responding to the color target. However, the typical Stroop interference is *reversed* whenever incongruent prime-target pairings (GREEN – *red*) occurred more frequently than congruent (RED- *red*) prime-target pairings. Namely, the reaction times are faster on incongruent than on congruent trials. One explanation of that reversed Stroop effect is that participants capitalize on the predictive information provided by the prime words (e.g., Merikle et al., 1995). Given that there are only two possible colors, participants learn to expect that the target color on each trial will be the color not named by the preceding prime word. Such a predictive strategy would facilitate performance on the incongruent trials and slow performance on the congruent trials, thus leading to a reversal of the typical Stroop interference. Interestingly, Merikle and colleagues have consistently reported that the predictive strategy in the two-color variant of the Stroop task is adopted by participants only when the prime words are consciously perceived. In contrast, when the words are not consciously perceived, the standard interference effect is rather found. For example, Merikle and Joordens (1997b, Experiment 1A) presented a single central prime word (RED or GREEN in grey color) for 33 ms, which was followed either immediately or after a delay of 134 ms (depending on the group of participants) by a pattern-mask (seven ampersands in grey color), thus resulting in a prime-mask stimulus onset asynchrony –SOA- of either 33 ms[2] or 167 ms, respectively. The masked prime word was always followed by a central target, which consisted of seven ampersands displayed –until response- in either red or green, with participants being required to make a button-press response to indicate the target color as quickly as possible. For both groups of participants, the prime-target SOA remained fixed at 300 ms. On 75% of the trials, the prime and the target were incongruent (RED-green), whereas on the remaining 25% of the trials, the prime and the target were congruent (RED-red). Merikle and Joordens found a significant crossover interaction between masking condition and prime-target congruency: With a delayed mask (i.e., a prime-mask SOA of 167 ms) the reaction times (RTs) were *faster* on incongruent than on congruent trials. Such a

[2] In a number of previous studies, Merikle and colleagues (e.g., Cheesman & Merikle, 1985; 1986; Merikle et al., 1995) have consistently found that a prime-mask SOA of 33 ms is below most participants' threshold for subjective awareness. This result has been confirmed in a series of recent experiments in our labs. Thus, when we required participants to make forced-choice binary tasks (e.g. an animate/inanimate judgment) on words presented for 33 ms and immediately masked, their performance was reliably better than chance (i.e., d' > 0). Yet, they reported a complete phenomenal lack of stimulus awareness, thus suggesting that words presented under immediate masking conditions were *below* a subjective threshold, but *under* an objective threshold for stimulus awareness.

reversed (i.e., facilitatory) Stroop effect suggests that participants consciously identified the prime word followed by a delayed mask, and then used the predictive information provided by the word to anticipate the target color (see also Logan, Zbrodoff & Williamson, 1984). On the contrary, with an immediate mask (i.e., a prime-mask SOA of 33 ms), a typical Stroop interference effect emerged, thus suggesting that participants experienced no awareness of the primes when they were presented for 33 ms and immediately followed by a mask. This crossover in performance (i.e., reversed Stroop vs. interference effect) provides an additional qualitative difference that distinguishes conscious from unconscious perception.

In a series of recent experiments (e.g., Daza & Ortells, 2000a; Daza, Ortells, & Fox, in press) using the Merikle and Joordens' task, we have replicated and extended the findings by these authors. We have consistently shown that behavioural consequences of perceiving a stimulus with or without phenomenological awareness are qualitatively different not only by the sign (i.e. interference vs. facilitation) of the corresponding Stroop priming effect, but also by the time course of each priming effect (see Figure 1).

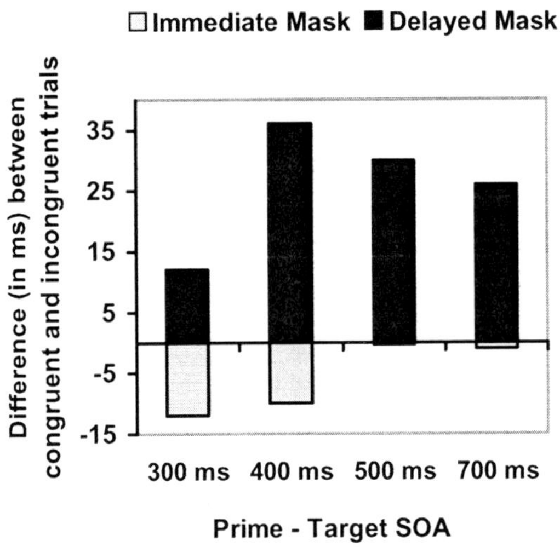

Figure 1. Stroop-priming effects from color words presented under immediate vs. delayed mask conditions as a function of the prime-target SOA (Daza, Ortells & Fox, in press).

When the prime word was immediately masked, such that participants reported a complete absence of conscious awareness of word's identity, a reliable Stroop interference was consistently found at a prime-target SOA of 300 ms, thus replicating the Merikle and Joordens' findings. Additionally, whereas such an interference effect was still significant at a prime-target SOA of 400 ms, it declined to zero at longer SOA intervals of 500 ms and 700 ms (Daza et al., in press, Experiments 2 and 3), thus suggesting a relatively fast decay over time for that supposedly "automatic" Stroop-priming effect. Regarding the prime words that were followed by a delayed mask, an opposite facilitation (i.e., reversed Stroop) effect was found, although this effect reached significance at prime-target SOAs of 400 ms or longer. At a SOA interval of 300 ms, we consistently failed to replicate the reversed Stroop reported by

Merikle and Joordens, regardless of whether our participants were either more or less actively encouraged for relying on the predictive information provided by the word to optimize their performance.[3] The fact that participants' self-reports in all our experiments indicated that they were aware of the word's identity on most (or many) of trials under the delayed masking condition, suggests that consciously perceiving a stimulus constitutes a necessary, though not sufficient condition, in order to that stimulus guides intentional actions. To the extent that a reversed Stroop mainly reflects the contribution of strategic processes, and on the basis of the literature suggesting that controlled (strategic) processes build up much more slowly (and are often more sustained) than automatic processes (e.g. De Groot, 1984; Neely, 1977; Ortells et al., 2001; Posner & Snyder, 1975a; 1975b; Shenaut & Ober, 1996; Shiffrin & Schneider, 1977), the Daza et al.' findings (see Figure 1) suggest that a prime-target SOA of 300 ms would not give most observers sufficient time to use the prime word in a strategic manner to anticipate the target color.[4]

How to explain, then, the fact that Merikle and Joordens (1997b) we able to observe a significant reversed Stroop with a delayed mask using a short SOA of 300 ms? The reasons for the discrepancies between their results and those by Daza et al. (in press) remain unclear. But an inspection of Figure 1 in their paper (p. 224) suggests that the overall reaction times in their study were faster (i.e., lower than 400 ms) than those reported by Daza et al. (i.e., greater than 400 ms). Whereas all participants in our experiments had no prior experience with this kind of task, it could be the case that some (or many) of the subjects in Merikle and Joordens' study had previously participated in other similar studies. If this were actually the case[5], and on the basis of prior evidence showing that the time necessary to implement strategic processes can be shortened by practice (e.g., Logan & Zbrodoff, 1982; see also Logan, 1985), it remains possible that in Merikle and Joordens' study, a seemingly "short" SOA of 300 ms could still represent enough time for strategy development for a great proportion of their participants.

In either case, the Daza et al.' findings of differential SOA functions for the "automatic" versus "strategic" components of the two-color word Stroop effects, provide further strength for the claim that these two Stroop effects observed under the immediate vs. delayed mask conditions, would indeed reflect different underlying processes.

In a number of additional experiments in our labs using other priming procedures, we have found a fairly similar differential time-course of effects stemming from conscious vs. unconsciously perceived stimuli (e.g., Daza & Ortells, 2000b; Daza, Ortells & Noguera, 2001). The basic procedure used in these experiments was as follows: On each trial an uppercase prime word (e.g. CAT) was displayed at the centre of the screen for 33 ms, and was followed either immediately or after a variable delay of either 34, 134, 234 or 334 ms (depending on the experiment) by a pattern mask. The mask offset was followed by a central

[3] The differential time-course showed by both types of Stroop-priming effects in Daza et al.'s experiments could not be attributed to strategy shifts across prime-target SOA conditions, as an identical result pattern emerged irrespective of the prime-target SOA was manipulated either between-participants or in a within-participant (random) design.

[4] Note, in fact, that less than one half of participants in each of the Daza et al.'s experiments did show a reversed (facilitatory) Stroop from consciously perceived stimuli at the shortest prime-target SOA (300 ms).

[5] We have recently known by both Phil Merikle (personal communication, July 16th, 2001) and Steve Joordens (personal communication, October 25th, 2001) that some of their participants were given considerable training to use the prime word to anticipate the target color. Furthermore, some of the participants in their study could

(lowercase) target word (e.g., hand) on which participants made a semantic categorization (animals vs. body parts) task (see Figure 2). On the 80% of the trials (unrelated condition) the prime and target words belonged to different semantic categories (e.g., CAT – mouth; NOSE – dog), whereas on the remaining 20% of the trials (related condition) the prime-target pairs were highly -associated words of the same semantic category (e.g. CAT – dog; NOSE – mouth).

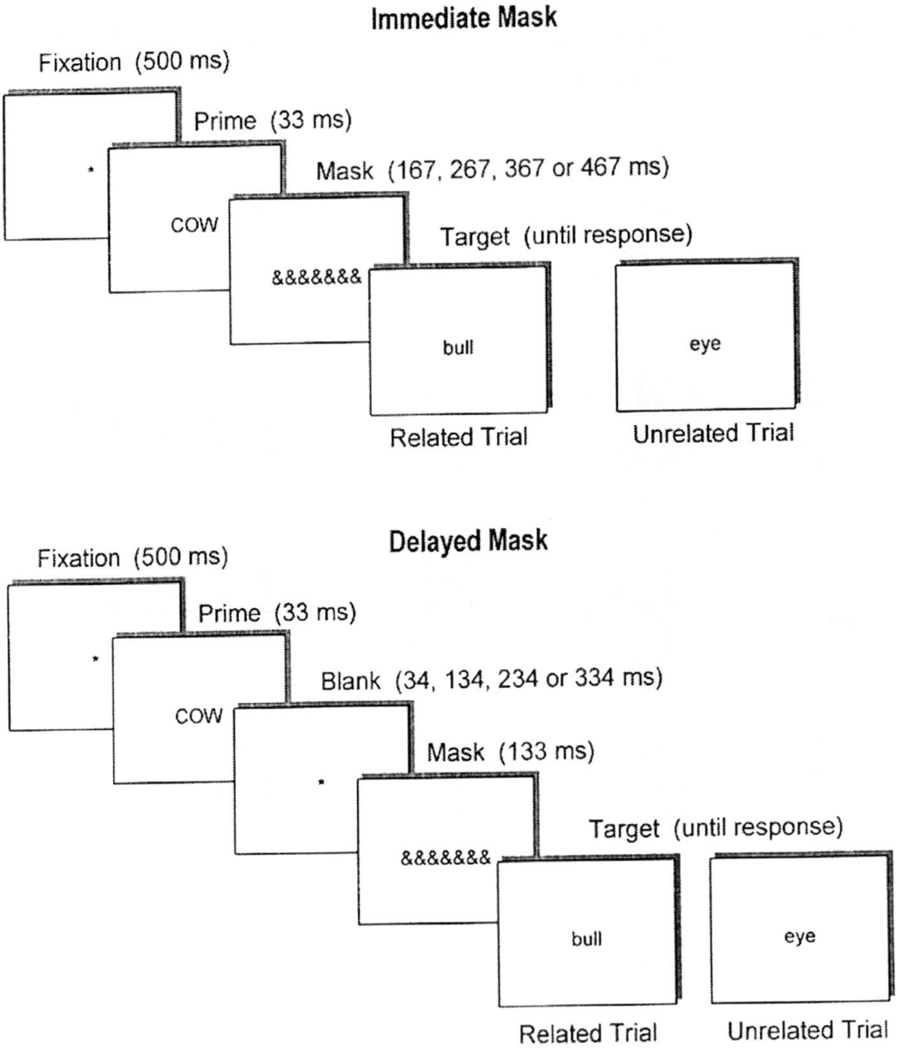

Figure 2. Sequence of events for trials in the priming procedure used by Daza & Ortells (2000b; see also Daza, Ortells & Noguera, 2001).

As in our Stroop-priming studies, irrespective of whether participants were either aware (delayed mask) or unaware (immediate mask) of the prime word, they were encouraged to use the predictive information provided by the prime word to optimise their categorization

have been in other "perception without awareness" experiments, experiments that gave them experience with briefly presented and masked items.

performance. So, given a particular prime word, they should expect that the upcoming target would belong to a different semantic category, as the unrelated trials were much more frequent than the related trials (see also Neely, 1977). The prime-target SOA was again manipulated at four levels: 200, 300, 400 and 500 ms.

The overall pattern of results mirrored to that found with the Stroop-priming task. Namely, when participants experienced no awareness of the prime word's identity (immediate mask), we obtained facilitatory semantic priming (i.e., faster reaction times on related than on unrelated trials). Such a supposedly "automatic" priming effect was significant at prime-target SOAs of 200, 300 and 400 ms, but it completely disappeared at the longer SOA of 500 ms (see Figure 3).

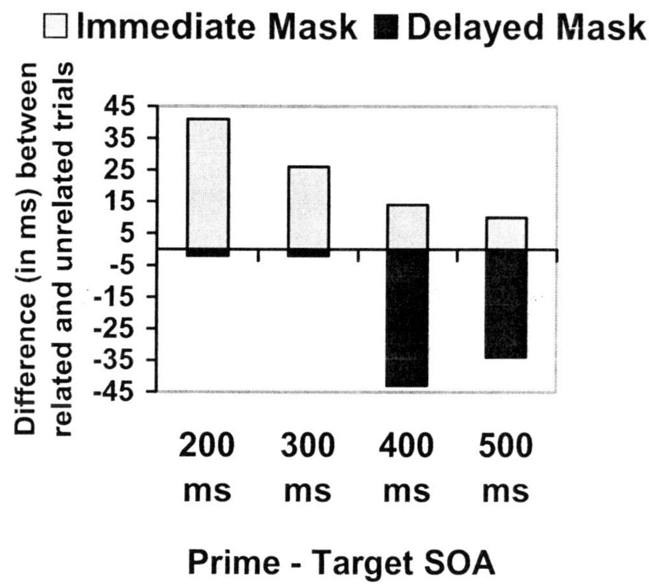

Figure 3. Semantic priming effects from masked words as a function of the masking (immediate vs. delayed) condition and the prime-target SOA (Daza & Ortells, 2000b).

On the other hand, when the prime word was followed by a delayed mask, such that participants were able to consciously perceive the word, a reversed semantic priming (i.e., interference) was rather found: reaction times were slower for related than for unrelated trials. Yet, as depicted in Figure 3, this strategic reversed priming was only found at SOAs of 400 ms or longer, with no priming effect being observed at the shorter SOAs of 200 and 300 ms. This latter result is clearly consistent with prior evidence from priming studies suggesting that obtaining a reliable strategy-dependent effect would usually require the use of time intervals of 400 ms or longer (e.g., De Groot, 1984; Di Pace, Longoni & Zoccolotti. 1991; Fuentes & Tudela, 1992; Neely, 1977; Ortells et al, 2001; Posner & Snyder, 1975a; 1975b; Shenaut & Ober, 1996).

One could argue that the lack of reversed priming effects in the delayed masking condition at the shorter SOAs (see also Daza et al., in press) it does not necessarily reflect the absence of strategic influences at that time intervals. Rather, it could be the case that the strategic processes contribute also immediately, but such an influence was overridden by an equally large automatic influence. As the automatic influence subsides (i.e., as the SOA interval is lengthened), the strategic influences would then predominate. To examine whether the reversed priming effects in our task mainly reflects the contribution of strategic (controlled) processes, or rather a combination of both, strategic and automatic influences, we have conducted an experiment in which the level of practice with the priming task (using a fixed prime-target SOA of 400 ms) was manipulated (Daza et al, 2001). If this last factor could influence to a greater extent the implementation of strategic (controlled) processes than that of automatic processes, and to the extent that the reversed priming from consciously perceived words mainly reflects the contribution of strategic influences, then we expected that such a priming effect was significant only when participants reached some level of practice with the task. In contrast, an unconsciously perceived prime word (immediate masking) should produce reliable automatic (facilitatory) priming even when participants have not enough practice with the task, with such a priming effect remaining similar across trial blocks. In other words, we expected that priming effects interacted with trial block with a delayed, but not with an immediate mask. The results were clearly consistent with our predictions (see Figure 4). Thus, an unconsciously perceived prime produced facilitatory priming, which reached significance since the first trial block. Conversely, with a delayed mask, the reversed priming effect started to be significant since the second block of trials, with no reliable priming being observed at the first trial block.

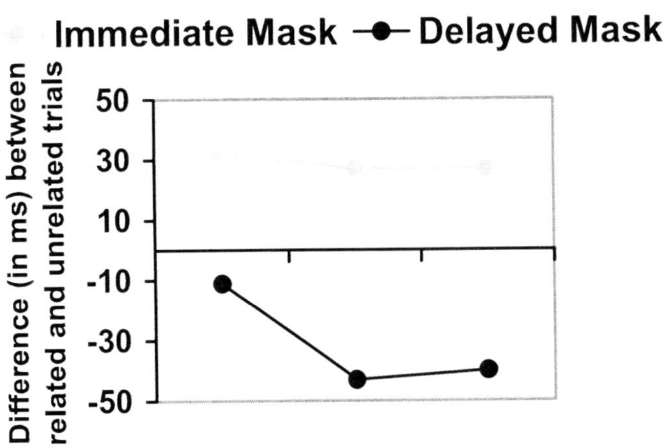

Figure 4. Semantic priming effects from words presented under immediate vs. delayed mask conditions as a function of the trial block (Daza, Ortells, & Noguera, 2001).

3. Conclusions

The question of whether perception can occur without awareness has been the focus of much research and controversy for many years. Traditional empirical approaches have generally aimed to find dissociations between performance on direct and indirect tests of perception. With the dissociation paradigm, perception without awareness is demonstrated whenever the indirect test shows sensitive to stimulus information on which the direct test shows null sensitivity. A weakness of studies using this dissociation paradigm, however, is that the conclusiveness of findings presumed to provide evidence for unconscious perception, depends entirely upon the adequacy of the measures used to define consciousness. Regardless of whether introspective (subjective) or behavioural (objective) measures of awareness are used, it is always possible to question whether that measure is actually successful (i.e., exhaustive) in guaranteeing a complete absence of *all* relevant conscious experiences. This would explain why traditional studies have not led to completely convincing evidence for unconscious perception, with the reported findings being open to alternative interpretations.

However, over the last two decades a number of alternative research strategies have been developed, which provide new methodologies for studying perception without awareness that do not require either the use of exhaustive measures of awareness or a convincing demonstration of a null direct effect. One particularly useful way of approaching to the study of unconscious perception is to focus on its distinction from conscious perception. That is, how do conscious and unconscious perceptual processes differ? As pointed out by Merikle and Daneman (1998; see also Merikle, 1992), the distinction between conscious and unconscious cognition is much more significant if both kinds of processes lead to qualitatively different consequences than if unconscious perception is simply a quantitatively weaker form of conscious perception.

Throughout the present review we have summarized a series of interesting qualitative differences in the behavioural consequences (e.g., cognitive and affective reactions) of perceiving a stimulus with or without awareness. The demonstration of qualitative differences in performance across aware and unaware conditions has considerable potential for at least two reasons. First, it constrains alternative interpretation of data (i.e., validating measures of awareness), thus providing more compelling evidence for the existence of unconscious perception than dissociations reported by traditional approaches. Secondly, the suggestion that conscious and unconscious perception can involve qualitatively different operations raises interesting questions about the nature of unconscious processing, and points to new insights and understandings regarding human cognition.

Acknowledgments

The writing of this chapter was supported by a Grant BSO2000-1411-C02-02 from Ministerio de Ciencia y Tecnología (Dirección General de Investigación) of Spain, to Juan J. Ortells, María T. Daza, Carmen Noguera and María J.F. Abad. Correspondence could be sent to Juan J. Ortells, Departamento de Neurociencia y Ciencias de la Salud. Facultad de Humanidades y Ciencias de la Educación. Universidad de Almería. 04120. Almería. Spain (E-mail: jortells@ual.es).

REFERENCES

Adams, J.K. (1957). Laboratory studies of behavior without awareness. *Psychological Bulletin*, *54*, 383-405.

Baker, L.E. (1937). The influence of subliminal stimuli on verbal behavior. *Journal of Experimental Psychology*, *20*, 84-100.

Balota, D.A. (1983). Automatic semantic activation and episodic memory. *Journal of Verbal Learning and Verbal Behavior*, *22*, 88-104.

Bar, M., & Biederman, I. (1998). Subliminal visual priming. *Psychological Science*, 9, 464-469.

Cheesman, J., & Merikle, P.M. (1984). Priming with and without awareness. *Perception and Psychophysics*, *36*, 387-395.

Cheesman, J. & Merikle, P.M. (1985). Word recognition and consciousness. In D. Besner, T.G. Waller, & G.E. McKinnon (Eds.). *Reading research: Advances in theory and practice*. New York: Academic Press.

Cheesman, J., & Merikle, P.M. (1986). Distinguishing Conscious from Unconscious Perceptual Processes. *Canadian Journal of Psychology*, *40*, 343-367.

Correa, A., Fox, E., Carmona, E., Noguera, C., Lupiáñez J. & Tudela P. (2002, April). The processing of emotional faces at two levels of awareness in subclinical anxiety. *Poster presented at the IV Conference of the Spanish Society for Experimental Psychology*, Oviedo, Spain.

Dagenbach, D., Carr, T.H., & Wihelmsen, A. (1989). Task-induced strategies and near-threshold priming: conscious influences on unconscious perception. *Journal of Memory and Language*, *28*, 412-443.

Dalrymple-Alford, E.C. (1972). Associative facilitation and interference in the Stroop color-wprd task. *Journal of Experimental Psychology*, *98*, 274-276.

Daza, M.T. & Ortells, J.J. (2000a, March). Procesamiento automático vs. controlado de palabras: relación entre atención y conciencia. [Automatic vs. controlled word processing: Relations between attention and consciousness]. *Paper presented at the III Conference of the Spanish Society for Experimental Psychology,* Barcelona, Spain.

Daza, M.T. & Ortells, J.J. (2000b, September). Curso temporal del procesamiento semántico automático (no-consciente) vs. controlado (consciente) de estímulos lingüísticos. [Time-course of automatic (unconscious) vs. controlled (conscious) semantic processing of verbal stimuli]. *Paper presented at the XII Conference of the Spanish Society for Comparative Psychology*, Granada, Spain.

Daza, M.T., Ortells, J.J., & Fox, E. (in press). Perception without awareness: Further evidence from a Stroop priming task. *Perception and Psychophysics*.

Daza, M.T., Ortells, J.J. & Noguera, C. (2001, September). Priming semántico consciente vs. no-consciente producido por palabras enmascaradas parafoveales. [Conscious vs. unconscious semantic priming from parafoveal masked words]. *Paper presented at the III Attention Meeting (RECA3)*, Almería, Spain.

Debner, J.A., & Jacoby, L.L. (1994). Unconscious perception: Attention, awareness, and control. *Journal of Experimental Psychology: Learning, Memory, and Cognition*, *20*, 304-317.

De Groot, A.M.B. (1984). Primed lexical decision: combined effects of the proportion of related prime-target pairs and the stimulus-onset asynchrony of prime and target. *Quarterly Journal of Experimental Psychology, 36A*, 253-280.

Dehaene, S., Naccache, L., Le Clec'H, G., Koechlin, E., Mueller, M., Dehaene-Lambertz, G., van de Moortele, P.F., & Le Bihan, D. (1998). Imaging unconsious semantic priming. *Nature*, *395*, 597-600.

Di Pace, E., Longoni, A.M. & Zoccoloti, P. (1991). Semantic processing of unattended parafoveal words. *Acta Psychologica*, *77*, 21-34.

Dixon, N.F. (1981). *Preconscious processing*. London: Wiley.

Dosher, B.A. (1998). The response-window regression method--some problematic assumptions: comment on Draine and Greenwald (1998). *Journal of Experimental Psychology: General*, *127*, 311-317.

Draine, S.C. & Greenwald, A.G. (1998). Replicable unconscious semantic priming. *Journal of Experimental Psychology: General*, *127*, 286-303.

Dunlap, K. (1900). The effect of imperceptible shadows on the judgment of distance. *Psychological Review*, *7*, 435-453.

Eriksen, C.W. (1960). Discrimination and learning without awareness: A methodological survey and evaluation. *Psychology Review*, *67*, 279-300.

Esteves, F., Dimberg, U., & Öhman, A. (1994). Automatically elicited fear: conditioned skin conductance responses to masked facial expressions. *Cognition and Emotion*, *8*, 393-413.

Esteves, F., & Öhman, A. (1993). Masking the face: recognition of emotional facial expressions as a function of the parameters of backward masking. *Scandinavian Journal of Psychology*, *34*, 1-18.

Farah, M. J. (2001). Consciousness. In B. Rapp (Ed.), *The Handbook of Cognitive Neuropsychology* (pp. 159-182). Psychology Press, UK.

Forster, K.I., & Davis, C. (1984). Repetition priming and frequency attenuation in lexical access. *Journal of Experimental Psychology: Learning, Memory and Cognition*, *10*, 680-698.

Forster, K.I., & Veres, C. (1998). The prime lexicality effect: form-priming as a function of prime awareness, lexical status and discrimination difficulty. *Journal of Experimental Psychology: Learning, Memory and Cognition*, *24*, 498-514.

Fox, E. (1996). Selective processing of threatening words in anxiety: The role of awareness. *Cognition and Emotion*, *10*, 449-480.

Fox, E., Russo, R., Bowles, R. & Dutton, K. (2001). Do threatening stimuli draw or hold visual attention in subclinical anxiety? *Journal of Experimental Psychology: General*, *130,* 681-700.

Fuentes, L.J., & Tudela, P. (1992). Semantic processing of foveally and parafoveally presented words in a lexical decision task. *Quarterly Journal of Experimental Psychology*, *45A*, 299-322.

Greenwald, A.G., & Draine, S. (1997). Do subliminal stimuli enter the mind unnoticed?. Tests with a new method. In J. Cohen & J. Schooler (Eds.). *Scientific approaches to consciousness* (pp. 83-108). Mahwah, NJ: Erlbaum.

Greenwald, A.G., Draine, S., & Abrams, R.L. (1996). Three cognitive markers of unconscious semantic activation. *Science*, *273*, 1699-1702.

Greenwald, A.G., Klinger, M.R., & Schuh, E.S. (1995). Activation by marginally perceptible ("subliminal") stimuli: Dissociation of unconscious from conscious cognition. *Journal of Experimental Psychology: General, 124*, 22-42.

Groeger, J.A. (1984). Evidence of unconscious semantic processing from a forced-error situation. *British Journal of Psychology, 75*, 305-314.

Groeger, J.A. (1988). Qualitatively different effects of undetected and unidentified auditory primes. *Quarterly Journal of Experimental Psychology, 40A*, 323-339.

Holender, D. (1986). Semantic activation without conscious identification in dichotic listening, parafoveal vision, and visual masking: A survey and appraisal. *The Behavioral and Brain Sciences, 9*, 1-66.

Jacoby, L.L. (1991). A process dissociation framework: Separating automatic from intentional uses of memory. *Journal of Memory and Language, 30*, 513-541.

Jacoby, L.L., Lindsay, D.S., & Toth, J.P. (1992). Unconscious influences revealed: Attention, awareness, and control. *American Psychologist, 47*, 802-809.

Jacoby, L.L., Toth, J.P., & Yonelinas, A.P. (1993). Separating conscious and unconscious influences of memory: Measuring recollection. *Journal of Experimental Psychology General*, Vol 122(2): 139-154.

Kemp-Wheeler, S.M., & Hill, A.B. (1988). Semantic priming without awareness: some methodological considerations and replications. *Quarterly Journal of Experimental Psychology, 40A*, 671-692.

Kunst-Wilson, W.R., & Zajonc, R.B. (1980). Affective discrimination of stimuli that cannot be recognized. *Science, 207*, 557-558.

Logan, G.D. (1985). Executive control of thought and action. *Acta Psychologica, 60*, 193-210.

Logan, G.D., & Zbrodoff, N.J. (1982). Constraints on strategy construction in a speeded discrimination task. *Journal of Experimental Psychology: Human Perception and Performance, 8*, 502-520.

Logan, G.D., Zbrodoff, N.J., & Williamson, J. (1984). Strategies in the color-word Stroop task. *Bulletin of the Psychonomic Society, 22*,135-138.

Mandler, G., Nakamura, Y. & Van Zandt, B.J.S. (1987). Nonspecific effects of exposure on stimuli that cannot be recognized. *Journal of Experimental Psychology: Learning, Memory, and Cognition, 13*, 646-648.

Marcel, A.J. (1980). Conscious and preconscious recognition of polisemous words: locating the selective effect of prior verbal context. In R.S. Nickerson (Ed.). *Attention and performance VIII*. Hillsdale, New Jersey: Erlbaum.

Marcel, A.J. (1983a). Conscious and unconscious perception: Experiments on visual masking and word recognition. *Cognitive Psychology, 15*, 197-237.

Marcel, A.J. (1983b). Conscious and unconscious perception: An approach to the relation between phenomenal experience and perceptual processes. *Cognitive Psychology, 15*, 238-300.

McCauley, C., Parmelee, C.M., Sperber, C.D., & Carr, T.H. (1980). Early extraction of meaning from pictures and its relation to conscious identification. *Journal of Experimental Psychology: Human Perception and Performance, 6*, 265-276.

Merikle, P.M. (1982). Unconscious perception revisited. *Perception and Psychophysics, 31*, 298-301.

Merikle, P.M. (1992). Perception without awareness: Critical issues. *American Psychologist, 47*, 792-795.

Merikle, P.M., & Cheesman, J. (1987). Current status of research on subliminal perception. In M. Wallendorf & P.F. Anderson (Eds.). *Advantages in Consumer Research, Vol. XIV*. Provo, UT: Association for Consumer Research.

Merikle, P.M., & Daneman, M. (1998). Psychological investigations of unconscious perception. *Journal of Consciousness Studies, 5*, 5-18.

Merikle, P.M., & Joordens, S. (1997a). Measuring unconscious influences. In J.D. Cohen, & J.W. Schooler (Eds.), *Scientific approaches to consciousness* (pp.109-123). Mahwah, NJ: Erlbaum.

Merikle, P.M., & Joordens, S. (1997b). Parallels between perception without attention and perception without awareness. *Consciousness and Cognition, 6*, 219-236.

Merikle, P.M., Joordens, S., & Stolz, J.A. (1995). Measuring the relative magnitude of unconscious influences. *Consciousness and Cognition, 4*, 422-439.

Merikle, P. M., & Reingold, E.M. (1991). Comparing direct (explicit) and indirect (implicit) measures to study unconscious memory. *Journal of Experimental Psychology: Learning, Memory and Cognition, 17*, 224-233.

Merikle, P.M., & Reingold, E.M. (1998). On demonstrating unconscious perception: comment on Draine and Greenwald (1998). *Journal of Experimental Psychology: General, 127*, 304-310.

Merikle, P.M., Smilek, D., & Eastwood, J. (2001). Perception without awareness: perspectives from cognitive psychology. *Cognition, 79*,115-134.

Miller, J.G. (1942). *Unconsciousness*. New York: Wiley.

Murphy, S.T., & Zajonc, R.B. (1993). Affect, cognition, and awareness: Affective priming with optimal and suboptimal stimulus exposures. *Journal of Personality and Social Psychology, 64*, 723-739.

Neely, J.H. (1977). Semantic priming and retrieval from lexical memory: Roles of inhibitionless spreading activation and limited-capacity attention. *Journal of Experimental Psychology: General, 106*, 226-254.

Nolan, K.A., & Caramazza, A. (1982). Unconscious perception of meaning: A failure to replicate. *Bulletin of the Psychonomic Society, 20*, 23-26.

Ortells, J.J., Abad, M.J.F., Noguera, C., & Lupiáñez,J. (2001). Influence of prime-probe stimulus onset asynchrony and prime precuing manipulations on semantic priming effects with words in a lexical decision task. *Journal of Experimental Psychology: Human Perception and Performance, 27*, 75-91.

Ortells, J.J. & Tudela, P. (1996). Positive and negative semantic priming of attended and unattended parafoveal words in a lexical decision task. *Acta Psychologica, 94*, 209-226.

Posner, M.I. & Snyder, C.R.R. (1975a). Attention and cognitive control. In R. Solso (Ed.). *Information Processing and Cognition: The Loyola Symposium*. Hillsdale. New Jersey: Erlbaum.

Posner, M.L. & Snyder, C.R.R. (1975b). Facilitation and inhibition in the processing of signals. In P.M.A. Rabbitt y S. Dornic (Eds.). *Attention and performance V*. New York: Academic Press.

Purcell, D.G., Stewart, A.L., & Stanovich, K.E. (1983). Another look at semantic priming without awareness. *Perception and Psychophysics, 34*, 65-71.

Reingold, E.M., & Merikle, P.M. (1988). Using direct and indirect measures to study perception without awareness. *Perception and Psychophysics*, *44*, 563-575.

Reingold, E.M., & Merikle, P.M. (1990). On the inter-relatedness of theory and measurement in the study of unconscious processes. *Mind and Language*, *5*, 9-28.

Searle, J.R. (1992). *The rediscovery of the mind*. Cambridge, MA: MIT Press.

Shenaut, G.K., & Ober, B.A. (1996). Methodological control for semantic priming in Alzheimer's disease. *Psychology and Aging*, *11*, 443-448.

Shiffrin, R.M. & Schneider, W. (1977). Controlled and automatic human information processing: II. Perceptual learning, automatic attending and a general theory. *Psychological Review*, *84*, 127-190.

Sidis, B. (1898). *The psychology of suggestion*. New York: Appleton.

Smith, S.D., & Merikle, P.M. (1999, June). Assessing the duration of memory for information perceived without awareness. *Poster presented at the 3rd annual meeting of the Association for the Scientific Study of Consciousness*, Canada.

Stroh, M., Shaw, A.M., & Washburn, M.F. (1908). A study in guessing. *American Journal Psychology*, *51*, 243-245.

Stroop, J.R. (1935). Studies of interference in serial verbal reactions. *Journal of Experimental Psychology*, *18*, 643-662.

Whalen, P.J., Rauch, S.L., Etcoff, N.L., Mclnerney, S.C., Lee, M.B., & Jenike, M.A. (1998). Masked presentations of emotional facial expressions modulate amygdala activity without explicit knowledge. *Journal of Neuroscience*, *18*, 411-418.

Williams, A.C. (1938). Perception of subliminal visual stimuli. *Journal of Psychology*, *6*, 187-199.

Chapter 3

EXPLAINING CONSCIOUSNESS AT MULTIPLE LEVELS

L. Andrew Coward
School of Information Technology
Murdoch University
Australia
lacoward@dijkstra.murdoch.edu.au

Ron Sun
Department of Computer Engineering and Computer Science
University of Missouri
Columbia, Missouri, USA
rsun@cecs.missouri.edu

ABSTRACT

This paper examines briefly the characteristics of successful theories in the physical sciences, and suggests that such theories create hierarchies of descriptions on many levels of detail in which causal relationships on one level are related to causal relationships on other levels. More detailed levels of description have higher information density but lower descriptive (or Kolmogorov) complexity. A similar hierarchy of descriptions (known as a functional architecture) is required in complex electronic systems to make it possible to build, repair and modify such systems. Human cognitive apparatuses have been constrained by natural pressures to adopt such functional architectures. A hierarchy of descriptions can be a good basis for a scientific theory of consciousness. Such a theory would describe the phenomenology of consciousness at a high level, then through a series of more detailed levels that could include computational models and physiology. The upper levels of a description hierarchy will then be constructed, from the highest level definitions of access and phenomenal consciousness, through implicit and explicit perception, implicit and explicit memory, and implicit and explicit skill learning. The capability of a range of existing computational cognitive models to provide descriptions of consciousness at a more detailed level but consistent with the causal relationships at high level is evaluated. Two models, CLARION and the RA, demonstrate some capabilities of this type, and they are compared in a number of aspects.

1. THE NATURE OF A SCIENTIFIC THEORY

What could a scientific theory of consciousness tell us, and what would be the limits of such a theory? We note that the physical sciences are presumably the paradigm for a rigorous scientific theory. Theories in this domain have achieved a high degree of sophistication that has developed steadily over the last 300 years or more. In thinking about what a theory of consciousness might look like, it is useful to analyze the nature of theories in the physical sciences. The development of physical science theory is illustrated in figure 1.

	Motion	**Gravity**	**Materials**
Direct Experience	Knowledge of movement for hunting, warfare, etc.	Rules for seasons, tides, intuitive knowledge that objects fall	Understanding of cooking, metal making, pottery
Classical	Newtonian mechanics	Newtonian gravity	Chemistry
19th Century	Hamiltonian dynamics		Atomic theory
Early 20th Century	Special relativity	General relativity	Bohr nuclear structure Quantum mechanics
Mid 20th Century	Quantum electrodynamics		Quantum electrodynamics Quantum chromodynamics Quantum electroweak theory
Late 20th Century		String Theory	String Theory

Figure 1. Development of theories in the physical sciences.

At the highest level of figure 1, domains of human experience that form the roots from which the physical sciences have developed are illustrated. Experience of movement in day to day life, in hunting, or in warfare is the domain that is described by the scientific theories of motion on successively deeper levels. The need to derive dates for planting crops from an understanding of the movements of stars and planets has led to theories of gravity. The study of metalworking and other material processes was the starting point for successively deeper sciences of matter and energy, from chemistry through atomic theory to quantum mechanics and on to current studies in string theory. The deeper theories provide an account that unifies multiple apparently independent domains at the higher level.

What are the differences between these theories? When a theory is described as deeper, what is the difference from "less deep" theories? A successful theory on any level creates entities at that level of descriptive detail and causal relationships between these entities that explain a certain range of data. If a causal relationship exists at the higher level, a causal relationship at the more detailed level must exist to explain it. Conversely, if a causal relationship exists at the lower level, it is often, though not always, manifested at the higher level. That is, entities and causal relationships in a theory at a deeper level should map into entities and causal relationships at a higher level of description.

In a successful theory at a deeper level causal relationships at the detailed level predict (sometimes unexpected but experimentally confirmed) causal relationships at the higher level. For example, pre-atomic or macro chemistry established entities (such as acids, alkalis, and salts), and consistent causal relationships (such as an acid plus an alkali produces a salt plus water). Atomic theory established entities such as H+ and OH- radicals and causal relationships between them that could be mapped into the entities and causal relationships of macro chemistry. Atomic theory also led to predictions of the existence of additional types of atoms that were subsequently discovered.

The key differences between levels are the information density of descriptions and the number of independent entities and causal relationships required for explaining phenomenology. In this context, information density means the total number of bits needed to support a description. The deeper the level, the higher the information density of a description of the same phenomenon. Descriptions at the level humans generally describe their experiences have the lowest information density. This statement may seem at first counter-intuitive. However, it is important to recognize that the comment refers to the informational content of descriptions of exactly the same phenomenon at different levels of detail. A description of cooking a meal at even the macro chemical level would require inclusion of hundreds or thousands of different chemicals interacting in their different ways. At the atomic level the description of the same phenomenon would require inclusion of each one of the order of 10^{24} different atoms. The quantum mechanical level would require inclusion of each of the individual electrons and nuclei, with each of these particles being described by a probability distribution rather than by a point position and velocity. A description of even the simplest macro-level phenomena would be painstakingly long and convoluted if fully constructed using more detailed theories at a deeper level.

In contrast to this higher information density, the number of causal relationships required to support descriptions is smaller and smaller at deeper levels. Thus theories at deeper levels are more "succinct". For example, the relationships amongst the vast number of possible acids, alkalis, and salts at the macro chemistry level (including all the organic acids, alcohols and esters) can be described at the atomic level using less than 100 elements and a simpler set of relations and equations. In other words, theories at deeper levels have lower description lengths, or lower Kolmogorov complexity. Kolmogorov complexity is a theoretical/mathematical measure of how many bits are needed on a theoretical model of computation (i.e., the Turing machine) to capture a process or a description (i.e., an algorithm; Li and Vitanyi 1997). It is a solid, though often neglected, foundation upon which we may theorize the relation between phenomenology and science. In a way, the whole scientific enterprise may be viewed as being aimed at creating descriptions (i.e., theories) of the phenomenological world at a lower and lower Kolmogorov complexity. This is a major, if not ultimate, goal of science. Shorter descriptive lengths of deeper theories leads to denser descriptions of phenomena.

In practice, though, full descriptions of high level phenomena as a human being experiences them are almost never created at the more detailed levels. For example, there has never been a description of cooking a recipe constructed at the quantum mechanical level. Detailed descriptions are created for a number of limited but precisely defined phenomena at the high level. If a large and representative sample of such limited high level phenomena can be consistently described at the detailed level, the detailed theory is considered to be valid. The validity is immediately questioned if a high level phenomenon which does not have a consistent detailed description is found. Physical scientists, like all human beings, can only handle a limited amount of information at one time, due to information density which we alluded to earlier. They must therefore use a high level theory for thinking about broader phenomena, and then focus in on smaller areas to apply the more detailed theory.

The paradoxical effect of deeper levels of theories, due to the combined effect of descriptive (Kolmogorov) complexity and information density, leads to interesting compromises --- multiple levels of descriptions may be used and complex relations among them are created, maintained, and explored. As an illustration of this process, consider the following quotation from a report from the front lines of theoretical physics (Greene 1999):

> "The universe is governed by the principles of quantum mechanics to fantastic accuracy. Even so, in formulating theories over the past half century, physicists have followed a strategy that, structurally speaking, places quantum mechanics in a somewhat secondary position. In devising theories, physicists often start by working in a purely classical language that ignores quantum probabilities, wave functions, and so forth - a language that would be perfectly intelligible to physicists in the age of Maxwell and even in the age of Newton - and then, subsequently, overlaying quantum concepts upon the classical framework. This approach is not particularly surprising, since it directly mirrors our experiences. At first blush, the universe appears to be governed by laws rooted in classical concepts such as a particle having a definite position and a definite velocity at any given moment in time. It is only after detailed microscopic scrutiny that we realize that we must modify such familiar classical ideas. Our process of discovery has gone from a classical framework to one that is modified by quantum revelations, and this progression is echoed in the way that physicists, to this day, go about constructing their theories."

The key point here is that physicists need to work with a hierarchy of theories in order to be able to manage the information. This approach can result in problems if the theories at different levels in the hierarchy are not fully consistent, but the solution is not to work only at the most detailed level, but to find a more consistent theory at the inconsistent level.

The issue for a scientific theory of consciousness is therefore as follows: is it possible to create a high-level phenomenological or psychological causal description of the phenomena of consciousness, and a consistent series of more detailed causal descriptions down to a description in terms of the deepest level which makes a difference to causal relationships at the highest level. This deepest level is generally taken to be an abstract (e.g., computational), or physiologically detailed, description of the operations of neurons and neurochemicals. Although it has been argued that quantum mechanics may change causal relationships at the psychological level of description of consciousness (Penrose 1994), thus far there is no convincing evidence that quantum mechanics makes any difference to causal descriptions at higher levels with regard to consciousness (for example, quantum mechanics makes no difference to causal descriptions at the chemical level), we will ignore this level for the time being.

2. Mechanistic Views of Consciousness

Such a theory would be a mechanistic (or processing-oriented) explanation. There would be no dualistic magic involved. What can be expected of such a theory of consciousness is an appreciation of how causal relationships at a deeper level give rise to the phenomena of consciousness at the phenomenological level, a better understanding of causal relationships at the psychological level, and a understanding of how silicon-based (electronic) systems might generate conscious phenomena.

A mechanistic explanation is generally taken to be computational in the broad sense of the term computational. Here the term "computation" is used to denote any process that can be realized computationally, ranging from chaotic dynamics (Freeman 1995) and Darwinian competition (Edelman 1989), to quantum mechanics (Penrose 1994; although we do not believe in its relevance). In terms of the sufficiency of such computational explanations, Jackendoff (1987) proposed the following hypothesis : "Every phenomenological distinction is caused by/supported by/projected from a corresponding computational distinction". This hypothesis remains viable, despite various assaults on it. These assaults (e.g. Edelman 1989, Freeman 1995, Damasio 1994, Penrose 1994, Searle 1980) failed to show that computation, in general, cannot account for the nature of consciousness, although they had some legitimate complaints about specific computational approaches and models.

A theory of consciousness which is analogous with a theory in physical sciences must first establish a set of entities or conditions X, Y, Z, etc. at a high level with consistent causal relationships such as the presence of X and Y results in Z. (Such causal relationships might include: perception of a red object and a verbal input of "What color is that" causes the perceiver to say "That is red"; or perception of a red object and a verbal input of "What are your feelings right now" causes the perceiver to say "I feel sad".) Then, the theory must also establish a set of entities or conditions at a lower level so that combinations of entities or conditions x, y, z, etc. correspond with the high level X, Y, Z etc. in such a way that if X plus Y causes Z at the high level of description, then x plus y causes z at the lower level of description. Descriptions at the high level contain less densely packed information than descriptions at a detailed level. However, descriptions at the more detailed level should have a lower Kolmogorov complexity. This means that the number of entities which are needed on the detailed level in general should be much smaller than the number of entities at the high level, and the number of different types of causal relationships at the detailed level generally should be smaller as well.

On the other hand, the lower density of descriptions at the higher level means that it is possible that many states at the detailed level could correspond with indistinguishably different states at the higher level. We may generalize the token identity theory concerning psychophysical correspondence (see e.g. Braddon-Mitchell and Jackson 1996): instances of a high level (psychological) state are instances of detailed level (physical) states (i.e. correspondence of categories across levels). However, consistent with the supervenience principle (no mental difference without a physical difference, see Kim 1993), no two different states at high level can correspond with the same state at a detailed level.

Given the wide difference in information content between psychological and physiological descriptions, it may be necessary to establish one or more intermediate levels of description to bridge the gap, in a manner analogous with the atomic level descriptions that bridge the gap between quantum mechanics and the acid, alkali etc. descriptions of chemistry as discussed earlier. In this context (i.e. correspondence of categories across levels) Jackendoff's hypothesis leads to the use of computational modeling to provide the intermediate levels of description. Computational modeling will need to map into both the psychological and physiological levels of description. According to our earlier characterization of levels, computational modeling could be at a level higher than the physiological theories, but could also be at a more detailed level. In either case, computational theories should map to physiological descriptions.

It is not always apparent that, for the design and management of complex electronic systems (e.g., computers), hierarchies of descriptions exist on many levels of detail. In such systems, a "functional architecture" exists which makes it possible to describe the operation of the system on many levels of detail. For example, carrying out a spell check within a word processing program could be described at the highest level as "check the spelling of all the words in this document". At a more detailed level, the description might include elements such as "load the spelling tool from hard disk to RAM". At yet more detailed levels the description would include software instructions such as "while x is true do y" and at an even deeper level would include assembly code instructions such as "jump 342". In such a functional architecture, the functionality of the system at the highest level is expressed as functional requirements which are separated into relatively independent modules. Each module is separated into sub-modules, sub-modules into yet smaller components and so on until the smallest functional elements of the system are reached. In a complex real time system (such as a telecommunications switch), there may be seven or eight levels of description from highest level functional requirements to executable machine code level, and a similar number of levels of description for the hardware, from system description to transistor, which interact at different levels with the software hierarchy. The modularization is strongly constrained: all modules on one level must perform roughly the same proportion of system operations, and although information exchange between modules is essential to coordinate functionality, such information exchange must be minimized. A functional architecture which meets these two constraints is called a "simple functional architecture".

Why does such a functional architecture exist in electronic systems? The reasons are that without such a functional architecture it would be extremely difficult to build, test, repair, or modify the functionality of the system, and independent design of modules by different engineers would also be impractical. To understand the operation of these constraints, imagine a design process in which (in a caricature of biological evolution) a large number of technicians were given a large number of devices and told to connect at random. Periodically the result would be tested, and eventually a system found which performed as required. There are some severe problems with such a system. There are no blueprints which can guide the process of building another copy: the only option is to duplicate the original, device by device, connection by connection. If an error were made in such a duplication process a functional problem would result, but there would be no easy way to use the knowledge of the problem to identify and correct the error. Similarly, if a device failed during operation it would be very hard to identify which device was defective. Finally, device level changes would generate complex and unpredictable functional changes, and if there was a need to modify the functionality in a controlled fashion there would be no way to identify what device level changes would produce the desired functional modifications (Coward 1999b). There must therefore be simple logical paths which can relate high level system functionality to the operations of transistors and fragments of code. The hierarchy of modules as described makes it possible for such paths to exist. Such a hierarchy of modules is effectively a hierarchy of descriptions of exactly the same functionality at different levels of detail.

Biological brains are not the result of a deliberate design process, but are constrained by the needs to be constructed from DNA "blueprints", recover from construction errors and damage, and learn new behaviors without introducing random, undesirable changes into existing behaviors. Coward (1990) has therefore argued that selection pressures deriving from the advantages of ease of construction, recovery from damage, and unconfused learning might result in brains being constrained to adopt simple functional architectures as defined earlier. Brains may therefore have an intrinsic "description" hierarchy which may link the phenomena of consciousness to those of physiology. Within such an architecture, high level descriptions of the phenomena of consciousness could be mapped through successively more detailed descriptions.

Marr (1982) argued that complete understanding of an information processing device required understanding on three levels: the levels of computational theory, representation and algorithm, and hardware implementation. However, Marr's position was that the three levels are only loosely coupled, since "The choice of algorithm is influenced for example, by what it has to do and by the hardware in which it must run. But there is a wide choice available at each level, and the explication of each level involves issues that are rather independent of the other two." Functional architecture based description hierarchies are rather different. The functionality described on each level is precisely the same, the difference is only in the information density (level of details) and the length of the description, and in the smaller number of entities needed by descriptions at the more detailed level. Causal relationships at one level will always have to relate to causal relationships at other levels.

In addition, note that, in contrast with Marr, we do not view algorithmic description as necessarily being at a higher level than physiological description (or hardware implementation). This is because computational theories (algorithms) may well provide a more succinct explanation than empirical generalizations of neuroscience which is consistent with physiological data. We have earlier alluded to the minimum description length principle -- why we need to condense information to obtain shorter explanations, based on what can maximally enhance generality (universality) of theories (Li and Vitanyi 1997). A succinct computational theory (not every computational theory) can reduce the number of causal relations needed to explain the world: some relations may be merged, some split, but all accounted for at a deeper level.

3. The Phenomena of Consciousness at the Highest Level

Below, we will employ a (rough and idiosyncratic) hierarchy to address the issue of consciousness at multiple levels simultaneously. Our hierarchy starts with a high level definition of consciousness, then describes the psychology of implicit and explicit perception, memory, and learning, and some of the causal relationships which can be observed. The next stage is to analyze various models of consciousness at a more detailed level. In some cases these models themselves have several levels of descriptions. The final stage is to discuss the ability of these models to map consistently into the neurophysiological level of description.

As a starting point, a description of the phenomena of consciousness at the highest level (the phenomenological level) is required. We are not going to address the full extent of this level in this paper, but only a key distinction: (1) conscious processes are directly accessible and (2) unconscious processes are inaccessible directly. Conscious processes are directly accessible (e.g. directly verbally expressible) without involving intermediate interpretive or transformational steps, which is a requirement prescribed and/or accepted by many others (see e.g. Clark 1993, Hadley 1995). Unconscious processes are, in contrast, inaccessible directly, thus exhibiting different properties (see, e.g., Heidegger 1927, Dreyfus and Dreyfus 1987, Berry and Broadbent 1988, Reber 1989).

The ways in which consciousness manifests itself can be separated into two types: access consciousness and phenomenal consciousness. Access consciousness refers to the direct availability of the mental content for access (e.g. verbal report), while phenomenal consciousness refers to the phenomenal content of the mental content, that is, what something feels like and the immediacy and vividness of such feeling, as Thomas Nagel brought out in "What is it like to be a bat" (Nagel 1974).

4. Access Consciousness

There have been various suggestions with regard to the functional role of access consciousness: for example, the veto view as suggested by Libet (1985), which states that the function of consciousness is to veto unconsciously initiated actions, or the counterbalance view as suggested by Kelley and Jacoby (1993), which is a generalization of the veto view. The question which these two views did not address is why one needs counterbalance, whether in the form of occasional veto or something else. On the other hand, Reber (1989), Stanley et al (1989), and others claimed that conscious and unconscious processes are each suitable for different situations, so that either a conscious or an unconscious process will be applied to a situation depending on which one is more suitable.

Similarly, the language/planning view of Crick and Koch (1990) suggests that the function of consciousness is to enable the use of language and (explicit) planning. However, the question remains why one should use language and planning (in an explicit and conscious way) on top of unconscious processes.

An alternative explanation (based on the representational difference view of consciousness of Sun 1999; more discussion later) is that the function of the conscious/unconscious distinction lies in the synergy that this distinction affords. As shown in Sun (1994, 1999), the interaction of the conscious and the unconscious (as two distinct processes, with different representations) can, in many common circumstances, lead to synergy in performance. This is the synergy view of consciousness proposed in Sun (1994, 1995, 1999).

Let us discuss this view in more detail. First of all, there is the question of the source of the synergy. As indicated by psychological data (e.g., on implicit learning and implicit memory), conscious processes tend to be more crisp and focused (selective), while unconscious processes tend to be more complex, broadly scoped (unselective), and context-sensitive (see Reber 1989, Berry and Broadbent 1988, and Seger 1994 regarding complexity; see Hayes and Broadbent 1988 regarding selectivity). Due to their vastly different characteristics, it should not come as a surprise that the interaction of the conscious and the unconscious leads to synergistic results. In the statistical literature, as in e.g. Breiman (1996) and Efron and Morris (1973), it is well known that combining diversified processes can improve performance. Taking this line of reasoning one step further, it is not farfetched to speculate that synergy was the reason evolution created consciousness (Sun 1999). Coward (1999b) has suggested a process by which natural pressures could incrementally give rise to such a synergistic functionality, starting from simple (pre-speech) signaling.

There is some psychological evidence directly in support of the synergy view. In psychological experiments, Willingham et al (1989) found that those subjects who acquired full explicit (conscious) knowledge (in a serial reaction time task) appeared to learn faster than those who did not have full explicit knowledge. Stanley et al (1989) reported that subjects' learning improved (in a dynamic control task) if they were asked to generate verbal instructions for other subjects along the way during learning. That is, a subject was able to speed up his/her own learning through the emphasis on explicit processes. In addition, in terms of learned performance, Willingham et al (1989) found that subjects who verbalized (while performing serial reaction time tasks) were able to attain a higher level of performance than those who did not verbalize. This phenomenon may also be related to the self-explanation effect reported in the cognitive skill acquisition literature (Chi et al 1989). In all of these cases, it can be the emphasis on conscious processes that helped performance. Finally, synergy has also been demonstrated through computational simulation, in the domains of commonsense reasoning (Sun 1994, 1995) and skill learning, including in all of the aforementioned psychological experimental tasks (Sun 1999, Sun et al 2000).

Explicitness of conscious processes also enables explicit control and manipulation, which allows meta-level processes (Nelson 1993). Such control and manipulation can include, for example, selecting a reasoning method, controlling the direction in which reasoning goes, enable/disable certain inferences, or evaluating the progress of reasoning. When meta-level processes get assimilated into regular processes, further meta-level processes can be developed on top of them. Thus, potentially, we can have many levels of self-control and self-manipulation of mental processes. (By no means we are claiming that unconscious processes cannot be controlled and/or manipulated at all, but that it is clearly more difficult to do so due to less accessible representation.) Meta-level processes may be another reason for synergy.

Therefore, conscious processes are characterized by explicit (i.e., localist or symbolic; Sun 1999) representations, as well as explicit meta-level regulation (i.e., control and manipulation of the processes operating on explicit representations). These two aspects together distinguish conscious processes from unconscious processes (as in, e.g., the CLARION model; see Sun et al 2000). Synergy results from the coexistence of the two different types of representations and consequently the coexistence of the two different types of processes operating on the two types of representations, respectively. A ``teleological" explanation of access consciousness follows directly from these two aspects. It should be emphasized that this dichotomous structure is minimal and necessary for understanding cognition, in the sense of a minimal explanatory construct at any level of description. There are qualitative and fundamental differences between the two types of processes. It is difficult to believe that one side of the dichotomy can be derived from the other ontogenetically, without some minimal structure to begin with. Thus, the distinction should be innate. Notice also that most of the high animal species are not capable of developing an elaborate and complete conceptual system (with symbolic processing abilities) while humans rarely fail to develop such a system. The constancy of this inter-species difference points to the innateness of such a difference and the innateness of the dichotomous structure. It is thus more convincing to hypothesize that the dichotomous structure is a given innate structure for humans, and should be incorporated into any architecture.

This explanation of consciousness also encompasses the other views concerning the functional role of consciousness. According to the synergy view, consciousness can certainly veto or counterbalance unconsciousness, given the right circumstances when such veto or counterbalance improves performance (that is, if they lead to synergy). Likewise, the synergy view can explain the situational difference view, in that, while in general both types of processes are present due to their synergistic effect, in some extreme cases, it may be advantageous to use only conscious or unconscious processes. For example, when a task is well practiced and therefore there is no longer a need for synergy, unconscious processes alone suffice, which leaves conscious processes for other tasks. (This phenomenon is known as automatization.) The synergy view also encompasses the language/planning view, because it explains why one should use (consciously) language/planning on top of unconscious processes: It is because of the possibility of improved performance through the interaction of both types of processes.

Another possible alternative is that the function of access consciousness is to expand the range of information available to determine the appropriate behavior in a given situation (Baars 1988, Coward 1990). To give a simple example, consider what happens when an object such as a tree is perceived. Direct sensory input generates some kind of mental activation which could be interpreted as alternative behavioral recommendations to walk around the tree, cut down the tree for firewood or furniture, lean on the tree for example. However, if the range of activity is expanded to include information recorded in the past when similar activations were present, it could include memories of past experiences in which trees have been a factor, including information recorded about the activities during those experiences. This function might be of limited value in generating behavior in relatively simple situations such as in response to a tree, but would be particularly important for generating behavior in extremely complex situations. Activation of information about a number of past situations with some resemblance to the current situation, including the behaviors adopted and their consequences, might make it possible to synthesize a more sophisticated behavior in the current situation.

The expansion of information range adds alternative behaviors in any situation in which it is invoked, and the actual behavior is selected from the total available. It is therefore advantageous to have the possibility of expanded information range, being able to adaptively adjust the detail of information accessed in coping with the world. It is clearly consistent with the synergy explanation, and can be viewed as one possible deeper level description of the synergistic functionality.

5. Phenomenal Consciousness

The functional role of phenomenal consciousness has been regarded as a far more difficult question. Let us examine the notion of qualia, which refers to the "phenomenal content" of conscious experience (Nagel 1974, Chalmers 1993, Block 1994). The major problem with the notion lies in the difficulty it poses for functionalism, the currently popular view in philosophy of mind that the defining feature of any mental state is the set of causal relations it bears to other mental states, environmental effect on the body, and behavior of the body. A functional organization capable of generating behavior alone is not a necessary and sufficient condition for consciousness (Searle 1980). If cognitive functioning can occur without qualia, then qualia may not have a functional role and functionalism cannot explain them.

It is possible that there could be a range of different states at a detailed level which generate the same externally observed behavior, some of which correspond with phenomenal consciousness and others do not (recall our earlier discussions of levels of description). We could then distinguish different possible functional structures capable of the same behavior, separating those that are capable of behavior but not the more intrinsic properties of the mind from those that are capable of both the behavior and the more intrinsic properties of the mind (Sun 1999). Thus, the search for phenomenal consciousness can, after all, be the search for a functional structure - the right functional structure.

Beyond functionalism, we also need to address the physical nature of phenomenal consciousness. A mechanistic explanation must be attempted, despite the failure that we have been seeing in such attempts so far. There are plenty of examples of dualistic "explanations" of difficult phenomena having evaporated after further explorations that led to a better mechanistic account. To see how the representational difference view of consciousness can account for phenomenal consciousness, let us examine some possibilities. Qualia might be accounted for by the totality of a multi-modal multi-level organization and its collective states (i.e., the total-states; Sun 1999). These total-states are of extremely high complexity involving a nexus of external perception (of many modalities), internal states, emotion, implicit and explicit memory, implicit and explicit representation, implicit and explicit decision making, and so forth (Sun 1999). This was termed the "manifold" by Van Gulick (1993), and the "superposed complex" by Lloyd (1995). In this approach (Sun 1999), a particular kind of phenomenal quality may be accounted for by a particular region of a total-state space (involving the totality of all the aforementioned aspects), which gives rise to the sense of what something is like (Nagel 1974) without explicit awareness of all the details. The complexity of organization may explain the "irreducibility" of phenomenal experience - the difficulty (or impossibility) of describing phenomenal qualities (qualia). Clearly, a region in a total-state space can only be formed on the basis of particular organizations of modules (Chalmers 1993) that support a space of total-states. Qualia are thus partially the result of functional architectures of cognitive apparatuses.

Are qualia epiphenomenal according to accounts like these? The answer is no, if we contrast the condition in which we are aware (with qualia) with the one in which we are not (without qualia). Qualia entail high-intensity, qualitatively distinguishing states which change the behavior of the agent experiencing them. Although it is possible to imagine that there are corresponding unconscious states (without qualia) that can lead to similar action on the part of the agent (Searle 1980), the activations which generate the subjective experience of qualia also serve useful functions (Nelson 1993).

In summary, at the highest level there is the conscious/ unconscious distinction. Some of the causal relationships which we focus on at this level are: conscious processes are accessible, but unconscious processes are not; conscious processes can cause vivid experiential states but unconscious processes cannot; both processes can cause behavior. A successful theory at a more detailed level must model two different types of processes in such a way that these causal relationships are maintained. An even more successful theory must also demonstrate that its processes can also be modeled at a physiological level of detail.

6. Psychology of Consciousness: A More Detailed Level

6.1. Implicit and Explicit Mental Processes

An important distinction is made in psychology between implicit and explicit mental processes (see for example the 1998 review edited by Kirsner et al). The distinction is recognized in the areas of perception, memory, and learning. In general, explicit and implicit phenomena have been distinguished by the presence or absence of access consciousness (Schacter and Graf 1986). The distinction between implicit and explicit had one of its origins in the observation of memory deficits (for example corresponding with the kinds of memory tasks amnesiacs can and cannot perform; Dunn 1998), explicit and implicit memory retrieval (Jacoby et al 1996), and in various learning tasks that cannot be easily learned in a conscious way (Reber 1989).

In amnesiacs such as Korsakovs patients (Butters 1984), the most obvious symptoms are the inability to create new declarative memories. Memories existing prior to onset are generally unaffected, although there can be loss of all memory from a section of time immediately preceding onset. However, general intelligence is unaffected, and there is an ability to learn skills even though there is no memory of the skill learning experience. For example, performance in solving the Tower of Hanoi problem steadily improves over a number of sessions, even though at the beginning of each session the patient has no memory of seeing the problem before (Cohen et al 1985).

6.2. Implicit and Explicit Perception

The key of implicit perception is that the memory of the perception is not accessible to conscious introspection. In the experiments of Triesman (1960), subjects were presented with two sections of very different text, one to each ear. They were asked to repeat aloud (to shadow) the text heard by one specific ear only. During presentation, the two texts were switched between ears part way through. The subjects tended to respond by (erroneously) switching to repeating the text from the other ear (i.e. the meaningful continuation). The implication is that the information presented to the unattended ear was being subjected to a significant degree of semantic analysis, although no information from the unattended ear was accessible to conscious recollection.

6.3. Implicit and Explicit Memory

Consider now the differences between explicit and implicit memory (Kirsner 1998). In laboratory testing of both normal and amnesiac subjects, testing of explicit memory is of two major types: free recall and recognition. In free recall, subjects are presented with lists of words or other objects and in a subsequent test phase asked to recall the items on the list. Recognition testing is similar except that in the test phase subjects are supplied with items and asked if they occurred in the original list. A typical test of implicit memory is repetition priming. In one form of repetition priming, subjects are presented with letter strings which may or may not be English words. The subject is asked to classify each letter string as a genuine word or meaningless string, and their reaction time to reach a decision is measured. In the first phase of the experiment a series of strings are presented. In the second phase both new strings and repeats from the first phase are presented. It is found that the reaction time for repeated strings is significantly shorter than for new strings.

There are a number of dissociations between explicit and implicit memory which indicate separate underlying mechanisms (Kirsner 1998). In explicit memory, unless subjects are asked to make old/new judgments immediately, accuracy is low and declines systematically as a function of time (Scarborough et al 1977; Tulving et al 1982). Explicit memory invokes conscious awareness, while implicit memory appears outside of consciousness, and is intact in amnesiac patients (Schacter and Graf 1986; Shimamura 1986). Performance in explicit memory testing is strongly affected by organizing material to be remembered, but such processes have minimal effect on implicit memory (Schacter et al 1989). Implicit memory is more strongly reduced by changes to the lexical form of words than explicit memory (Forbach, Stanner and Hochhaus 1974). The magnitude of the implicit memory effect is also much lower for words with a high natural language frequency. In general, explicit memory involves conscious awareness, while implicit memory appears to be outside this awareness.

6.4. Implicit and Explicit Skill Learning

Cognitive skill acquisition is widely studied in cognitive science (e.g., VanLehn 1995), in particular, learning to solve problems in intellectually oriented tasks such as arithmetic, puzzle solving, elementary geometry, and LISP programming (Anderson 1982, 1993, VanLehn 1995, Rosenbloom et al 1993). However, skills vary in complexity and degree of cognitive involvement. They range from simple motor movements and other routine tasks in everyday activities to high-level intellectual skills.

The inaccessibility of procedural knowledge is accepted by most researchers and embodied in most computational models that capture procedural skills, including Anderson (1983, 1993) and Rosenbloom, Laird, and Newell (1993). Dreyfus and Dreyfus (1987) proposed the distinction of analytical and intuitive thinking, and believed that the transition from the former to the latter is essential to the development of complex cognitive skills (on the basis of phenomenological analyses of chess playing at different stages of learning chess). This transition is very similar to the declarative-to-procedural transition as advocated by Anderson (1983, 1993), although they are not exactly identical.

Procedural knowledge can be highly efficient once it has been developed and can work independently without involving declarative knowledge in many cases (e.g., Anderson, 1983, 1993; Dreyfus and Dreyfus, 1987). However, declarative knowledge is also advantageous in many situations: Declarative knowledge can speed up the learning process when constructed on-line during skill learning (Mathews et al 1989, Sun et al 1996), declarative knowledge can facilitate the transfer of a skill (Willingham et al 1989, Sun and Peterson 1999) by speeding up learning in new settings, and declarative knowledge can help in the communication of knowledge and skills to others.

Several lines of research demonstrate that individuals may learn to perform complex skills without first obtaining a large amount of explicit declarative knowledge (e.g., Berry and Broadbent 1988, Stanley et al 1989, Lewicki et al 1992, Willingham et al 1989, Reber 1989, Schacter 1987, Schraagen 1993, Posner et al 1997, Sun et al 2000). In research on implicit learning, Berry and Broadbent (1988), Willingham et al (1989), and Reber (1989) expressly demonstrate a dissociation between explicit knowledge and skilled performance, in a variety of tasks including dynamic control tasks (Berry and Broadbent 1988), artificial grammar learning tasks (Reber 1989), and serial reaction tasks (Willingham et al 1989). There are indications that explicit knowledge may arise from procedural skills in some circumstances (Stanley et al 1989). Using a dynamic control task, Stanley et al. (1989) finds that the development of declarative knowledge paralleled but lagged behind the development of procedural knowledge.

Even in high-level cognitive skill acquisition (VanLehn 1995), Schraagen (1993) reports that learning in the domain of designing psychological experiments often involves generalizing specific knowledge to form generic schemas/rules in addition to specializing general knowledge to specific knowledge.

There can be inconsistencies between procedural and declarative knowledge of a skill as revealed by studies of specialist expertise. "In some areas of expertise, there is a dissociation between what experts say they do and what they do ..... [An expert verbal] description typically bears only a superficial relationship to the expertise" (Speelman 1998). Verbal descriptions by experts describing their expertise often correspond with beginner methods rather than actual methods, and requiring a verbal description can even result in the expert reverting to the less effective beginner method (Bainbridge 1977; Berry 1987).

7. Models of Consciousness

7.1. Various Models

There have been a variety of explanations for the distinction between conscious and unconscious processes. These explanations can be regarded as attempts to describe the phenomena discussed in the previous section at a more detailed level. A major distinction at the highest level such as the distinction between implicit and explicit mental processes could plausibly be expected to reflect one or more major functional separations needed at the more detailed level. (An example of this would be the ways in which the major functional separation between memory and processing in conventional computer systems results in two types of processes at the highest level: reading/writing memory and execution). Many existing cognitive models can be categorized on the basis of the following types of functional separations: (1) differences in knowledge organization between two subsystems; (2) differences in knowledge content between two subsystems; (3) differences in knowledge representation between two subsystems; (4) differences in knowledge processing between two subsystems.

For instance, Anderson (1993) proposed in his ACT* model that there are two types of knowledge: declarative knowledge is represented by semantic networks, and is consciously accessible; procedural knowledge is represented by production rules, and is inaccessible. The difference thus lies in the different way of organizing knowledge: whether knowledge is organized in an action-centered way (procedural knowledge) or in an action independent way (declarative knowledge). Both types of knowledge are implemented symbolically (using either symbolic semantic networks or symbolic production rules). The model does not explain the fundamental, qualitative phenomenological differences between the conscious and the unconscious (e.g. in terms of conscious accessibility).Although the knowledge organization is apparently different between semantic networks and production rules (with different degrees of action-centeredness), the difference is insufficient to account for the qualitative phenomenological difference, since both are symbolically represented and fundamentally the same. The difference in conscious accessibility is assumed in the model rather than intrinsic. To put this another way, there is nothing in the model which demonstrates at its own level and in its own terms that conscious processes cause speech (verbal reports) but unconscious processes do not. Causal relationships at higher levels do not follow from the detailed model.

Hunt and Lansman's (1986) model is almost the exact opposite of Anderson's model, although the emphasis is on knowledge access (as opposed to knowledge organization). In their model, the "deliberate" process of production matching and firing, which is serial, is assumed to be a conscious process, while the spreading activation (Collins and Loftus 1975) in semantic networks, which is massively parallel, is assumed to be an unconscious process. Despite the different emphases, the problem with this model is the same, the differences between conscious and unconscious processes in the model do not on their own level and in their own terms generate the necessary causal relationships of conscious and unconscious processes. (A number of other views had the same problem, such as Anderson 1993, Bower 1996, Logan 1988, and so on.)

There have also been various proposals in neurobiology that there are different processing pathways in the brain, some of which lead to conscious awareness while others do not. For example, Milner and Goodale (1995), Damasio et al (1990), and LeDoux (1992) proposed various versions of this view. Likewise, Schacter (1990) and Revonsuo (1993) suggested, based on neuropsychological data, that multiple modules coexist in the brain, each of which performs specialized processing (without incurring conscious awareness), with the exception of one module that is solely responsible for conscious awareness. Each of the specialized modules sends its output to the conscious module and thus makes the output consciously accessible. The problem of these biologically motivated two-system views is that, although there is ample biological evidence that indicates the existence of multiple pathways (in visual, language, and other processing modes), some of which are correlated with conscious awareness while some others are not, there is no explanation of why some result in consciousness while others do not, that is, what is different, mechanistically or computationally, between these different pathways. Mere association with different neural pathways does not constitute an explanation.

Yet another two system view is based on the representational difference. As proposed in Sun (1994, 1995, 1999), different representational forms (in different subsystems) may be used to explain the qualitative phenomenological difference between the conscious and the unconscious. According to connectionist theorizing (Sun 1995), in localist (or symbolic) representation, one distinct entity (e.g., a node in a connectionist model) represents a concept. Therefore, the representation is easily accessible. In distributed representation, a non-exclusive set of entities (e.g., a set of nodes in a connectionist network) are used for representing one concept and the representations of different concepts overlap each other; that is, a concept is represented as a pattern of activations over a set of entities (a set of nodes). Therefore, the representation is not easily accessible (relatively speaking). The mechanistic difference in accessibility between the two types of representation accounts for the phenomenological difference in accessibility between conscious and unconscious. To put this in a more simplistic way, verbal reports (speech) can be generated more easily from a localist representation. This view has been implemented in the CLARION model, which has two components using localist and distributed representations respectively (see Sun et al 2000).

Coward (1990, 1999a, 2000) proposed another two subsystem based model in which both knowledge representation and processing differ between the two subsystems. The knowledge in one subsystem is partial and ambiguous information about individual perceptual events, in the other it is behavioral interpretations of the information recorded in clustering. Knowledge recorded in the latter subsystem is behavior specific and not accessible to another behavior. The difference between conscious and unconscious processes resides partly the inability of speech behaviors to access information in the latter subsystem and partly in the much wider range of information active in the former subsystem during conscious processes. A primary difference from CLARION is that there are no unambiguous symbolic representations.

In contrast to these two-system views, there are also views that insist on the unitary nature of the conscious and unconscious; that is, they hold that the conscious and the unconscious are the different manifestations of the same underlying system or process (Dennett 1991). The difference in this case is between different processing modes in the same system. For example, as suggested by Baars (1988) and many others, some sort of coherence in the activities of the mind (or the brain) gives rise to consciousness. The emphasis is on internal consistency which supposedly produces consciousness. The distinction of the conscious and the unconscious is linked to the distinction between coherent and incoherent activities in the mind/brain. For example, Mathis and Mozer (1996) suggested that being in a stable attractor of a dynamic system (a neural network in particular) leads to consciousness. Crick and Koch (1990), on the other hand, suggested that synchronous firing of neurons (at 35-75 Hz in particular) leads to conscious awareness. There is also the variety that insists on the reverberation of information flows in various cortical and sub-cortical areas that leads to consciousness, as suggested by Damasio (1990). The difficulty with these views is that there is no explanation why coherence (whether it is in the form of attractors, reverberation, or synchronous firing) leads to consciousness; that is, what is qualitatively different about coherence (in any of these above forms) that can account for the qualitative, phenomenological difference in consciousness.

Notably, some of these views are based on biological data (i.e., concerning biological mechanisms, processes or systems; for example, the synchronous firing view, the reverberation view, and the two-pathway view). However, being biological does not automatically constitute an explanatory sufficiency, especially not in terms of underlying mechanisms, unless causal relationships at the biological level can be demonstrated to explain the appropriate causal relationships at the psychological level. The issue is, if it is assumed that coherence at the detailed level correlates with consciousness at the high level, what are the causal relationships between coherence and, for example, speech generation, and the corresponding absence of the possibility of causal links between states lacking coherence and speech generation. Such an explanation of causal relationships is absent for the discussion of one-system models.

7.2. Issues with these Models

What are the key issues here involved in establishing mechanistic/computational explanations of consciousness? First of all, we want a mechanistic (computational) explanation, which is the bottom line. Without mechanistic explanations, we can never claim to have achieved a scientific understanding of the nature of consciousness.

Furthermore, our mechanistic explanations should account for the qualitative phenomenological difference between the conscious and the unconscious, in mechanistic terms. The explanatory sufficiency, generality, and parsimony of our mechanistic constructs need careful attention also. Our explanations should have sufficient explanatory power to account for all major aspects, in parsimonious way and with generality (that is, with low Kolmogorov complexity). One way to achieve such explanatory power is through theoretical reduction, as opposed to mere empirical verification/confirmation (e.g. Crick and Koch 1990). Reduction means mapping conceptual entities in one theoretical framework to some other entities in another that are more fundamental, more tangible, or better understood, and thereby grounding the former level in the latter. For example, identifying biological substrates of consciousness such as done in the two-pathway view is important, but it by itself does not explain the phenomenological distinction between the conscious and the unconscious. It is an empirical verification and confirmation of the distinction, but not a reduction and not a sufficient mechanistic explanation of the phenomenological distinction.

Likewise, implementing conscious accessible and inaccessible knowledge in semantic networks and production rules, respectively, is useful in that it instantiates a cognitive model and enables the simulation of cognitive data, but it again does not constitute a reduction and does not explain the phenomenological. In contrast, the representational difference views do attempt to explain or reduce the phenomenological distinction in mechanistic terms. In other words, these views demonstrate causal relationships between their functions which correlate with the causal relationships observed at the phenomenological level. We will therefore focus on the models based on representational difference and discuss how these models can account for the more detailed phenomena described in section 6.

7.3. Description of the CLARION Model

7.3.1. The High Level Architecture

The CLARION model contains three major components as illustrated in figure 2. The top level and the bottom level both produce behavioral recommendations in response to an input condition, and a separate subsystem selects which behavior to carry out.

Both top and bottom levels receive the same sensory inputs, generate behavioral recommendations, and receive feedback on the consequences of accepted recommendations. The bottom level encodes procedural rules using distributed representations. Each module within the level corresponds with a behavior, and is generally uninterpretable in symbolic terms. Modules are evolved by consequence feedback (i.e. by reinforcement learning; e.g. Watkins 1989). The top level of the architecture encodes propositional rules using symbolic representations. Each unit or module within the level has a clear conceptual meaning/interpretation (i.e. a semantic label). Conceptual units are established by selecting an input condition which resulted in generation of a successful behavior by the bottom level, and coding it as a propositional rule leading to the same behavior. Such propositional rules can then also generate behaviors and can be evolved independently of modules in the bottom level. The two levels thus in time generate independent behavioral recommendations, and the selection subsystem selects a behavior from the range of alternatives generated by the two levels.

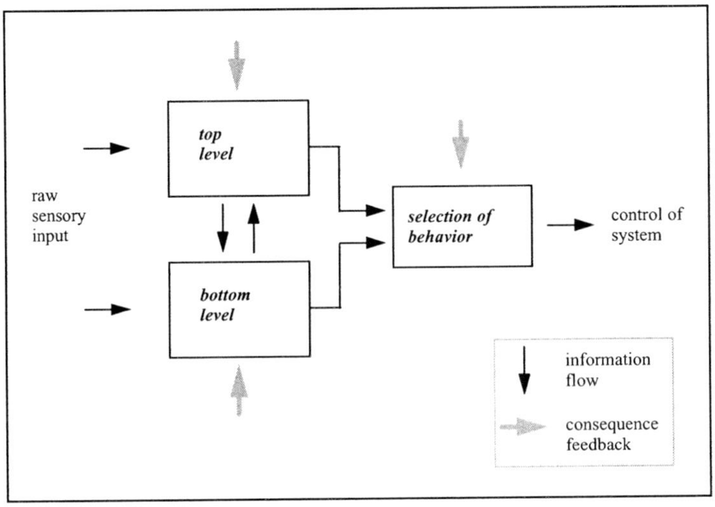

Figure 2. High level view of the CLARION architecture.

7.3.2. The Bottom Level

Modules within the bottom level correspond with possible alternative behaviors. In response to an input condition, the output from each module is a numeric value which can be interpreted as the relative value of the behavior corresponding with current input conditions. Modules are implemented as three layer backpropagation networks. The I/O mappings are acquired by supervised or reinforcement learning algorithms. Use of a reinforcement learning algorithm such as Q-learning (Watkins 1989) allows sequential behaviors to emerge. The system can learn to take into account future steps in longer and longer sequences. The bottom level is more sensitive to subtle or complicated forms of information. Processes operating at this level may be aptly described as associative, complex and fuzzy. The similarity of two items, defined as the similarity between their representational patterns, plays a major role, because connections between patterns direct the processing.

7.3.3. The Top Level

Modules within the top level consist of propositional rules. Each rule is implemented as a localist connectionist network. Rule learning occurs in response to every input condition. If some action decided by the bottom level is successful then the level extracts a rule that corresponds to the decision and adds the rule to the rule network. Then, in subsequent interactions with the world, the agent verifies the extracted rule by considering the outcome of applying the rule: if the outcome is not successful, then the rule should be made more specific and exclusive of the current case (shrinking); if the outcome is successful, the agent may try to generalize the rule to make it more universal (expansion). If two rules give similar results, they are merged. If a rule needs to be shrunk but cannot be without making it unresponsive to all input conditions, it is deleted. Activations above 0.9 are considered 1, and activations below 0.1 are considered 0. So rules are discretized and thus crisp/binary.

One input condition may correspond with multiple rules. In these circumstances, one rule is chosen from the matching set (by voting or randomly). Processes in the top level are more crisp and discrete, and thus more precise, reliable and selective. It is possible to follow exact rules. The top level allows explicit control and manipulation (due to explicit, localist representation) for deciding, altering and controlling reasoning methods. Each module has a clear conceptual meaning/interpretation (i.e. a semantic label).

7.3.4. Selection of Behavior

The different characteristics of the two levels makes the combination of the outcomes from the two levels advantageous. This subsystem makes the final decision of which action to take by incorporating outputs from the two levels. The corresponding values for an action from the two levels are combined in a weighted sum. The top level indicates that its recommended action has an activation value, which is 0 or 1 because rules are binary, and the bottom level indicates that its recommended action has a numeric activation value. Stochastic decision making (with Boltzmann distribution) based on the weighted sums is then performed to select an action. Relative weights of the two levels can be automatically set based on the probability matching of the relative performance of the two levels (which is commonly observed in animal behavior). That is, if the success rate of the decisions made by the top level is sb and the success rate of the bottom level is st, then the weights are st / (sb + st) for the top level and sb / (sb + st) for the bottom level. At a particular moment (and in a particular task), whether the top level, the bottom level, or both are used may be altered by a number of factors (as known from the psychological literature) e.g. instructions, or situational demands such as complexity or multiplicity of tasks, and so on (Sun et al 2000).

7.3.5. General Properties of the Model

As a preliminary to a more detailed comparison of the model with the psychological phenomena described in section 6, the processes in the model which correspond with four major psychological processes are described in general terms. These processes are: acquisition of a skill; recognition that a perceived object or condition is familiar (i.e. has been perceived before); reminiscence of a past object or condition; and generation of verbal reports describing a mental state.

In the CLARION model, skills can be acquired by two different types of learning modes: top-down learning and bottom-up learning. In top-down learning, top-level knowledge is learned first and then assimilated into the bottom level. In bottom-up learning explicit knowledge is extracted from implicit knowledge in the bottom level which is learned through trial and error first. Depending on the circumstances there might be mixed learning (both top-down and bottom-up), separate learning (learning within each level separately), or no learning at all. Recognition that an object or condition is familiar depends upon the existence of a large number of rules in the top level which correspond with the condition, and knowledge in the neural networks in the bottom level. Recognition that a new condition closely resembles a particular past condition can be performed, first of all, at the bottom level, which is similarity-based and associative, but also through the interaction of the two levels which leads to the explicit recognition of what a new condition is similar to by bottom-up partial activation of top-level nodes. Reminiscence of a past object or condition could be implemented by some top level nodes recommending activation of other nodes even in the absence of the input condition for those nodes, and speech could be implemented by some top level nodes recommending generation of speech elements.

7.4. Description of RA (Recommendation Architecture) Model

7.4.1. The High Level Architecture

The RA has a major functional separation into three subsystems as illustrated in figure 3.

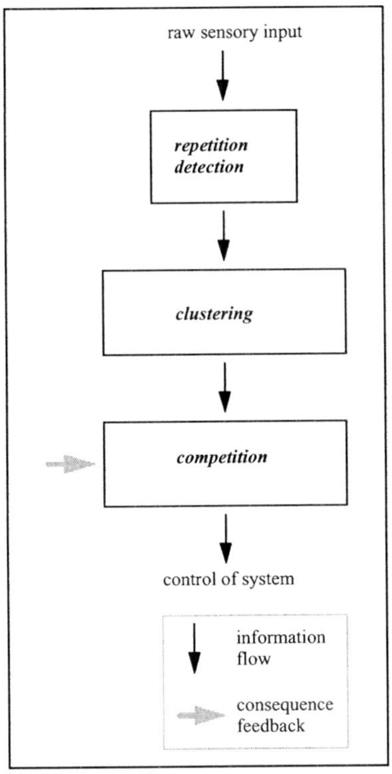

Figure 3. High level view of the recommendation architecture.

The repetition detection subsystem takes raw sensory inputs and generates as outputs selected combinations of those inputs. Each output is determined by a set of inputs, and the nature of one output is not affected by other outputs, so the interaction between different processes within this subsystem is minimal. There must be an initial construction process for this function, but after that initial process the outputs for a particular set of inputs do not change with experience. The clustering subsystem selects and permanently records combinations of inputs from the repetition detection subsystem and indicates by an output that they are occurring, and also indicates if a sufficiently similar repetition occurs in the future by generating exactly the same output. The recorded repetitions are organized into a hierarchy. The clustering subsystem compresses a potentially huge input space into a relatively small output space which can more easily be used to determine behavior. The competitive subsystem interprets the outputs of the clustering subsystem as alternative behavioral recommendations and selects one behavior. The probability of the clustering function generating a similar behavior in response to a similar combination of clustering outputs is modulated by feedback of consequences (although the clustering outputs themselves are not directly affected by consequence feedback). Separate competitive subsystems exist for different types of behavior, and these separate competitive subsystems compete for control of system behavior.

7.4.2. The Clustering Subsystem

The clustering subsystem is presented with inputs generated by the repetition detection subsystem from the current system environment, and selects and permanently records combinations of these inputs. The first recording and any subsequent repetition of a selected combination is indicated by a signal. The subsystem is organized into successive level of devices. The first level records combinations of inputs, the second level records combinations of first level combinations, and so on until a final level records potentially very complex combinations of combinations. The signals generated by the final level are the outputs from clustering to competition. Recording is permanent: once a combination has been recorded at any level and a signal produced, if the same combination repeats the same signal will be generated.

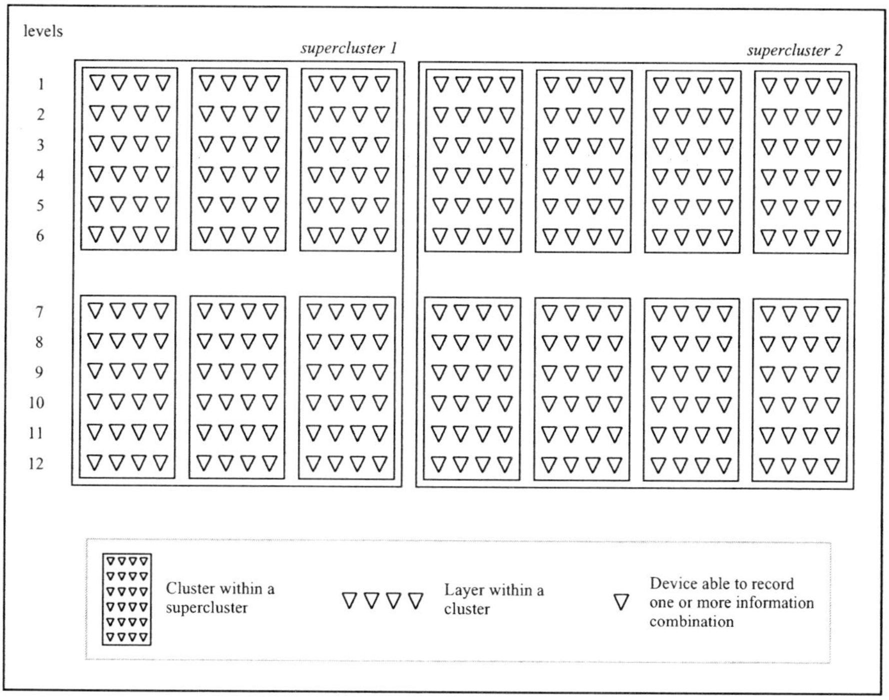

Figure 4. The structure of the clustering subsystem which is overlaid on the primary sequence of device levels recording information combinations of steadily increasing complexity. Clusters detect broad types of repetition, and layers within clusters determine whether the current input condition is sufficiently similar to the cluster type to justify producing an output by recording of additional combinations. Layers also determine if there is a requirement for additional clusters. The upper layer of clusters detect the presence of combinations of the order of complexity of objects and the lower layer combinations of the order of complexity of groups of objects (Coward, 2001). Superclusters detect repetitions appropriate for different major types of behavior.

When presented with a set of inputs derived from the current external and internal conditions, any already recorded combinations generate their signals, and additional combinations are recorded on every level until an output is produced from the final level, at which point recording stops. The clustering subsystem thus always finds the repetition of some information combinations which it indicates to the competitive subsystem. However, because the system heuristically defines the combinations to be recorded, the combinations only correlate partially with functionally relevant patterns and categories in the inputs. The competitive subsystem uses the presence of these repetitions to determine the appropriate behavior. The output space must be a major compression of the input space, but discriminate between input conditions for which different behaviors are appropriate. Structures within the clustering subsystem use some combinations of repetitions to determine when new combinations are to be recorded on different levels to achieve such compression (see figure 4). Devices, layers of devices, clusters, layers of clusters and superclusters as shown in figure 4 form the functional hierarchy of the recommendation architecture.

7.4.2. The Competitive Subsystem

The competitive system interprets the functionally ambiguous outputs of the clustering subsystem into a system behavior using a form of reinforcement learning. Devices within the competitive subsystem are neural networks. They can have both excitatory and inhibitory inputs with different input weights. In its simplest form the competitive subsystem is made up of a series of parallel pipes. Each pipe corresponds with one type of behavior, and the pipe with the strongest output is the behavior in response to the current input condition. In a pipe there are two layers of devices. The first layer receives excitatory inputs from the clustering subsystem, the second layer receives excitatory inputs from the first layer of the same pipe and inhibitory inputs from the first layer of other pipes. The threshold of all devices in all pipes is lowered until there is an output from at least one pipe. The strongest output is taken as the behavior, if there are several outputs of equal strength, one is selected at random. If the consequences of the selected behavior are good, all recently active excitatory input weights are increased and inhibitory weights reduced, and vice versa for bad consequences. Over a period of experience such a subsystem converges on a behavioral interpretation of the clustering subsystem outputs.

The separation between clustering and competitive subsystems is essential in any system which learns to perform a complex combination of functions. In such a system, modules handling different functional components must heuristically determine which outputs from other modules will be taken as inputs. Once a module begins to produce an output, it cannot change that output (for example in response to feedback of consequences) without influencing a potentially wide range of other modules which may play no role in the action which generated the consequences. Hence the hierarchy of modules which manages the interaction of all the different functions must be separated from the subsystem which uses reinforcement learning and selects behavior at each point in time (Coward 2000).

7.4.4. General Properties of the RA Model

We will overview how the model handles four psychological processes in general. The acquisition of a skill in the RA model is the process of weight adjustment in the competitive subsystem in response to feedback of consequences. Different combinations of repetitions indicated by the clustering subsystem are interpreted as different behaviors, although the repetition definitions themselves are not changed by the interpretation process. The recognition that a perceived object or condition is familiar derives from the degree to which the recording of new information combinations is required in order to generate an output in response to the condition. In reminiscence, the competitive subsystem interprets the outputs of the clustering subsystem as a recommendation to activate combinations of repetitions which are not currently active but have often been active on past occasions when the current clustering outputs were present. Generation of speech results from one interpretation of the clustering subsystem outputs.

8. Accounting for Psychological Phenomena at the Highest Level

The question is how the two models account for the important distinction in psychology between implicit and explicit mental processes. One central issue is how to account for the differences in the ease with which implicit and explicit processes can be expressed verbally.

One approach, taken by the CLARION model, is to associate the difference with differences in the ease with which a representation can be transformed into a corresponding conceptual label, which is taken as a prerequisite for verbal reporting and explicit thinking. One mathematically grounded way of defining explicitness as employed in CLARION is based on how easy a representational item can be transformed into a corresponding conceptual label, and thus a proper basis for defining explicitness. The transformation can be viewed as an algorithmic process - an algorithm transforms representation into corresponding labels. Thus, the explicitness of a representation can be defined based on the descriptive complexity of the simplest possible algorithm that can be used to transform it. Explicitness of a representation is therefore defined as the inverse of Kolmogorov complexity of its transformation -- a theoretical measure of the algorithmic complexity (Li and Vitanyi 1997), which we used earlier to define multiple levels of descriptions. Without getting into the technical details of this definition, it suffices to point out that the transformation of localist (symbolic) representation has lower Kolmogorov complexity compared with the transformation of distributed representation. This definition of explicitness gives us a solid foundation for claiming, based on connectionist theorizing, that localist representation is more accessible than distributed one, and thus for the hypothesis that the difference in representational accessibility can explain the difference in phenomenological accessibility. There can also be a host of additional factors. In the RA, one additional factor is whether or not the expanded information activation is present in response to the outputs which generated the behavior.

CLARION Model

In general, implicit processes are modeled by processes in the bottom level, and explicit processes are modeled by processes in the top level. The major distinction at the psychological level corresponds with a major functional separation at a more detailed level of description, as well as the difference in representation (localist vs. distributed). The reason the functional separation exists is that generation of behaviors by two alternative processes and selecting between the alternatives results in a better performing behavior than either process separately (that is, synergy). Representations in the top level are more explicit than representations in the lower level, as defined above, and therefore can be more easily expressed in speech. In process terms, because processes in the top level are on functional elements which correspond closely with semantic concepts, processes in this level can readily be communicated by speech.

RA Model

The primary source of the distinction between implicit and explicit processes at the psychological level is the separation between clustering and competition: information in the competitive subsystem is not accessible to the processes generating speech within the clustering subsystem. The reason the separation exists is the requirement to learn a complex combination of functions. A second source of the distinction is whether a recommendation to activate an expanded information set is accepted during the mental process. If such an activation is not present, the mental states making up the process are less rich.

9. Accounting for Consciousness at a More Detailed Level: Comparison of Models

It is not enough to account for a general difference between conscious and unconscious processes, or even to account in detail for one specific area such as skill learning. A viable theory must demonstrate the capability to account for a broad range of psychological phenomena, and preferably make some predictions on both psychological and physiological levels. To illustrate this process, we will compare the ability of the two models to account for phenomena across a wider range, and discuss some of the predictions which might be made.

9.1. Implicit and Explicit Perception

In implicit perception, the perception is not accessible to conscious introspection, as illustrated by the experiments of Triesman (1960), as discussed earlier.

CLARION Model

In CLARION, the bottom level can process information from different sources simultaneously, due to its modular structures. In order for information to be consciously accessible, however, it has to be represented at the top level. In the top level of CLARION, some conceptual nodes perform meaning analysis and generate speech recommendations. Other nodes generate recommendations biasing the relative probability of acceptance of alternatives from the bottom levels, which processes all inputs but generates results which may not be accessible without the top level. Yet other nodes respond to external instructions. The shift of text shadowing from one ear to the other following meaning analysis rather than explicit instructions, as in the experiments of Triesman (1960), can be modeled as the top-level nodes responsible for meaning analysis overruling nodes for following external instructions, which results in only information of consistent continuity, after processing by the bottom level, reaching the top level nodes.

RA Model

In implicit perception, outputs from the clustering subsystem are generated, but these outputs do not result in expanded information activation. So any repetitions imprinted during implicit perception will not incorporate information from the expanded information space, and will therefore not be activated in the future by such activations. The perception is therefore inaccessible to conscious introspection. In the experiments of Triesman (1960), input to both ears generates clustering outputs corresponding with behavioral recommendations. These recommendations include both recommendations for expanded information activation and "pay attention" recommendations which determine the input accepted for more complex processing within the clustering subsystem (for example, which outputs generated in the top layer of clusters in figure 4 are allowed to proceed to the bottom layer). The input from the ear for which the text is shadowed initially results in accepted recommendations. When the text continuity shifts, the recommendation to pay attention to the unshadowed ear will now be accepted because of the continuity of information with that currently being held active from the recent past.

9.2. Dissociations between Implicit and Explicit Memory

The next question is how the alternative models can account for differences between explicit and implicit memory, as illustrated for example, by implicit and explicit recognition of letter strings as described earlier.

CLARION Model

In CLARION, explicit memory may be driven by processes in the top level, while implicit memory may be driven by processes in the bottom level. The faster time decay for explicit memory results from the fact that top level has a high decay parameter than bottom level. The effect of organization of material on performance for explicit but not implicit memory is accounted for because the top level learning mechanism is more crisp, and thus more sensitive to the way the materials are presented, while in bottom level neural network learning is generally not affected as much by the input presentation. The performance for words with high natural language frequency can be understood on the basis that for those words, behavioral recommendations are well established in both levels, and since response speed is correlate with the degree of learning, there will be little change in response times for such words.

RA Model

In the RA, words operate by activating input level repetitions similar to those present in the past when, for example, both the word and the object indicated by the word were present. This mechanism means that there is a set of connections which activate a "mental image" of the object associated with the word when combinations recorded in the past when the word was heard repeat (Coward 1990). Response time for implicit memory depends on the weights in favor of a verbal response which have been assigned to already existing repetitions indicating activity in that set of connections. These repetitions will be complex combinations of information very specific to the particular word.

In explicit memory, recall of words on a previously presented list requires that expanded information activation during the original experience resulted in recording of some information combinations which include information from both the test environment and words on the list. In later testing, expanded information activation deriving from being reminded of the test environment must generate enough of these combinations that speaking a word on the list becomes an available behavior. Recognition that a word was on the previously presented list requires detection of the presence of already recorded combinations which include information from both the original test environment and the word.

The faster decay time for explicit memory results from decline in expanded activation probability of recorded information combinations which rarely or never repeat. Semantic organization can affect the type of information recorded in explicit memory. Lexical changes will change the information combinations used for implicit memory, and the different combinations will have smaller weights. Words with high natural language frequency will generate such a large activation of repetitions in the image activation path that implicit memory response will be rapid, and the effect of tuning input weights on future performance will be small.

9.3. Implicit and Explicit Skill Learning

As discussed earlier, some of the key phenomena which should be modeled at a deeper level include: learning skills which vary in complexity and degree of cognitive involvement, from simple motor movements to high-level intellectual skills; distinctions between the two types of knowledge, and transitions from one to the other in the development of complex cognitive skills; explicit knowledge may arise from procedural skills in some circumstances; the ability to learn and operate skills without involving explicit declarative knowledge; in some cases explicit declarative knowledge can speed up learning in new settings; a dissociation between explicit knowledge and skilled performance;

CLARION Model

The CLARION model was originally developed in the context of skill learning, and allows learning of skills of varying complexity. It provides a model for initial implicit learning shifting to explicit learning with a bottom-up learning mechanism that extracts explicit rules from neural networks (Sun et al 2000), and vice versa. Inconsistencies between explicit and implicit knowledge could arise through learning differences between the two levels (that is, reinforcement learning vs. rule learning). It provided a process by which declarative knowledge can accelerate the acquisition of implicit knowledge and vice versa through simultaneous use of two sets of learning and performance mechanisms (Sun et al 2000).

RA Model

For a skill being learned implicitly at first, then explicitly, groups of repetition conditions become associated with skill related behaviors, but initially these groups are not associated with speech behaviors. Explicit access to the operation of the skill occurs if similar groups of repetition conditions then become associated with speech behaviors on the topic of the skill. Inconsistencies can arise if such speech behaviors become associated with different groups of repetition conditions from the actual skill. On the other hand, explicit learning followed by implicit occurs if the initial groups of repetitions applied to the skill behavior are existing speech related groups. Reinforcement learning in the competitive subsystem causes the groups associated with the skill to diverge from the speech related groups.

9.4. Detailed Models of Specific Psychological Deficits

In amnesias such as those of Korsakov's patients, the most obvious symptoms are inability to create new declarative memories. Memories existing prior to the onset of the symptom are generally unaffected, although there can be loss of all memory from a section of time immediately preceding onset. However, general intelligence is unaffected, and there is an ability to learn skills even though there is no memory of the skill learning experience.

CLARION Model

In CLARION, acquiring new skills in the absence of declarative memory by amnesiacs is modeled as loss of capability to modify nodes in the top level while retaining the ability to modify nodes in the bottom level, given the functional separation between the two levels. The retention of existing information is accounted for by the unchanged (and unchangeable to some extent) existing top-level nodes, and by the bottom-level nodes. Acquiring new skill is possible because the bottom level is undamaged. To account for the loss of a "time slice" of memory, we hypothesize that damage to the top level remove nodes that are relatively recent (and thus less well established), and thus remove access to a recent slice of memory.

RA Model

In the RA model, loss of ability to acquire new declarative memories results from loss of the ability to record new information combinations in the clustering subsystem. Such a loss could, however, be consistent with generating outputs in response to repetition conditions already recorded. In an adult these repetition conditions would be extensive. In a novel situation such as learning Tower of Hanoi, if thresholds in the clustering subsystem were lowered sufficiently, some of these existing repetition conditions would generate outputs in response to the new situations and groups of such repetitions become associated with the new behavior. No new information would be recorded, and there would therefore be no explicit memory of the learning process. If local damage removed pathways to recently recorded information, a "time slice" of memory would be lost.

9.5. Functional and Information Plausibility

One issue is whether the models can scale to handle complex combinations of functions. A specific aspect is acquiring new learning without damaging past learning. A second issue is the need to work with high volumes of noisy input data which does not correlate easily with functionally useful cognitive categories.

CLARION Model

The CLARION model has demonstrated its ability to heuristically organize binary input spaces containing up to 41 inputs (i.e., in the order of 10^{12} states). The system can also handle analog input, with noise (Sun and Peterson 1998). The ability to control actions in tasks such as navigating through a minefield has been demonstrated.

The preservation of prior learning is achieved through modular structuring of the model (Sun and Peterson 1999), which clusters tasks according to their overall similarity to each other. This modularization not only speeds up reinforcement learning (in the bottom level) (Sun and Peterson 1999), but also ensures, to some degree, past learning is preserved.

RA Model

The RA has demonstrated its ability to heuristically organize binary input spaces containing up to 10,000 inputs, or 10^{3000} different input states (Coward 2001). When presented with a sequence of very noisy input conditions which are all different but which have enough similarities, a system can heuristically define repetition conditions in a sequence of such inputs. The output space compresses the input space by a factor which increases with the size of the input space. With a binary input space dimension of 10,000 the compression is by a factor of several hundred.

It has been demonstrated that new categories can be learned without interference with the recognition of existing categories (Coward 2000). The reasons are that new information combinations are added to clustering without changing older combinations, and associating of different skills with combinations of repetitions is separated into different modules for each skill within the competitive subsystem, so the learning of different skills is to some degree decoupled.

9.6. Biological/Physiological Plausibility

The issues here are whether a model has the potential ability to map into known physiology. This includes such questions as whether physiological structures in biological brains correspond with expected functional separations of models, whether computational mechanisms correspond with neuronal properties, and whether biological processes resemble model processes.

The human brain demonstrates some major structural differentiations. It is separated into major structures (such as cortex, basal ganglia, thalamus, hippocampus and many others). The cortex, regarded as a primary structure for higher cognitive functions, is a thin sheet made up of about six distinguishable layers, and the cortex is also separated into large numbers of columns which extend vertically through the sheet (Calvin 1995).

The basal ganglia (BG) has been identified to be the locus of reinforcement learning (Schultz et al 1997). The dopamine neurons in the BG likely predict reinforcement to be received and thereby drive the learning process using the discrepancy between the prediction and the actual reinforcement received (Houk et al 1995, Schultz et al 1997).

CLARION Model

The central role of modularity in motor control learning has been argued from the neurophysiological standpoint (e.g., Wolpert and Kawato 1998). In relation to the methods of modularization used in CLARION, it has been argued that such modules reside mainly in the supplementary motor area, while the selection of modules is likely done in the basal ganglia (see e.g. Bapi and Doya 1999). However, Houk et al (1995) suggested that action modules resided in the basal ganglia in the form of matrix regions. The learning (both the updating of modules and the updating of the module selection mechanism) is likely to be controlled by processes in the pre-supplementary motor area (Nakamura et al 1998).

There has been indications that in the human brain, the temporal lobe is often correlated with consciously controlled sequence learning (Keele et al 1998). In implicit memory research, however, it is found that explicit recall of sequences is often correlated with frontal lobe activities (Posner et al 1997). Thus, it may well be the case that the temporal lobe and/or the frontal lobe are both responsible for the explicit processing corresponding to the top level of CLARION. The bottom level of the model, responsible for implicit processing, is likely distributed across a wide range of brain regions (Keele et al 1998), including the pre-supplementary motor area, the supplementary motor area, and the motor cortex (Nakahara et al 1997).

RA Model

This model predicts that there will be a major separation in biological brains between clustering and competitive subsystems. There may be multiple subsystems of each type, but neuronal mechanisms in the two types are radically different. The cortex may correspond with a set of clustering subsystems, and such structures as the basal ganglia and cerebellum correspond with competitive subsystems (Coward 1990).

The model also requires functional separations into clusters and layers within the clustering subsystem which may correspond with columns and layers in the cortex. Because explicit knowledge is recorded in all the multiple clusters concerned with different broad types of behavior, local damage to the clustering subsystem may affect one type of behavior but will not in general remove explicit memories. This is consistent with the classic work of Lashley (1950) on the effects of local damage to the cortex in rats. The identification of the basal ganglia (BG) as the locus of reinforcement learning is consistent with the RA.

10. Conclusions

A definition of a scientific theory of consciousness as a hierarchy of descriptions of the same phenomena on many levels, in which the deeper levels have higher information density but lower descriptive complexity, has been proposed based on analogy with physical sciences. It is argued that biological brains have been constrained into simple functional architectures (successive levels of isomorphic descriptions) by the needs to be constructed, to recover from damage, and to learn, and that the hierarchy of descriptions implicit in such architectures can be the basis for such a scientific theory.

The outlines of some attempts at such a theory at high levels have then been constructed beginning with access and phenomenal consciousness as the highest level description, and implicit and explicit mental processes as the next level of detail.

It has then been demonstrated that many current cognitive models do not have the capabilities to provide the next deeper levels of descriptive detail of the higher level phenomena, and are therefore inadequate as the basis for a scientific theory. Two current models have been demonstrated to have some capabilities of this type: CLARION and RA. The capabilities of these models to describe implicit and explicit mental processes and to account for the experimental dissociations between the processes were then compared. The consistency of the two models with the physiological level was also compared.

The comparison shows the ability of both models to account for many of the observed phenomena. CLARION has the advantage of greater simplicity, RA has the advantage of a stronger potential to scale to very complex combinations of functionality. A further point is that simulation of a range of implicit and explicit mental processes to a deeper level of detail which makes it possible to compare such models based on detailed experimental results is needed.

REFERENCES

Anderson, J. R. (1983). *The Architecture of Cognition*, Harvard University Press, Cambridge, MA.

Anderson, J. R. (1993). *Rules of the Mind.* Lawrence Erlbaum Associates. Hillsdale, NJ.

Baars, B. (1988). *A Cognitive Theory of Consciousness*. Cambridge University Press.

Bainbridge, L. (1977). Verbal reports as evidence of the process operator's knowledge. *International Journal of Man-Machine Studies*, 11, 411-436.

Bapi, R. and Doya, K. (1999). MFM: Multiple forward model architecture for sequence learning. In: L. Giles and R. Sun, (eds.) *Proceedings of the IJCAI'99 Workshop on Neural, Symbolic, and Reinforcement Methods for Sequence Learning.*

Berry, D.C. (1987). The problem of implicit knowledge. *Expert Systems*, 4, 144-151.

Berry, D. and Broadbent D. (1988). Interactive tasks and the implicit-explicit distinction. *British Journal of Psychology*. 79, 251-272.

Block, N. (1994). On a confusion about a function of consciousness. *Brain and Behavioral Sciences*.

Bower, G. (1996). Reactivating a reactivation theory of implicit memory. *Consciousness and Cognition*, 5, 1/2, 27-72.

Braddon-Mitchell, D. and Jackson, F. 1996: *Philosophy of Mind and Cognition*, Oxford, Blackwell.

Breiman, L. (1996). Bagging predictors. *Machine Learning*, 24, 123-140.

Butters, N. (1984). Alcoholic Korsakov's Syndrome: An Update, *Seminars in Neurology* 4, 2, 226-244.

Chalmers, D. (1993). *Towards a Theory of Consciousness*. Ph.D Thesis, Indiana University.

Calvin, W.H., (1995). Cortical Columns, modules, and Hebbian cell assemblies. In M.A. Arbib, Ed., *The Handbook of Brain Theory and Neural Networks*, Cambridge, MA: Bradford Books/MIT Press.

Chi, M., Bassok, M., Lewis, M., Reimann, P. and Glaser, P. (1989). Self-explanation: how students study and use examples in learning to solve problems. *Cognitive Science.* 13. 145-182.

Clark, A. (1992). The presence of a symbol. *Connection Science* 4, 193-205.

Cohen, N.J., Eichenbaum, H., Deacedo, B.S., and Corkin, S. (1985). Different memory systems underlying acquisition of procedural and declarative knowledge. In D.S. Olton, E. Gamzu, and S. Corkin (Eds.), *Memory dysfunction: an integration of animal and human research from preclinical and clinical perspectives* (54-71). New York: New York Academy of Sciences.

Collins, A. and Loftus, J. (1975). Spreading Activation theory of semantic processing, *Psychological Review*, 82, 407-428.

Coward, L. A. (1990). *Pattern Thinking*, New York: Praeger.

Coward, L.A. (1999a). A physiologically based approach to consciousness, *New Ideas in Psychology*, 17 (3), 271-290.

Coward, L.A. (1999b). A physiologically based theory of consciousness, in Jordan, S. (ed.), *Modeling Consciousness Across the Disciplines*, 113-178, Maryland: UPA.

Coward, L.A. (2000). A Functional Architecture Approach to Neural Systems. *International Journal of Systems Research and Information Systems*, 9, 69 - 120.

Coward, L.A. (2001). The Recommendation Architecture: lessons from the design of large scale electronic systems for cognitive science. *Journal of Cognitive Systems Research* 2(2), 111-156.

Crick, F. and Koch, C. (1990). Toward a neurobiological theory of consciousness. *Seminars in the Neuroscience*, 2, 263-275.

Damasio, A. (1994). *Decartes' Error.* Grosset/Putnam, New York.

Dennett, D. (1991), *Consciousness Explained.* Little Brown.

Dreyfus, H. and Dreyfus, S. (1987). *Mind Over Machine: The Power of Human Intuition*, The Free Press, New York.

Dunn, J. (1998). Implicit Memory and Amnesia, in Kirsner, K., Speelman, C., Maybery, M., O'Brien-Malone, A., Andersen, M., and MacLeod, C. (Eds). *Implicit and Explicit Mental Processes*, pp 99-117, New Jersey: Erlbaum.

Edelman, G. (1989). *The Remembered Present: A Biological Theory of Consciousness.* Basic Books, New York.

Efron, B. and Morris, C. (1973). Combining possibly related estimation problems. *J. Royal Statistical Society.* 35, 379-421.

Forbach, G.B., Stanners, R.F., and Hochhaus, L. (1974). Repetition and practice effects in a lexical decision task. *Memory and Cognition*, 2, 337-339.

Freeman, W. (1995). *Societies of Brains.* Lawrence Erlbaum Associates, Hillsdale, NJ.

Greene, B., (1999). *The Elegant Universe*, New York: Norton.

Hayes, N. and Broadbent, D. (1988). Two modes of learning for interactive tasks. *Cognition*, 28, 249-276.

Hadley, R. (1995). The explicit-implicit distinction. *Minds and Machines* 5, 219-242.

Heidegger, M. (1927). *Being and Time.* English translation published by Harper and Row, New York. 1962.

Houk, J. Adams, J. and Barto, A. (1995). A model of how the basal ganglia generate and use neural signals that predict reinforcement. In J. Houk, J. Davis, and D. Beiser, (eds.) *Models of Information Processing in the Basal Ganglia*, MIT Press, Cambridge, MA.

Hunt, E. and Lansman, M. (1986). Unified model of attention and problem solving. *Psychological Review*. 93 (4), 446-461.

Jackendoff, R. (1987). *Consciousness and the Computational Mind*. MIT Press, Cambridge, MA.

Jacoby, L. , Toth, J. and Yonelinas, A. (1993). Separating conscious and unconscious influences of memory: measuring recollection. *Journal of Experimental Psychology: General*, 122 139-154.

Keele, S., Ivry, R., Hazeltine, E., Mayr, U. and Heuer, H. (1998). *The cognitive and neural architecture of sequence representation*. Technical report No.98-03, Institute of Cognitive and Decision Sciences. University of Oregon.

Kelley, C. and Jacoby, L. (1993). The construction of subjective experience: memory attribution. In: *Consciousness*. eds. M. Davies and G. Humphreys. Blackwell, Oxford, UK.

Kim, J. (1993) *Supervenience and Mind*, Cambridge University Press

Kirsner, K. (1998). Implicit Memory, in Kirsner, K., Speelman, C., Maybery, M., O'Brien-Malone, A., Andersen, M., and MacLeod, C. (eds), *Implicit and Explicit Mental Processes*, pp 13-36, New Jersey: Erlbaum.

Lashley, K. S.(1950). In search of the engram. *Symposia of the Society for Experimental Biology* 4, 454-482.

LeDoux, J. (1992). Brain mechanisms of emotion and emotional learning. In: *Current Opinion in Neurobiology*. Vol.2, No.2, 191-197.

Lewicki, P., Hill, T. and Czyzewska, M. (1992). Nonconscious acquisition of information. *American Psychologist*. 47, 796-801.

Li, M. and Vitanyi, P. (1997). *An Introduction to Kolmogorov Complexity and Its Applications*. Springer, Heidelberg, Germany.

Libet, B. (1985). Unconscious cerebral initiative and the role of conscious will in voluntary action. *Behavioral and Brain Sciences*. 8, 529-566.

Lloyd, D. (1995). Consciousness: a connectionist manifesto. *Minds and Machines*. 5: 161-185.

Logan, G. (1988). Toward a theory of automatization. *Psychological Review* 95, (4), 492-527.

Marr, D. (1982). *Vision*, W.H. Freeman: New York.

Mathews, R. Buss, R. Stanley, W. Blanchard-Fields, F. Cho, J. and Druhan, B. (1989). Role of implicit and explicit processes in learning from examples: a synergistic effect. *Journal of Experimental Psychology: Learning, Memory and Cognition*. 15, 1083-1100.

Mathis, D. and Mozer, M. (1996). Conscious and unconscious perception: a computational theory. *Proceedings of 18th Annual Conference of Cognitive Science Society*, 324-328. Erlbaum, Mahwah, NJ.

Milner, D. and Goodale, N. (1995). *The Visual Brain in Action*. Oxford University Press, New York.

Nagel, T. (1974). What is it like to be a bat? *Philosophical Review*. 4, 435-450.

Nakahara, H., Doya, K., Hikosaka, L. and Nagano, S. (1997). Reinforcement learning with multiple representations in the basal ganglia loops for sequential motor control. *International Joint Conference on Neural Networks*, 1553-1558.

Nelson, T. (Ed.) (1993). *Metacognition: Core Readings*. Allyn and Bacon.

Penrose, R. (1994). *Shadows of the Mind*. Oxford University Press. Oxford, UK.

Posner, M., DiGirolamo, G. and Fernandez-Duque, D. (1997). Brain mechanisms of cognitive skills. *Consciousness and Cognition*. 6, 267-290.

Reber, A. (1989). Implicit learning and tacit knowledge. *Journal of Experimental Psychology: General*. 118 (3), 219-235.

Revonsuo, A. (1993). Cognitive models of consciousness. In M. Kamppinen (ed.), *Consciousness, Cognitive Schemata and Relativism*. 27-130. Kluwer, Dordrecht, Netherland.

Rosenbloom, P., Laird, J. and Newell, A. (1993). *The SOAR papers: Research on Integrated Intelligence*. MIT Press, Cambridge, MA.

Scarborough, D.L., Cortese, C., and Scarborough, H.S. (1977). Frequency and repetition effects in lexical memory. *Journal of Experimental Psychology: Human Perception and Performance*, 3, 1-17.

Schacter, D.L., Bowers, J., and Booker, J. (1989). Intention, awareness and implicit memory: The retrieval intentionality criterion. In S. Lewandowsky.

Schacter, J.C. and Graf , P. (1986). Preserved learning in amnesic patients: Perspectives from research on direct priming. *Journal of Clinical and Experimental Neuropsychology*, 8, 727-743.

Schacter, D. (1987). Implicit memory: History and current status. *Journal of Experimental Psychology: Learning, Memory, and Cognition*, 13, 501-518.

Schacter, D. (1990). Toward a cognitive neuropsychology of awareness: implicit knowledge and anosagnosia. *Journal of Clinical and Experimental Neuropsychology*. 12 (1), 155-178.

Schraagen, J. (1993). How experts solve a novel problem in experimental design. *Cognitive Science*. 17, 285-309.

Schultz, W. Dayan, P. and Montague, R. (1997). A neural substrate of prediction and reward, *Science*, 275, 14, 1593-1603.

Searle, J. (1980). Minds, brains, and programs. *Brain and Behavioral Sciences*. 3, 417-457.

Seger, C. (1994). Implicit learning. *Psychological Bulletin*. 115 (2), 163-196.

Shimamura, A. (1986). Priming effects in amnesia: Evidence for a dissociable memory function. *Quarterly Journal of Experimental Psychology*, 38A, 619-644.

Speelman, C. (1998). Implicit Expertise: Do We Expect Too Much from Our Experts. In Kirsner, K., Speelman, C., Maybery, M., O'Brien-Malone, A., Andersen, M., and MacLeod, C. (Eds.), *Implicit and Explicit Mental Processes*, pp 135-147, New Jersey: Erlbaum.

Stanley, W. , Mathews, R., Buss, R. and Kotler-Cope, S. (1989). Insight without awareness: on the interaction of verbalization, instruction and practice in a simulated process control task. *Quarterly Journal of Experimental Psychology*, 41A (3), 553-577.

Sun, R. (1994). *Integrating Rules and Connectionism for Robust Commonsense Reasoning*. John Wiley and Sons, New York, NY.

Sun, R. (1995). Robust reasoning: integrating rule-based and similarity-based reasoning. *Artificial Intelligence*. 75, 2. 241-296.

Sun, R. (1999). Accounting for the computational basis of consciousness: A connectionist approach. *Consciousness and Cognition*, Vol.8, pp.529-565. December, 1999.

Sun, R. and Bookman, L. (eds.) (1994). *Computational Architectures Integrating Neural and Symbolic Processes*. Kluwer Academic Publishers. Norwell, MA.

Sun, R., Peterson, T. and Merrill, E. (1996). Bottom-up skill learning in reactive sequential decision tasks. *Proceedings of 18th Cognitive Science Society Conference*, 684-690. Lawrence Erlbaum Associates, Hillsdale, NJ.

Sun, R. and Peterson, T. (1998). Autonomous learning of sequential tasks: experiments and analyses. *IEEE Transactions on Neural Networks*, Vol.9, No.6, pp.1217-1234.

Sun, R. and Peterson, T. (1999) Multi-agent reinforcement learning: weighting and partitioning. *Neural Networks*, 12, 4-5, 127-153.

Sun, R. , Merrill, E. and Peterson, T. (2000). From implicit skills to explicit knowledge: a bottom-up model of skill learning. *Cognitive Science*, in press.

Taylor, J. (1997). The relational mind. in: A. Browne, (ed.) *Neural Network Perspectives on Cognition and Adaptive Robotics*. IOP, Bristol, UK.

Triesman, A.M. (1960). Contextual clues in selective listening, *Quarterly Journal of Experimental Psychology*, 12, 242-248.

Tulving, E., Schacter, D.L., and Stark, H.A. (1982). Priming effects in word-fragment completion are independent of recognition memory. *Journal of Experimental Psychology: Learning, Memory, and Cognition*, 8(4), 336-342.

Van Gulick, R. (1993). Understanding the phenomenal mind. In: *Consciousness*. eds. M. Davies and G. Humphreys. Blackwell, Oxford, UK.

VanLehn, K. (1995). Cognitive skill acquisition. In: J. Spence, J. Darly, and D. Foss, (eds.) *Annual Review of Psychology*, Volume 47, Annual Reviews Inc. Palo Alto, CA.

Watkins, C. (1989). *Learning with Delayed Rewards*. Ph.D Thesis, Cambridge University, Cambridge, UK.

Willingham, D., Nissen, M. and Bullemer, P. (1989). On the development of procedural knowledge. *Journal of Experimental Psychology: Learning, Memory, and Cognition*. 15, 1047-1060.

Wolpert, D. M. and Kawato, M. (1998). Multiple paired forward and inverse models for motor control. *Neural Networks*, 11, 1317-1329.

Chapter 4

PRECISION OF WARM'S WEIGHTED LIKELIHOOD ESTIMATION OF ABILITY FOR A POLYTOMOUS MODEL IN CAT

Shudong Wang*
Harcourt Assessment Systems
Tianyou Wang
ACT, Inc.

ABSTRACT

The purpose of this Monte Carlo (MC) study was to evaluate the relative accuracy of Warm's weighted likelihood estimate (WLE) in comparison with maximum likelihood estimate (MLE), expected a posteriori estimate (EAP), and maximum a posteriori estimate (MAP) using the generalized partial credit model under a variety of computerized adaptive testing conditions. The results of this study have provided a first indication that WLE is more accurate than MLE when a polytomous IRT model for fixed test length CAT is used in the CAT environment, and are consistent with previous findings that WLE and MLE have smaller bias but larger SE than both EAP and MAP. EAP is more accurate than MAP in a variety of CAT conditions. Although there are differences of conditional RMSE among four methods, this study showed that no statistically significant mean differences of RMSE among all four existing methods. Also, both EAP and MAP had advantages over WLE and MLE in terms of fidelity and test efficiency. These results also suggest that the test termination rule had more impact on the accuracy of ability estimation methods than did the size of item banks with similar item characteristics.

Keywords: computerized adaptive testing, ability estimation methods, polytomous responses, item response theory.

* All correspondence should be sent to: Shudong Wang, Suite 300, Three Bala Plaza West, Bala Cynwyd, PA 19004-3481, (610) 617-5178, Email: swang@HARCOURT.com

INTRODUCTION

Currently, computerized adaptive testing (CAT) is of considerable interest to the measurement and research community because of its advantages over the traditional paper-pencil tests (Lord, 1977; Kingsbury & Weiss, 1983; McBride & Martin, 1983; Urry, 1977; Wainer et al., 1990). So far, almost all real CAT applications have been based on dichotomous items. CATs with polytomous models can be applied to items with rating scales, or in some situations with multiple-choice items. However, polytomous CATs are not widely used in the educational testing settings because machine scoring of polytomous items is still difficult to achieve. Bennett, Steffen, Singley, Morley, and Jacquemin's (1997) research in computer scoring of open-end format items has shown new hope for polytmous item-based CAT.

The ability estimation procedure is one of the most important components in a CAT system. The accuracy of ability estimation methods used in CAT has a significant impact on the quality of CAT testing because it affects not only the final score reported, but also the item selection and test termination. The properties of various ability estimation methods have been studied extensively for CAT with dichotomous IRT models (e.g., Book & Mislevy, 1982; Wang & Vispoel, 1998; Wang, Hanson & Lau, 1999). Among the methods that have been studied, Warm's (1989) weighted likelihood estimate (WLE) method was found particularly promising because it not only reduces the bias of the MLE method as it originally purported to do, but reduces its standard error (SE) as well computerized adaptive testing for fixed test length CATs. The Bayesian methods, such as expected a posteriori estimate (EAP) and maximum a posteriori estimate (MAP), though having smaller SEs than the non-Bayesian methods, all have large bias toward the prior means. Samejima (1998) expanded the WLE method to include the general polytomous IRT models. Warm claimed the WLE removed the first-order bias term from MLE. In general, it is desirable to have unbiased parameter estimation. Reducing bias in ability estimation can be important in several situations (Wang, Hanson, & Lau, 1999). One such situation is when comparability between CAT and paper-pencil versions is needed; unbiased ability estimation methods may help to reduce the need to equate the CAT and paper-pencil versions (Eignor & Schaeffer, 1995; Segall, 1995; Segall & Carter 1995; Wang & Kolen, 1997). Another situation is in domain-reference testing or certification and licensure testing where cut scores are set on the ability scales. In that situation, bias can systematically affect the precision of the cut score and consequently the validity of the classification decisions. However, bias may be of little concern when the main purpose of a CAT test is to rank order examinees. The purpose of this study is to empirically investigate the properties of the WLE in CAT conditions using a generalized partial credit model, and compare it with WLE, EAP, and MAP.

THE GENERAL APPROACH TO REDUCING THE BIAS OF MLE

Firth (1993) identified two approaches that reduce the MLE bias. One is a corrective approach, and the other is a preventive approach.

The corrective approach includes two methods that have been extensively studied in the literature; one encompasses computationally intensive procedure, such as jackknife and bootstrap methods (Quenouille, 1949,1956); the other is simply to subtract, as estimate of the first-order bias, $\text{Bias}_1(\text{MLE}(\hat{\theta}))$ from the MLE estimate; the bias-corrected estimate is then expressed as

$$\hat{\theta}_{CorrectBias} = \hat{\theta}_{MLE} - Bias_1(MLE(\hat{\theta})) \,. \tag{1}$$

Both of these methods may succeed in removing term $\text{Bias}_1(\text{MLE}(\hat{\theta}))$ (Warm, 1989). A common feature of the two methods is that they are 'corrective' in nature; that is, the MLE $\hat{\theta}$ is first calculated, and then corrected. For that reason, both methods require the existence of a finite $\hat{\theta}$.

The preventive approach (Firth, 1993) to reduce MLE estimate bias, on the other hand, modifies the score function before the MLE estimate is calculated. In general, the MLE is derived as a solution to the score equation $S(\theta)$:

$$S(\theta) = \frac{\partial}{\partial \theta} l(\theta) = 0 \,, \tag{2}$$

where $l(\theta)=\ln L(\theta)$ is the log likelihood function for any given model.

The bias in $\hat{\theta}$ can be reduced by introducing a small bias (first-order) into the score function (Firth, 1993). For given bias $B(\theta)$, score function $S(\theta)$ can be modified to be $S^*(\theta)$ by simple triangle geometry as illustrated in Figure 1. Since the Fisher Information function is defined as

$$I = I(\theta) = -E\left[\frac{\partial^2}{\partial \theta^2} \ln L(\mathrm{u} \mid \theta)\right] = -E\left[\frac{\partial^2}{\partial \theta^2} l(\theta)\right] = -E\left[\frac{\partial}{\partial \theta} S(\theta)\right], \tag{3}$$

which is the expectation of the negative value of $S(\theta)$ tangent at $\hat{\theta}$, or $-I(\theta) = S'(\theta)$. Then for any given bias $B(\theta)$, the score function can be shifted by the amount of $I(\theta)B(\theta)$. This defines a modified score function

$$S^*(\theta) = S(\theta) - I(\theta)B(\theta) \tag{4}$$

and hence a modified estimate θ^*, is given by solving $S^*(\theta) = 0$. It can be seen that, in general, the $O(n^{-1})$ bias may be removed from the maximum likelihood estimator by introduction of an appropriate bias term into score function. It is not an assumption of this procedure that bias reduction is always desirable. The merits of bias reduction in any particular problem will depend on a number of factors (Copas, 1988).

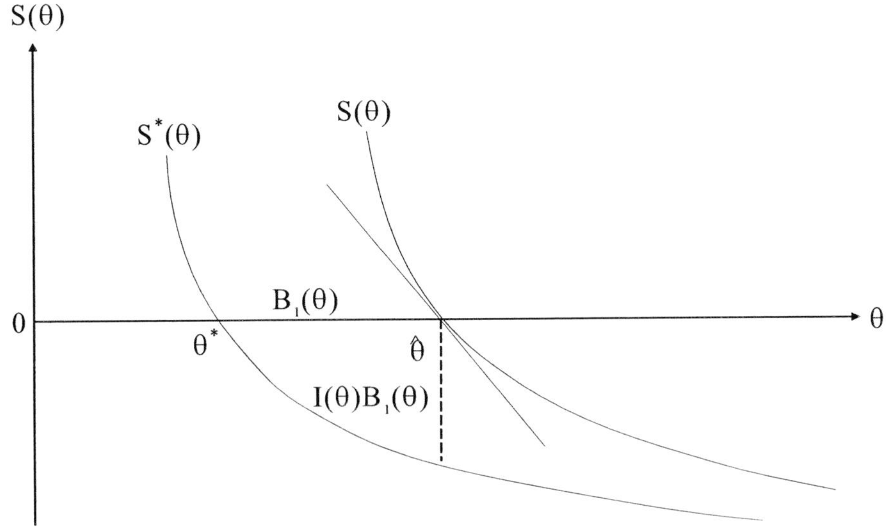

Figure 1. Modification of the Score Equation.

Although Warm (1989) did not have a general rationale in deriving his WLE method to reduce the bias of MLE, his method is actually a special case of Firth's (1993) preventive appraoch (Wang, Hanson & Lau, 1999). Warm derived the WLE method based on a conjecture from Lord's (1983, 1986) bias functions of MLE and Bayse Modal estimation method. His final result, however, is equivalent to the general result in Equation 4.

EXPANSION OF WLE FOR POLYTOMOUSLY SCORED RESPONSES

Samejima (1998) expanded Warm's weighted likelihood estimation method to include polytomously scored responses. Samejima used the right adjusting term for reducing bias (i.e., the term in Equation 4) without being aware of the general rationale formulated by Firth (1993). In order to express this adjusting term, Samejima (1993a, 1993b) first generalized Lord's MLE bias function in 3PL for polytomous responses, then based on the bias function for polytomously scored items, applied Warm's WLE to the polytomous responses.

The MLE bias function for general discrete responses is given by

$$Bias(MLE(\theta;\hat{\theta}_v)) \cong E[\hat{\theta}_v - \theta \,|\, \theta] \cong -\frac{1}{2[I(\theta)]^2}\sum_{j=1}^{n}\sum_{k=0}^{m_j}\frac{\frac{\partial}{\partial\theta}P_{jk}(\theta)\frac{\partial^2}{\partial\theta^2}P_{jk}}{P_{jk}(\theta)}, \qquad (5)$$

where $P_{jk}(\theta)$ is the probability of a response in category k to item j with the assumption that $P_{jk}(\theta)$ is, at least, five times differentiable with respect to θ. When $P_{jk}(\theta)$ is modeled by Samejima's Graded Response Model (1969), the bias function (Equation 5) can be expressed as Equation A1 (see Appendixes).

When $P_{jk}(\theta)$ is modeled by Muraki's Generalized Partial Credit Model, this bias function can be expressed as Equation A4 (see Appendixes). A straightforward expansion of Warm's WLE for 3PL to general discrete response can be expressed as

$$\frac{\partial}{\partial\theta}\ln L(\mathrm{v}\mid\theta)+\frac{\partial}{\partial\theta}\ln w(\theta)=\sum_{k_j\in v}\frac{\partial}{\partial\theta}\ln P_{jk}(\theta)-Bias(MLE(\theta,\hat{\theta}))I(\theta)\equiv 0 \tag{6}$$

where $I(\theta)$ is the test information function. Muraki and Bock (1999) provided a computer program to calculate WLE for both the generalized partial credit model and the graded response model.

METHOD AND DATA

A Monte Carlo (MC) simulation method was used to evaluate the ability estimation method in this study. Specifically, 21 true ability values ranging from –4.0 to 4.0 by increments of 0.4 were used for the GPCM. CAT was simulated for 500 examinees at each of the 21 true ability parameter points. Both descriptive methods and inferential procedures were used to analyse the results from the simulation. The descriptive methods provide global summaries of the study results. Inferential procedure, on the other hand, can overcome some of the deficiencies of the descriptive methods by conceptualizing this MC study as a statistical sampling experiment (Harwell, 1997; Spence, 1983).

Independent Variables

(1) Ability Estimation Methods

The primary categorical independent variable examined in this study is ability estimation method. Four level categories of ability estimation methods for the generalized partial credit model are MLE, WLE, EAP, and MAP. The relationships among this independent variable and other independent variables (described below) are considered. For each method, the values of the true ability parameter used in this study are equally spaced across a fixed range on the θ scale.

(2) Test Termination Rules

The second categorical independent variable is test termination rule. The types of test termination rules to be investigated for the GPCM and the GRM are:

(a) Fixed test length: The CAT tests will be terminated after a certain number of items has been administered. Four test lengths used in this study are 5, 10, 15, and 20 items. Previous studies (Koch & Dodd, 1985; DeAyala, 1992; Dodd et al., 1989) employed fixed test lengths from 10 to 30 items. To search for the smallest acceptable test length, a test length of 5 items is also included in this investigation.

(b) Fixed test reliability: The tests will be terminated when certain values of estimated reliability are reached.

Since the relationship between reliability (ρ) and the standard error of ability estimation (SE) can be expressed as $\rho = 1- SE^2$ given the θ variance is 1, estimated reliability and estimated SE have the same effect on the CAT termination. Both MLE and WLE use the square root of the reciprocal of test information as SE. Both EAP and MAP Bayesian methods use the standard deviation of the posterior distribution as SE. The three values of reliability to be employed are 0.7, 0.8, and 0.9, which correspond to SEs of 0.55, 0.45, and 0.32 and test information values of 3.33, 5, and 10. A maximum test length of 33 items will be used to terminate the test for a given exaimee if the prespecified levels of reliability cannot be reached. This maximum test length is used because the smallest bank size is 33 items.

(3) Sizes of Item Bank

The third independent variable is the size of item bank. Although some research has found that item banks with 30 items may be sufficient for accurate θ estimation with few nonconvergence problems (Dodd, 1987, 1993; Dodd & DeAyala, 1994; Dodd et al., 1989; Koch & Dodd, 1989), these findings do not imply that any item bank composed of 30 or more items will be sufficent for polytomous CAT (Dodd et al., 1995).

The real-life item bank 1GP, which consist of 263 polytomously scored 1996 NAEP science items that have 3 to 5 categories for grades 4, 8, and 12, will be used for this study. The item parameters for the generalized partial credit model were calibrated by the Educational Testing Service, which used a very complex sampling design and scaling methodology for the NAEP Science Assessment. The details are described in the NAEP 1996 Technical Report (Allen, Carlson, & Zelenak, 1999), which is too complex to describe here. In summary, it used a matrix sampling procedure called the balanced incomplete block (BIB) spiraling, which required the examinees to take different subsets of items from the item bank. The item subsets were carefully balanced and spiraled among the examinees in test administrations. The item parameters were calibrated using a special version of combined BILOG/PARSCALE programs. The dimensionality analyses reported in the technical report indicated that the response data were close to being unidimensional (Allen, Carlson, & Zelenak, 1999, pp. 191-192). The accuracy of item parameter estimation and the dimensionality issues of the response data are not of great concern to this study because we used simulation techniques which treat the unidimensional IRT model as true and treat the item parameters as if they were true parameters. However, when the results of this study are applied to the real world testing situations, these issues have to be taken into account.

There are three sizes of item banks (1GP, 2GP, and 3GP) for the GPCM. The 66 items in the bank named 2GP and the 33 items in the bank named 3GP are randomly drawn from the bank named 1GP using the proportional stratified random sampling method (Gall et al., 1996), based on the same proportions of items with different category numbers in the 1GP bank. Table 1 shows the summary of descriptive statistics for the item parameters of these three item banks.

Table 1: Summary of Descriptive Statistics for the Estimates of Item Parameters of the Three Item Banks, 1GP, 2GP, and 3GP, under the Generalized Partial Credit Model.

Name/ Size	Parameter	Mean	Median	S.D.	Min.	Max.
1GP (263)	a	0.549	0.522	0.229	0.105	1.871
	b_1	0.713	0.720	2.011	-6.972	11.746
	b_2	1.270	1.264	2.640	-17.381	13.926
	b_3	1.034	1.004	2.371	-6.369	7.187
	b_4	0.822	0.822	2.546	-3.159	4.924
2GP (66)	a	0.539	0.527	0.171	0.171	1.200
	b_1	1.066	1.000	1.728	-3.204	7.399
	b_2	1.679	1.491	2.519	-2.665	13.926
	b_3	1.832	1.412	1.656	-0.856	5.506
	b_4	4.270	4.270	0.535	0.535	4.925
3GP (33)	a	0.560	0.523	0.190	1.90	1.055
	b_1	0.752	0.631	1.384	-2.738	3.437
	b_2	1.695	1.684	2.495	-3.638	7.293
	b_3	1.467	1.680	3.480	-6.369	7.187
	b_4	2.000	2.000	0.000	2.000	2.000

Dependent Variables

Based on the previous literature on CAT ability estimation, five criterion variables (or their log transformations) were used in this study: the biases, standard errors (SEs), root mean square errors (RMSEs), fidelity (correlation of the estimated and true ability parameters, (Wang & Vispoel, 1998), and administrative efficiency (the mean numbers of items needed to reach a criterion SE level). These criteria are used to examine the effects of the manipulated independent variables described in the last subsection because they provide complementary evidence. The bias, SE, and RMSE can be assessed both conditionally and overall (average) across an entire ability distribution. The conditional indexes are computed at each θ point, and overall indexes are computed by taking the absolute values of the conditional indexes and integrating them over a normally distributed θ for a population of examinees using the numerical integration method (Wang & Vispoel, 1988). These formulas are:

Conditional Indexes

$$Bias(\hat{\theta}) = \sum_{r=1}^{N} (\hat{\theta}_r - \theta) , \qquad (7)$$

$$SE(\hat{\theta}) = \sqrt{\frac{1}{N}\sum_{r=1}^{N}\left(\hat{\theta}_r - \frac{\sum_{t=1}^{N}\hat{\theta}_t}{N}\right)^2}, \qquad (8)$$

$$RMSE(\hat{\theta}) = \sqrt{\frac{1}{N}\sum_{r=1}^{N}(\hat{\theta}_r - \theta)^2}, \qquad (9)$$

where θ is the true ability of simulees, which was used to generate responses in the simulation, $\hat{\theta}_r$ is the estimated ability for the *r*th replication, and N is the number of replications. The number of replications in this MC study is the analogue of sample size. Because the primary goal is to assess the relative accuracy of ability estimation methods, the significance of a statistic is tested, and the empirical sampling distributions for the statistics are generated. In order to minimize the sample variance and increase the power to detect the effects of interest, a large number of replications are desired. In this study, relative accuracy is assessed by comparing the differences between the ability parameter estimates and the true ability across replications. In such a study, 500 replications are considered sufficient (Stone, 1993).

The RMSE can be separated into two components, Bias and SE ($RMSE^2 = Bias^2 + SE^2$).

Overall Indexes

$$\text{AVERAGE}_{\text{Bias}} = \sum_{i=1}^{21} |\text{Bias}(\hat{\theta})|\,|\theta_i * \text{weight}(\theta_i), \qquad (10)$$

$$\text{AVERAGE}_{\text{SE}} = \sqrt{\sum_{i=1}^{21} \text{SE}^2(\theta)|\theta_i * \text{weight}(\theta_i)}, \qquad (11)$$

$$\text{AVERAGE}_{\text{RMSE}} = \sum_{i=1}^{21} \text{RMSE}(\hat{\theta})|\theta_i * \text{weight}(\theta_i), \qquad (12)$$

where the weight(θ_i) are quadrature weights based on the standard normal distribution, and the θ_i are the 21 equally spaced true ability levels that range from –4 to 4 in increments of 0.4.

On the basis of the assumption, $\hat{\theta} = c\theta + e$, where c is a constant depending on bias, and e is the random error (If no bias, c=1). Wang (1995) defined the fidelity correlation as

$$r_{\theta\hat{\theta}} = \frac{\text{cov}(\theta,\hat{\theta})}{\sigma_\theta\sigma_{\hat{\theta}}} = \frac{c\sigma_\theta^2}{\sigma_\theta\sigma_{\hat{\theta}}} = \frac{c\sigma_\theta}{\sigma_{\hat{\theta}}} = \frac{c\sigma_\theta}{\sqrt{c^2\sigma_\theta^2 + \sigma_e^2}} = \frac{\sigma_\theta}{\sqrt{\sigma_\theta^2 + \frac{\sigma_e^2}{c^2}}}. \qquad (13)$$

Descriptive statistics are provided for all conditional indexes and overall indexes, Graphical presentations are proved for the conditional indexes. Inferential statistics are used only for the overall indexes.

Experimental Design

Two experimental designs were used in the analyses of the overall indexes. For the fixed-length tests, a 4 x 3 x 4 completely crossed ANOVA design was used. There are 4 ability estimation methods (MLE, WLE, EAP, and MAP), 3 item bank sizes (261, 66, 33), and 4 test lengths (5, 10, 15, and 20). For the fixed reliability tests, a 4 x 3 x 3 completely crossed ANOVA design was used: 4 ability estimation methods, 3 item bank sizes, and 3 reliability levels (0.7, 0.8, and 0.9). Effect sizes were computed and tabulated with significance level statistics.

RESULTS

Because of limited space, only partial results from the complete study are presented here. Plots for conditional indexes are presented at selected test lengths 5 and 10 items, and at selected reliabilities 0.7 and 0.9. Interested readers may refer to Wang (1999) for a more complete set of results.

Results Based on the Conditional Indexes

Fixed Test Length

Figures 2 through 7 show the bias, SE, and RMSE of four ability estimation methods (WLE, MLE, EAP, and MAP) with different ability levels for four fixed test lengths based on bank size 263. It is clear from figures 2 and 3 that the WLE has the smallest bias over most range of ability among all four estimation methods for all test lengths. This result also matches the dichotomous model case (Wang, Hanson, & Lau, 1998). Both WLE and MLE have considerably less bias than the two Bayesian methods. It can be seen that the MLE has "outward bias", which means the bias of MLE is positively correlated with θ; while WLE almost has no bias. The biases of EAP and MAP have "inward bias", which means the bias is negatively correlated with θ.

From Figures 4 and 5, we can see that the WLE has less SE than MLE at almost all ability levels for both fixed test lengths, although the differences are small. This means that WLE not only reduces MLE's bias, but also reduces MLE's SE. It can be also seen that for both ability extremes, the Bayesian methods of EAP and MAP have far less variability than WLE and MLE; this reduction of variability is at the expense of increased bias. This finding is also consistent with previous findings that Bayesian methods had lower SE but higher bias when compared with MLE and WLE for dichotomous models (Wang, Hanson, & Lau, 1998; Warm, 1989).

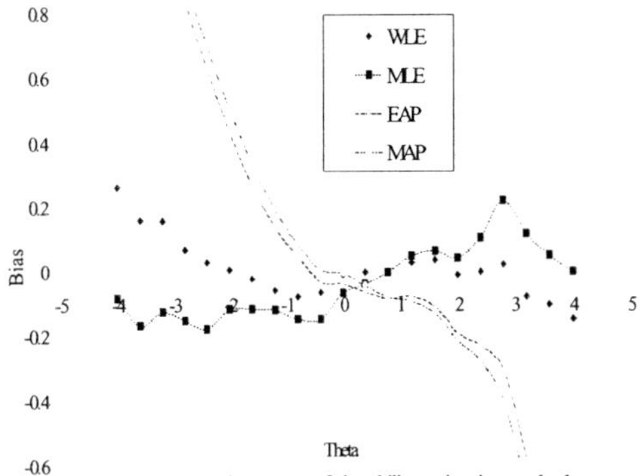

Figure 2. Bias curves of the ability estimation methods, test length = 5, bank size = 263.

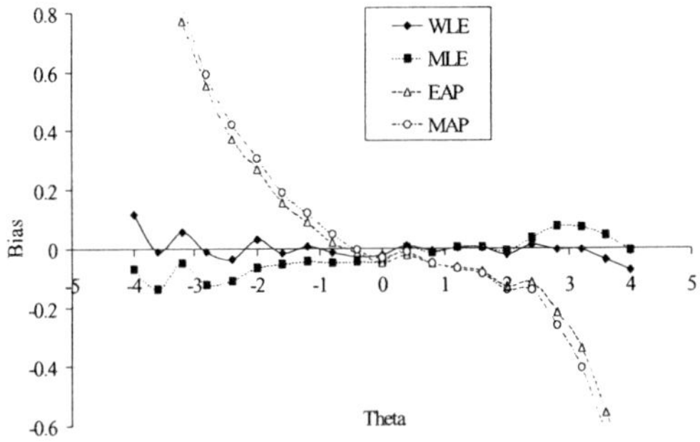

Figure 3. Bias curves of the ability estimation methods, test length = 10, bank size = 263.

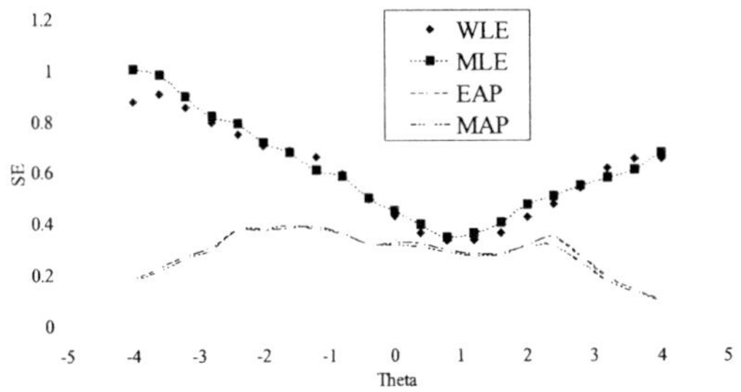

Figure 4. SE curves of the ability estimation methods, test length = 5, bank size = 263.

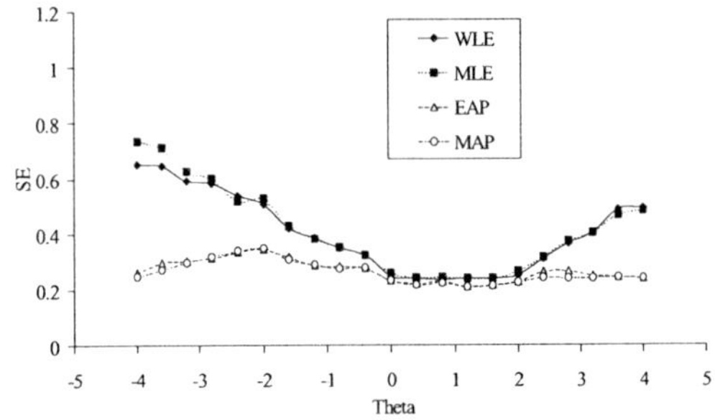

Figure 5. SE curves of the ability estimation methods, test length = 10, bank size = 263.

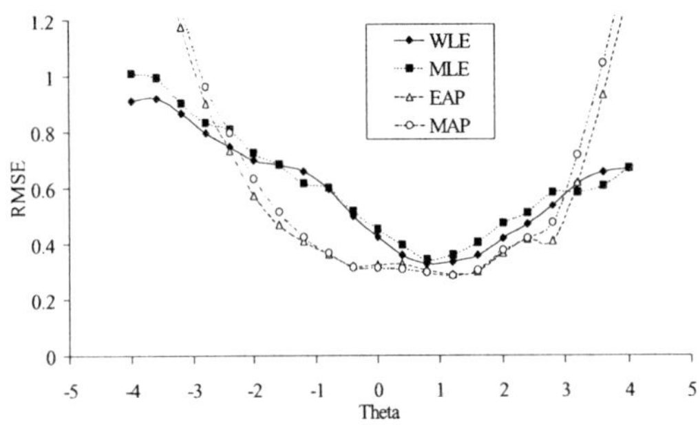

Figure 6. RMSE curves of the ability estimation methods, test length = 5, bank size = 263.

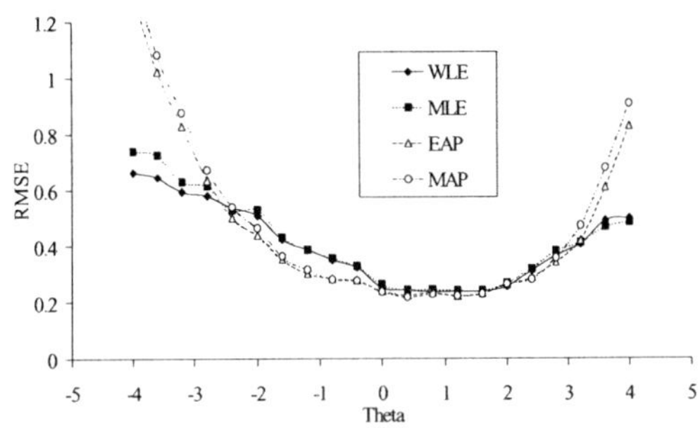

Figure 7. RMSE curves of the ability estimation methods, test length = 10, bank size = 263.

Figures 6 and 7 show that for all test lengths, the RMSE for all four estimation methods is similar along all ability levels, and WLE had lower RMSEs than MLE. All bias, SE, and RMSE decrease as the test length increases.

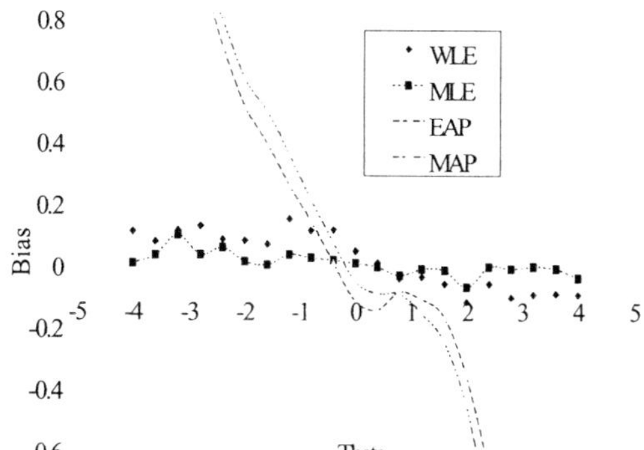

Figure 8. Bias curves of the ability estimation methods, reliability = 0.7, bank size = 263.

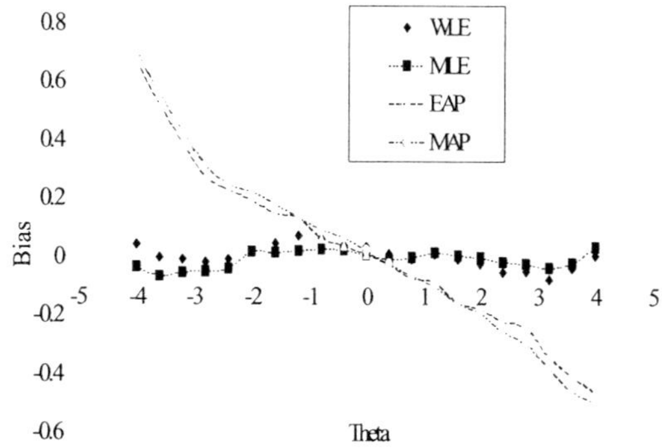

Figure 9. Bias curves of the ability estimation methods, reliability=0.9, bank size=263.

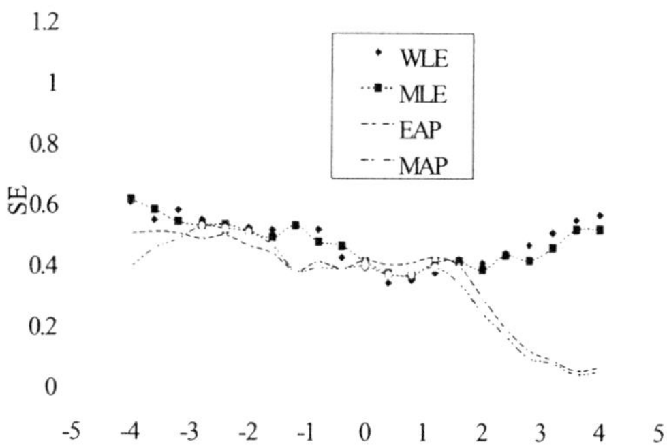

Figure 10. SE curves of the ability estimation methods, reliability = 0.7, bank size = 263.

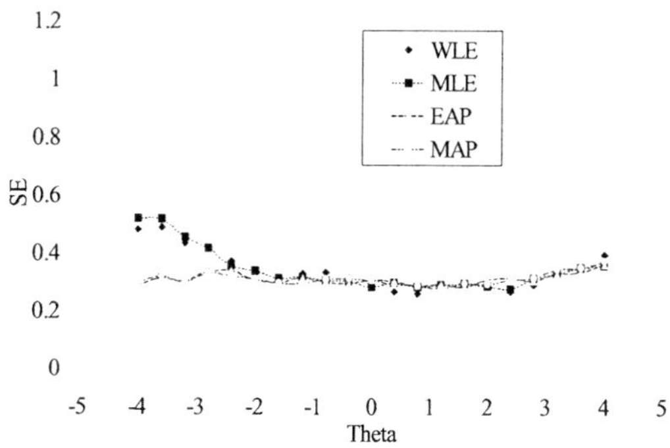

Figure 11. SE curves of the ability estimation methods, reliability = 0.9, bank size = 263.

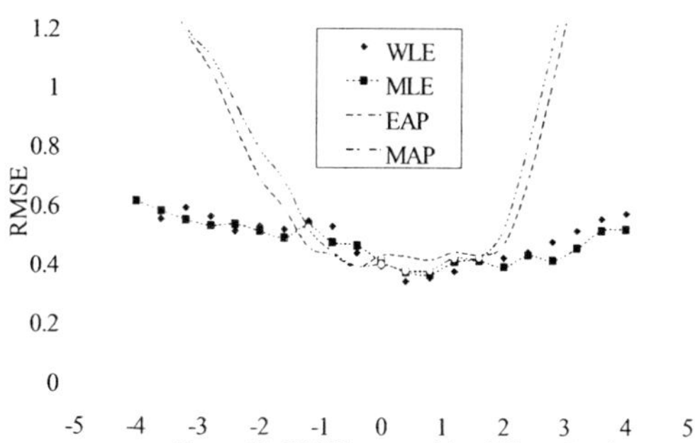

Figure 12. RMSE curves of the ability estimation methods, reliability = 0.7, bank size = 263.

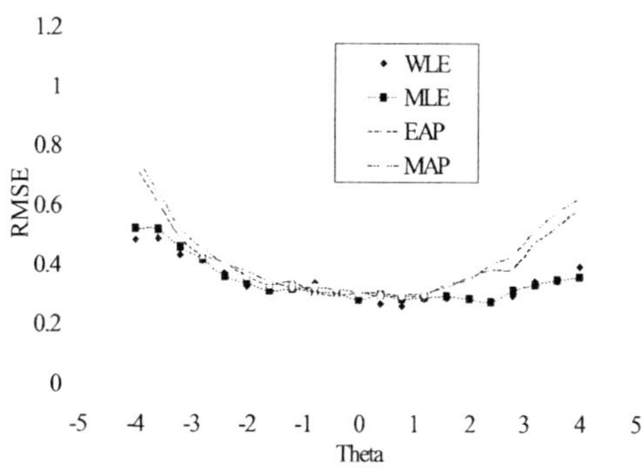

Figure 13. RMSE curves of the ability estimation methods, reliability = 0.9, bank size = 263.

Fixed Reliability

Figures 8 to 9 show the effects of fixing reliability at 0.7 and 0.9 on the biases of the four ability estimation methods based on item bank size 263. The fixed reliability changed the bias direction of MLE from "outward" to slightly "inward". The WLE further increased that "inward" bias, which is opposite with the situation of fixed test length. This means under a fixed reliability rule, WLE fails to reduce the bias of MLE.

Both MLE and WLE have remarkably smaller bias than the Bayesian methods. As indicated in Figures 10 to 11, both WLE and MLE have larger SE than the Bayesian methods, and WLE has smaller SE, than MLE. Figures 12 though 13 show that the non-Bayesian methods have lower RMSEs than the Bayesian methods at extreme ability levels, but have higher RMSEs than Bayesian methods at middle level of ability. WLE performed slightly better than MLE. All bias, SE, and RMSE decrease as the test reliability increases.

Figures 14 to 15 show the mean numbers of items required for achieving reliabilities 0.7 and 0.9 by using different ability estimation methods. For average ability level examinees, the number of items required for MLE and WLE are approximately twice those for EAP and MAP; for extreme ability level examinees, the numbers of items required for MLE and WLE are tripled. The sizes of item bank have slight effects on the all conditional indexes.

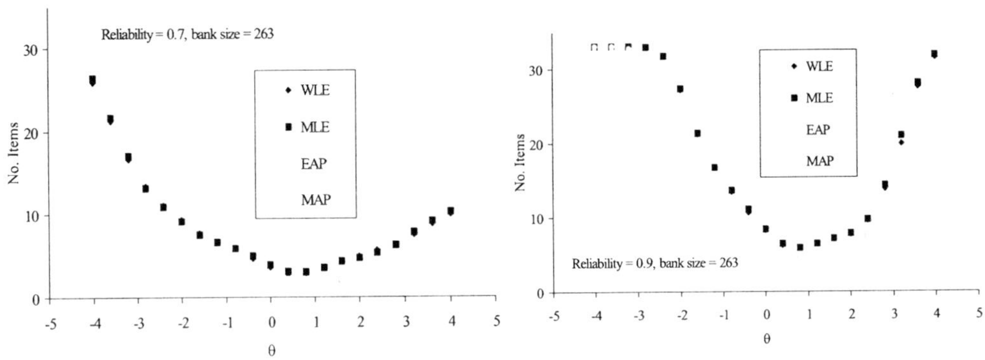

Figure 14-15. Mean numbers of items required for terminating the test.

Results Based on the Overall Indexes

Table 2 shows the results of the three-way ANOVA of absolute bias, average SE, and average RMSE for the fixed test length termination rule, and Table 3 shows the results of the three-way ANOVA for fixed test length reliability rule.

In general, the results for overall indexes further support the results of conditional indexes. The factor of ability estimation method had the most influence on absolute bias; it accounted for 54.0% of the total variance of absolute bias for the fixed test length termination rule and 74.0% of the total variance of absolute bias for the fixed reliability termination rule across the three sizes of item banks. The test termination rule also played a important part in estimated absolute bias; it represented 15.4% of the total variance for the fixed test length and 15.4% of the total variance for the fixed test reliability test. Both WLE and MLE show significantly less absolute bias than the Bayesian methods, and WLE performed best. On the other hand, the SE was affected mostly by the factor of test termination rule. The fixed test length accounted for 65.9% of the total variance in average SE, while the fixed test reliability accounted for 59.1%. The second most influential factor for SE varies by termination rule. For the fixed test length termination rule, it is the ability estimation method that accounted for 16.7% variance in SE; and for the fixed test reliability termination rule, it is the size of item bank that accounted for 13.9% variance in SE. The SE of both WLE and MLE are significantly larger than the SEs of the Bayesian methods. WLE had significantly lower SE than MLE, and EAP had significantly smaller SE than MAP. RMSE (log of RMSE) was affected mostly by test termination rule. The fixed test length accounted for 31.2% of the total variance of RMSE, and the three-way interaction of method x size x length accounted for 16.1%. The fixed test reliability accounted for 87.6% of the total variance of RMSE, and the ability estimation accounted for 6.3%. There were no significant differences in RMSE among ability estimation methods except the difference between WLE and MLE.

CONCLUSION

The precision of ability estimation method used in CAT has significant impact on the quality of CAT testing, because it not only affects the final score reported, but also affects which items are selected for particular examinee. In general, for all four ability estimation methods (WLE, MLE, EAP, and MAP), both conditional and overall indexes (bias, SE, RMSE, fidelity) of dependent variables decrease as the values of independent variables (test length, test reliability, and item bank size) increase. The magnitudes of the differences among those dependent variables decrease as the values of independent variables increase.

WLE out performed MLE in terms of all the dependent variables studied, and WLE performed better than the Bayesian methods in terms of bias. MLE had less bias than both Bayesian methods. Both EAP and MAP showed more favorable results with SE and fidelity than either the WLE or MLE did; EAP did a better job than MAP for almost all conditions. Different test termination rules had significant impact on those dependent variables for given ability estimation methods, especially for WLE and MLE methods. Although the quality of item banks has vast effects on the conditional distribution of bias, SE, RMSE, and test efficiency (Wang & Vispoel, 1998), the size of item bank had less impact on the differences among the dependent variables than test termination rules. This study confirms Warm's conclusions that (a) WLE is unbiased to first order, while MLE, EAP, and MAP are biased, and (b) the WLE method has small variance over entire range of θ for fixed test length CAT testing.

A limitation of this study is that in the CAT simulations, the item parameters were regarded as true item parameters and free of error. van der Linden and Glas (2000) have suggested that errors in item parameter estimates can have a dramatic effect on CAT ability estimation. Further research on the impact of item parameter error on the accuracy of polytomous CAT ability estimation is needed.

APPENDIX

The MLE bias function for the Samejima's Graded Response Model is

$$B(\theta;\hat{\theta}_v) \cong E[\hat{\theta}_v - \theta \mid \theta] \cong -\frac{1}{2[I(\theta)]^2}\sum_{j=1}^{n}\sum_{k_j}\frac{\frac{\partial}{\partial\theta}P_{jk}(\theta)\frac{\partial^2}{\partial\theta^2}P_{jk}(\theta)}{P_{jk}(\theta)}$$

$$= -\frac{1}{2[I(\theta)]^2}\sum_{j=1}^{n}\sum_{k_j}\left\{\frac{Da_j\exp[Da_j(\theta-b_{jk})]}{\{1+\exp[Da_j(\theta-b_{jk})]\}^2}-\frac{Da_j\exp[Da_j(\theta-b_{jk+1})]}{\{1+\exp[Da_j(\theta-b_{jk+1})]\}^2}\right\} *$$

$$\left\{\frac{D^2a_j^2\left(\frac{\exp[Da_j(\theta-b_{jk})]-\exp\{2[Da_j(\theta-b_{jk})]\}}{\{1+\exp[Da_j(\theta-b_{jk})]\}^3}-\frac{\exp[Da_j(\theta-b_{jk})]-\exp\{2[Da_j(\theta-b_{jk+1})]\}}{\{1+\exp[Da_j(\theta-b_{jk+1})]\}^3}\right)}{P_{jk}(\theta)}\right\} \quad \text{(A1)}$$

where I(θ) is the test information:

$$I(\theta) = E[I_v(\theta)\mid\theta] = \sum_{v} I_v(\theta)P_v(\theta) = \sum_{j=1}^{n} I_j(\theta) = \sum_{j=1}^{n}\sum_{k_j} I_{k_j}(\theta)P_{k_j}(\theta) \quad \text{(A2)}$$

and $I_j(\theta)$ is Item information

When $P_{jk}(\theta)$ modeled by the Muraki's Generalized Partial Credit Model, the MLE bias function become

$$I_j(\theta) = \sum_{k=0}^{m_j}\frac{[P'_{jk}(\theta)]^2}{P_{jk}(\theta)} - 0 = \sum_{k=0}^{m_j}\frac{[P'_{jk}(\theta)]^2}{P_{jk}(\theta)} = \sum_{k=0}^{m_j}\frac{[P^{*\prime}_{jk} - P^{*\prime}_{j,k+1}]^2}{[P^{*}_{jk} - P^{*}_{j,k+1}]} \quad \text{(A3)}$$

$$B(\theta;\hat{\theta}_v) \cong E[\hat{\theta}_v - \theta \mid \theta] \cong -\frac{1}{2[I(\theta)]^2}\sum_{j=1}^{n}\sum_{k_j}\frac{\frac{\partial}{\partial\theta}P_{jk}(\theta)\frac{\partial^2}{\partial\theta^2}P_{jk}}{P_{jk}(\theta)}$$

$$= -\frac{1}{2[I(\theta)]^2}\sum_{j=1}^{n}\sum_{k_j}D^3a_j^3P_{jk}\left[\mathrm{k} - \sum_{c=0}^{m}cP_{jk}\right]\left[\mathrm{k}^2 - 2k\sum_{c=0}^{m}cP_{jk} + 2\left(\sum_{c=0}^{m}cP_{jk}\right)^2 - \sum_{c=0}^{m}c^2P_{jk}\right] \quad \text{(A4)}$$

$$I_j(\theta) = \sum_{j=1}^{m_j} P_{k_j}(\theta) I_{k_j}(\theta) = \sum_{k=1}^{m_j} P_{k_j}(\theta)\left[-\frac{\partial^2}{\partial\theta^2}\log P_{k_j}(\theta)\right] = \sum_{k=1}^{m_j} P_{k_j}(\theta)\left\{\left[\frac{[P'_{jk}(\theta)]}{P_{jk}(\theta)}\right]^2 - \frac{P''_{jk}(\theta)}{P_{jk}(\theta)}\right\}$$

$$= D^2 a_j^2 \sum_{k=1}^{m_j} P_{k_j}(\theta)\left\{\sum_{c=1}^{m_j} T_c^2 P_{jc}(\theta) - \left[\sum_{c=1}^{m_j} T_c P_{jc}(\theta)\right]^2\right\} = D^2 a_j^2\left\{\sum_{c=1}^{m_j} T_c^2 P_{jc}(\theta) - \left[\sum_{c=1}^{m_j} T_c P_{jc}(\theta)\right]^2\right\}$$

$$= D^2 a_j^2 \sum_{c=1}^{m_j} [T_c - \overline{T}_j(\theta)]^2 P_{jk}(\theta) \qquad \text{(A5)}$$

Where

T_c is a linear integer scoring function; that is, $T_c = (1,2,\ldots,c,c+1,\ldots,m_j)$,

$$\overline{T}_j(\theta) = \sum_{c=1}^{m_j} T_c P_{jc}(\theta).$$

Where I(θ) is the test information and $I_j(\theta)$ is the item information, the relationship between I(θ) and $I_j(\theta)$ is described as A2. The item information $I_j(\theta)$ become

REFERENCES

Allen, N. L., Carlson, J. E., & Zelenak, C. A. (1999*). The NAEP 1996 Technical Report, NCES 1999-452.* Washington, DC: National Center for Educational Statistics.

Bennett, R. E., Steffen, M., Singley, M. K., Morley, M., & J acquemin, D. (1997). Evaluating an automatically scorable, open-ended response type for measuring mathematical reasoning in computer-adaptive tests. *Journal of Educational Measurement, 34,* 162-176.

Copas, J. B. (1988). Binary regression models for contaminated data. *J. R. Statist. Soc. B 50,* 225-65.

Dodd, B. G., Koch, W. R., & De Ayala, R. J. (1989). Operational characteristics of adaptive testing procedures using the Graded Response Model. *Applied Psychological Measurement, 13*, 129-143.

Eignor, D. R., & Schaeffer, G. A. (1995, April*). Comparability studies for GRE General CAT and the NCLEX using CAT.* Paper presented at the annual meeting of the National Council on Measurement in Education, San Francisco CA.

Firth, D. (1993). Bias reduction of maximum likelihood estimates [Correction: 95V82 p667]. *Biometrika, 80*, 27-38.

Gall, M. D., Borg, W. R., & Gall, J. P. (1996). *Educational Research: An introduction.* New York: Longman

Harwell, M. R. (1997). Analyzing the results of Monte Carlo studies in item response theory. *Educational and Psychological Measurement, 57*, 266-279.

Kingsbury, G. G., & Weiss, D. J. (1983). A comparison of IRT-based adaptive mastery testing and a sequential mastery testing procedure. In D. J. Weiss (Ed*.), New horizons in* *testing: Latent trait test theory and computerized adaptive testing* (pp. 257-238). New York: Academic Press.

Koch, W. R., & Dodd, B. G. (April, 1985). *Computerized adaptive attitude measurement.* Paper presented at the annual meeting of the American Educational Research Association, Chiago.

Lord, F. M. (1977). A broad-range tailored test of verbal ability. *Applied Psychological Measurement, 1*, 95-100.

Lord, F. M. (1983b). *Memorandum for: Ms. Stocking, Ms. M. Wang, Ms. Wingersky. Subject: Sampling variance and bias for MLE and Bayesian estimation of θ.* August 26, 1983 (Internal Memorandum). Princeton, NJ: Educational Testing Service.

McBride, J. R., & Martin, J. T. (1983). Reliability and validity of adaptive ability tests in a military setting. In D. J. Weiss (Ed.), *New horizons in testing: Latent trait test theory and computerized adaptive testing* (pp. 224-236). New York: Academic Press.

Muraki, E. (1992). A generalized partial credit model: application of an EM algorithm. *Applied Psychological Measurement, 16*, 159-176.

Muraki, E. & Book, D. (1999). *PARSCALE: Parameter scaling of rating data (Version 3.5).* Lincolnwood, IL: Scientific Software International.

Quenouille, M. H. (1949). Approximate tests of correlation in time-series. *J. R. Statist. Soc. B 11*, 68-84.

Quenouille, M. H. (1956). Notes on bias in estimation. *Biometrika 43*, 353-360.

Samejima, F. (1969). Estimation of latent ability using a response pattern of graded scores. *Psychometrika Monograph No. 17, 34*, (4, Pt. 2).

Samejima, F. (1993a). An approximation for the bias function of the maximum likelihood estimate of a latent variable for the general case where the item response are discrete. *Psychometrika, 58*, 119-138.

Samejima, F. (1993b). The bias function of the maximum likelihood estimate of ability for the dichotomous response level. *Psychometrika, 58*, 195-209.

Samejima, F. (1998). *Expansion of Warm's weighted likelihood estimator of ability for three-parameter logistic model to general discrete responses.* Paper presented at the Annual Meeting of the National Council on Measurement in Education (San Diego, April).

Segall, D. O. (1995, April). *Equating the CAT-ASVAB: Experiences and lessons learned.* Paper presented at the annual meeting of the National Council on Measurement in Education, San Francisco CA.

Segall, D. O., & Carter, G. (1995, April). *Equating the CAT-GATB: Issues and approach.* Paper presented at the annual meeting of the National Council on Measurement in Education, San Francisco CA.

Spence, I. (1983). Monte Carlo simulation studies. *Applied Psychological Measurement, 7*, 405-425.

Stone, C. A. (1993, July). *The use of multiple replications in IRT based Monte Carlo research.* Paper presented at European Meeting of the Psychometric Society, Barcelon.

Urry, V. W. (1977). Tailored testing: A successful application of latent trait theory. *Journal of Educational Measurement, 14*, 181-196.

van der Linden, W. J. & Glas, C. A. W. (2000). Capitalization on item calibration error in adaptive testing. *Applied Measurement in Education, 13*, 35-53.

Wainer, H., Dorans, N. J., Flaugher, R., Green, B. F., Mislevy, R. J., Steinberg, L., and Thissen, D. (1990). *Computerized adaptive testing: A primer.* Hillsdale, NJ: Lawrence Erlbaum Associates.

Wang, S. (1999). *The accuracy of ability estimation methods for computerized adaptive testing using the generalized partial credit model.* Unpublished doctoral dissertation, University of Pittsburgh.

Wang, T. (1995). *The precision of ability estimation methods in computerized adaptive testing.* Unpublished doctoral dissertation. Iowa City, IA: University of Iowa.

Wang, T., Hanson, B. A., & Lau, C. M. (1999. Reducing bias in CAT trait estimation: A comparison of approaches. *Applied Psychological Measurement, 23* 263-278.

Wang, T., & Vispoel, W. P. (1998). Properties of ability estimation methods in computerized adaptive testing. *Journal of Education Measurement*, *35*, 109-135.

Warm, T. A. (1989). Weighted likelihood estimation of ability in the item response theory. *Psychometrika, 54*, 427-450.

Chapter 5

COMPUTER PREPAREDNESS: A STUDY OF COMPUTER USE SELF-EFFICACY AMONG DECA STUDENTS IN KENTUCKY

Carol A. Decker*

Lincoln Memorial University, TN, USA

ABSTRACT

The unrestraint of computer technology and its ever present dependency in the workplace present continuous challenges for the business educator to produce skilled individuals. This study examined the preparedness of noted secondary business education students through their perceptions of computer use self-efficacy. Computer use attributes or computer education components are identified for their influence upon business student performance or preparation for work. A descriptive survey design was used employing the computer self-efficacy and self-efficacy of computer technologies scales. Data was gathered from regional Distributive Education Clubs of America (DECA) student representatives (n = 132) of 22 counties in southeastern Kentucky. Results suggest that students were ill prepared in specific technologies such as database management, word processing, educational software, programming, spreadsheets, and statistical processing. In addition, computer education that incorporates frequent use, student flexibility in preparing others, and exposure to statistical processing produces positive self-efficacy (performance). Recommendations include that business educators should implement new technologies more timely to peak student inspiration and curiosity for learning.

The unrestraint of computer technology and its ever present dependency in the workplace present continuous challenges for the business educator to produce computer skilled individuals into the 21st century. One method employed is participation in an organization known as Distributive Education Clubs of American (DECA) that provides business students, with interests in careers in management, marketing, or entrepreneurship, the opportunities to

* For correspondence: Lincoln Memorial University, 320 Debusk School of Business, Cumberland Gap Parkway, Harrogate, TN 37752; (W) (423) 869-6263 and/or 2227 Highway 11N, Niota, TN 37826; (H) (423) 337-6410; E-mail: Deckca@aol.com

obtain actual work experience. These students, under the advisory of business teacher educators, collaborate with education and business communities to gain a career foundation supplemented with leadership, etiquette, and ethical responsibility exposure. With a 180,000 student membership, over 110,000 secondary education students participate in competitive events based upon the student's occupational and chapter goals. Student competitive participation is often based upon the student's classroom subject performance, personable skills, and awareness of DECA concerns. These events are intended to be learning activities that evaluate student competence for entry or advancement in future careers in marketing, management, or entrepreneurship, as specified and validated by qualified occupational practitioners (DECA: An association of marketing students, 1999). While competency is an objective, business educators must ensure that such methods instill industry performance or preparedness.

Today's technically demanding work environments require individuals to possess both a confident belief and applied capability in order to be prepared and use progressive computer technology in a proficient manner. Business educators have overlooked this belief as the driver of applied capability, thus, performance preparedness (Decker, 1998; Decker, 1999a). Mager and Bandura (as cited by Wolf, 1997) agree that one's confident belief in their ability affects the degree of action taken. Devastating demands and ill experience with computers are likely to erode one's confidence to perform as well as any desire to undertake any computer-related activities.

Computer self-efficacy perceptions are reflected in the performance and proficiency realized by industry (Decker, 1998; Decker, 1999aa). Workplace performance and the student/employee's willingness to learn about computers and their related tasks is hindered by low self-efficacy levels (Bandura, 1977; Gist, 1987); consequently, attention to providing and ensuring that computer education results in self-efficacious computer interaction is a necessity.

SELF-EFFICACY FRAMEWORK

A representative definition of self-efficacy is an individual's belief in their ability to perform a particular task (Bandura, 1977; Bandura, 1982; Gist, 1989). Beginning some 20 years ago, it was posited that performance was a result of human interactions, mental cognition, and perceived self-efficacy. Although this framework produced performance, the degree of activity varied from proficient to reactive. As a result, the level of self-efficacy one achieves through human interactions and mental cognition is the focus believed to generate positive and skilled transfer to work environments (Decker, 1998; Decker, 1999aa). Bandura (1977) stated that information needed to make self-efficacious judgments are obtained by (a) performance accomplishments or enactive mastery, (b) vicarious learning experiences, (c) verbal persuasion, and (d) physiological arousal. Performance accomplishments make a cognitive register in which self-thought compares the possibilities of future task attainment. Bandura (1982) said

> People register notable increases in self-efficacy when their experiences disconfirm misbeliefs about what they fear and when they gain new skills to manage threatening activities. If in the course of completing a task, they discover something that appears intimidating about the

undertaking, or suggests limitations to their mode of coping, they register decline in self-efficaciousness despite their successful performance (p.125-126).

In the same study, he referred to verbal persuasion as the use of conversation and collaboration to reach a level of self-efficacy. Physiological arousal, conversely, acts as the responsible agent for changing emotions to fit a self-efficacy judgment mode. Vicarious learning experiences occur by watching and absorbing the struggles and successes of others (Popper & Lipschiz, 1993).

Gist (1989), and Gist and Mitchell (1992) noted self-efficacy as the basis for choosing what to do, sustaining amount of effort needed for attainment, and preserving experiences. Bandura (1977) found self-efficacy judgments to result in four types of behavior: (a) performance, (b) coping, (c) arousal, and (d) persistence of situations individuals choose for themselves.

Bandura (1982) noted self-efficacy as a determinant of choice behavior because it influences thechoice of behavioral settings. When individuals recognize coping as inadequate for addressing threatening situations, those situations are avoided. On the other hand, if people feel coping is acceptable, they may participate in that kind of situation. Bandura (1988) wrote that threeelements: (a) skills matched with basic competencies, (b) complex skills broken down into subskills, and (c) learning to apply skill competencies and subskills with different people and circumstances result in positive performance. He emphasized new skills are rarely used for long unless they are applied in work situations which provide an environment of experience for building personal confidence. Accordingly he said self-efficacy is not consistent, but instead requires acquisition and organization of skills into courses of action and such self-efficacy judgment influences activities and environmental paths chosen. Bandura (1977, 1982) contended that training programs can influence sources of information that, in turn, result in self-efficacy judgments to reliably measure performance in program objectives.

Self-efficacy theory is applicable in numerous settings which require performance or where performance is expected. In this study, self-efficacy is considered within the context of computers. Murphy, Coover, and Owen (1988) defined computer self-efficacy as an individual's belief that they can perform a specific computer task. Narrowing the definition of self-efficacy, Kinzie and Delcourt (1991) recognized fluctuating levels of self-efficacy with regard to specific technologies and derived the term self-efficacy of computer technologies. They defined self-efficacy of computer technologies as an individual's belief in their ability to use a specific computer technology such as word processing, electronic mail, and CD-ROM data bases.

To summarize, computer self-efficacy is enhanced by an individual's opportunity to perform computer tasks themselves, watch others perform computer tasks, listen to others experiences with computers, and to cognitively assess their own personal ability with computers. Lacking these situations in a computer education environment can result in reduced impact through eventual non-productivity. Individuals with higher levels of self-efficacy are likely to be more productive with computer-related tasks and organizational requirements.

REVIEW OF LITERATURE

First of all, previous literature is restrained by its failure to combine studies of secondary business students, computer use self-efficacy, and DECA preparation. Consequently, examining secondary business student performance preparedness via computer use self-efficacy is limitly researched. However, many studies attempt to establish connections between individual, organizational, and educational characteristics and performance as measured by computer self-efficacy. Measuring computer self-efficacy to determine one's preparation for work and future accomplishments or actions should be noted as a distinguishing objective among mostself-efficacy studies. Several existing studies utilize service and sales organization employees such as teachers, nurses, clerical, administrative or professional, and sales representatives to determine factors that explain performance measured by self-efficacy (Barling, & Beattie, 1983; Chowdhury, 1993; Decker, 1998; Decker, 1999aa; Earley, 1994; Henry & Stone, 1994; Hill, Kellenberger, 1996; Lee & Gillen, 1989; Parker, 1993; Smith, & Mann, 1987; Wolf, 1997). Compeau and Higgins (1995) examined the self-efficacious performance influence of "encouragement of others" upon academicians and management professionals. They also indicated that more computer use paralled higher self-efficacy measures. Decker (1999b) indicated that some required actions toward more self-efficacious technology performance is providing computer assisted instruction, allowing computer skilled individuals performance independence during training, and providing teaching opportunities for the skilled computer operators. Utilizing students and industry employees, some studies recognized computer self-efficacy as a component of user acceptance in advanced technology (Christoph, Schoenfeld, & Tansky, 1998; Ellen, Bearden, & Sharma, 1991; Hill, et al., 1987). Decker (1998, 1999a, 1999b) supported hands-on education as having an impact on self-efficacy and technology use. Several studies used a variety of university employees and state government employees to analyze how aspects of work affected general self-efficacy (Decker, 1998; Decker, 1999a; Jex & Gudanowski, 1992; Latham & Frayne, 1989; Prieto & Altmaier, 1994).

Computer self-efficacy and self-efficacy of computer technology studies historically have utilized students as a focal population (Kinzie & Delcourt, 1991; Murphy et al, 1988; Prieto & Altamaier, 1994). In an educational setting, Murphy et al. (1988), in validating the CSE instrument used in this study, found that the variable "previous computer use" was significant in predicting advanced computer skills. Further analysis by the authors found that previous computer use was not significant in predicting mainframe computer skills. Consequently, the researchers also found that the individual's level of education explained a significant portion of the variability in computer self-efficacy in advanced and mainframe computer skills. Also in an educational setting, Kinzie and Delcourt (1991) found that experience in computer coursework was the most significant predictor of self-efficacy. Significance occurred in all subscales of computer technologies; word processing, electronic mail, and CD-ROM Data Bases. Other studies, which again used mostly student populations, revealed significant types of education and educational situations and their impact upon self-efficacious performance (Earley, 1994; Eyring, Honson, & Francis, 1993; Garavan & McCracken, 1993; Kellenberger, 1996; Mathieu, Martineau, & Tannenbaum, 1993; Mitchell, Hopper, Daniels, George-Falvy, & James, 1994; Phillips & Russell, 1994; Prieto & Altmaier, 1994; Stevens, Bavette, & Gist, 1993; Torkzadeh & Koufteros, 1994).

Further evidence reveals information that computer instruction techniques promote self-efficacy's association with increased performance. Decker (1999b), Karsten and Roth (1998), and Zhang and Espinoza (1998) used students from computer and non-computer courses at the university level to find that instructors enhance self-efficacy performance through environmental settings and their own awareness of computer literacy differences among students. Specifically, Karsten and Roth (1998) suggested more relevant computer experience supersedes quantity of computer experience in gaining high performance. Also, they posited that specific computer self-efficacy is significantly related to computer performance rather than other measured areas of self-efficacy. Qutami and Abu-Jaber (1997) illustrated decreased self-efficacy performance due to a lack of computer or training experience and scientific orientation. Sing-Fai (1996) supported this notion of imposing computer complexity to entreat performance through the study of disabled individuals. Providing tasks of increased difficulty were shown to increase rehabilitation performance.

Purpose and Research Questions

The purpose of this study was to examine the computer preparedness of recognized secondary business education students through their perceptions of computer use self-efficacy. Computer use attributes or computer education components are investigated for their influence upon business student computer performance or preparation for computer work. The following research questions were addressed by the study:

1. What are the differences in computer preparedness among DECA students with regard to frequency of computer use, specific classroom computer use, home use, and computer assistance, as measured by computer self-efficacy.
2. What are the differences in computer preparedness among DECA students with regard to frequency of computer use, specific classroom computer use, home use, and computer assistance, as measured by self-efficacy of computer technologies.

Methodology

Population

A Computer Self-Confidence Assessment, distributed at a regional DECA conference in southeast Kentucky, was employed to gather data from a population (N = 132) of signified secondary business students. Using a purposive sampling method and recognizing the small population, a sample size of 132 was felt tenable by the researcher. The total sample of 132 responded, providing an overall response rate of 100%. Of the respondents, 58 (44%) were 17 years of age, 27 (21%) were 16 years of age, 24 (18%) were 18 years of age, 17 (13%) were 15 years of age, 4 (3%) were 14 years of age, and 2 (1%) were 19 years of age. With regard to gender, 86 (65%) respondents were female, while 46 (35%) were male. In terms of grade level, 59 (45%) were seniors, 44 (35%) were juniors, 19 (15%) were sophomores, and 7 (5%) were freshmen.

Instrument

The Computer Self-Confidence Assessment consisted of a background segment, the computer self-efficacy scale (CSE), and the self-efficacy of computer technologies scale (SCT). The CSE scale measured the degree of confidence people have about computer knowledge and skills after a education experience (Murphy et al., 1988). Scale items consisted of 32 activity statements developed from course designs for teaching graduate students and practicing professionals how to use micro and mainframe computers. The CSE asked individuals to rate their degree of confidence in performing the listed computer activities on a 5-point Likert scale (1 = Very Little Confidence, 5 = Quite a Lot of Confidence). Reliability was determined by the Cronbach alpha internal consistency coefficient of .95 (Torkzadeh & Koufteros, 1994).

The SCT scales measured self-efficacy of specific computer technologies such as word processing, electronic mail, and CD-ROM data bases (Kinzie & Delcourt, 1991). The scale contained a total of 27 activity items with 10 subscale items related to word processing, 10 items related to electronic mail, and seven items related to compact disc (CD-ROM) data bases. Study participants rated their degree of confidence by strongly disagreeing or strongly agreeing with the confidence statements on a 4-point Likert scale (1 = Strongly Disagree, 4 = Strongly Agree). The scale was validated by a pilot study utilizing a panel of 17 experts consisting of computer technology instructors, measurement experts, educational consultants, and graduate students. A critique of the questionnaire was also provided by a university instrument review committee. After the pilot study, a Cronbach alpha of .84 was determined. A present study Cronbach reliability test was conducted of the Computer Self-Confidence Assessment resulting in an estimated coefficient of .90.

The Background Information section of the assessment asked the students for the independent variables of frequency of computer use in four subscales (never, seldom, frequently, and always), previous classroom computer use in six subscales (database management, e.g., Dbase, AskSam; word processing, e.g., WordPerfect; statistical processing, e.g., SPSSX, SAS; spreadsheets, e.g., Lotus, Excel; programming, e.g., Basic, COBOL; educational software, e.g. Teaching Typing), home use, and computer assistance.

Analysis

With multiple continuous dependent variables and multiple discrete independent variables the research objectives were best addressed using multivariate analysis of variance (MANOVA). According to Tabacknik and Fidell (1983), MANOVA is the best test when the question is how groups differ among the dependent variables in combination. Also, using the MANOVA procedure protects against a Type I error when two-tailed tests of significance are used (Huck, Cormier, and Bounds, 1974). The MANOVA procedure was used to determine significant differences ($p < .05$) of frequency of computer use, specific classroom computer use, home use, and computer assistance on computer use self-efficacy (CSE & SCT in combination). As a follow-up test, the Wilks' lambda multiple comparison procedure was performed on each test of the independent variables on the dependent variables to detect significant mean differences. For significant differences found by the Wilks' lambda, the

Tukey HSD post hoc test addressed all pairwise comparisons among the independent variable subscales.

The dependent variables were CSE and SCT scores while specific independent variables were frequency of computer use (never, seldom, frequently, and always), specific classroom computer use (database management, word processing, statistical processing spreadsheets, programming, and educational software), home use (yes or no), and computer assistance (yes or no).

FINDINGS

Research Objective 1

- What are the differences in computer preparedness among DECA students with regard to frequency of computer use, specific classroom computer use, home use, and computer assistance, as measured by computer self-efficacy?

As shown in Table 1, the MANOVA disclosed significant differences in computer self-efficacy among DECA students with regard to frequency of computer use, home use, and computer assistance. Consequently, there was no significant differences found in computer self-efficacy pertaining to specific classroom computer uses in database management, word processing, statistical processing, spreadsheets, programming, and educational software.

Table 1: Multivariate Analysis of Variance for CSE Scores of Respondents Classified by Independent Variables

Source	df	SS	F	ϴ
CSE				
Frequency of Computer Use	4	4755.17	2 .97	.04*
Specific Classroom Computer Use				
Database Management	1	122.06	.23	.63
Word Processing	1	84.66	.16	.69
Statistical Processing	1	1861.69	3.49	.07
Spreadsheets	1	77.41	.15	.70
Programming	1	59.37	.11	.74
Educational Software	1	12.20	.02	.88
Home Use	2	6800.56	6.37	.00*
Computer Assistance	2	4536.20	8.50	.01*

* $p < .05$

Research Objective 2

- What are the differences in computer preparedness among DECA students with regard to frequency of computer use, specific classroom computer use, home use, and computer assistance, as measured by self-efficacy of computer technologies?

Table 2 demonstrates a significant difference in the self-efficacy of computer technologies among DECA students with regard to frequency of computer use, specific classroom computer use in statistical processing, and computer assistance. Similar to CSE results, no significant differences were found in SCT scores with respect to specific classroom computer use in database management, word processing, spreadsheets, programming, and educational software. The Hotellings T^2 multiple comparison follow-up procedure, as shown in Table 3, revealed significant differences in composite computer self-efficacy and self-efficacy of computer technologies scores of employees with respect to frequency of computer use, specific classroom computer use in statistical processing, home use, and computer assistance.

Table 2: Multivariate Analysis of Variance for SCT Scores of Respondents Classified by Independent Variables

Source	df	SS	F	θ
SCT				
Frequency of Computer Use	4	3092.58	4 .05	.01*
Specific Classroom Computer Use				
Database Management	1	141.80	.56	.46
Word Processing	1	300.93	1.18	.28
Statistical Processing	1	1073.23	4.22	.04*
Spreadsheets	1	132.71	.52	.47
Programming	1	89.65	.35	.56
Educational Software	1	167.81	.66	.42
Home Use	2	651.48	1.28	.28
Computer Assistance	2	988.30	3.89	.04*

* p < .05

Table 3: Multivariate Analysis of Variance for Mean Scores of Respondents Classified by Independent Variables

	Hotellings	T^2	df	F	θ
Frequency of Computer Use 3.04	1,131	397.79	.00*		
Specific Classroom Computer Use					
Statistical Processing		3.04	1,131	397.79	.00*
Home Use	3.04	1,131	397.79	.00*	
Computer Assistance	3.04	1,131	397.79	.00*	

* p < .05

Tukey's HSD all pairwise procedure, as shown in Table 4, revealed Tukey HSD post hoc significant differences in CSE mean scores among those participants using computers always and those using computers seldom and never. Other significant differences were noted among those participants using computers frequently and those using computers never and seldom. The Tukey HSD post hoc also revealed significant differences in CST mean scores among those participants using always and those participants using computer frequently, seldom, and never. Other significant differences were noted among those participants using computers frequently and never.

Table 4: Tukey's Honestly Significant Difference (HSD)[a] Procedure (Simultaneous Confidence Intervals) for Comparisons of Subscale Means for Variable of Frequency of Computer Use

	Frequency of Computer Use Subscales	Lower Confidence Limit	Mean	Upper Confidence Limit
CSE				
	A[b] - F[b]	-.54	18.58	37.69
	A - S[b]	15.04	35.69	56.18*
	A - N[b]	23.54	58.00	92.46*
	F - S	3.28	17.11	30.95*
	F - N	8.46	39.42	70.39*
	S - N	-9.52	22.31	54.14
SCT				
	A[b] - F[b]	2.90	16.12	29.34*
	A - S[b]	7.62	21.79	39.96*
	A - N[b]	18.75	42.58	66.42*
	F - S	-3.90	5.67	15.24
	F - N	5.04	26.46	47.88*
	S - N	-1.22	20.79	42.81

[a] Controlling for Type I experimentwise error rate at .05 level of significance
[b] Frequency of Computer Use: A = Always, F = Frequently, S = Seldom, N = Never
* Group means are significantly different
* $p < .05$

CONCLUSIONS AND DISCUSSION

Using computer self-efficacy measurement scales as a baseline for perceptions of confidence in performance, the study provided indications of DECA student capabilities. Results illustrated computer use attributes and computer education components to be considered when designing education or instruction to prepare the business student for the work world. With no indication of changes in self-efficacy, DECA students perceived their own general performance inadequacies in specific computer technologies such as database management, word processing, educational software, programming, spreadsheets and statistical processing. Clearly, these students perceived the classroom use of these technologies failed to make them prepared for work. These results leave one to speculate twofoldly. First, of all, students feel these technologies were not emphasized enough in the classroom. Secondly, students may feel these technologies, while frequently instructed, were not taught broadly enough to make them confidently perform in the workplace. Such known facts denote that DECA students may be redundantly exposed to the same uses or tasks of the technologies rather than being given the opportunity to explore expanded capabilities of the technologies. However, having examined DECA students in the self-efficacy of specific computer technologies, they indicate more confidence in preparation for work when they are able to use statistical processing in the classroom. These perceptions show that DECA students may become overly familiar with some tasks within the technologies of database management, word processing, educational software, programming, and spreadsheets and thus, become indifferent to performing the task, thereby, reducing their self-efficacy or performance confidence. However, when DECA students were presented with challenging, yet, less familiar tasks, such as statistical processing their confidence increases due to the inspiration of possibly performing different assignments as well as the technologies' perceived relevance.

Additionally, home use of computers produce higher confidence levels of performance. DECA students recognize that having a computer at home prepares them for work but unless they know how to apply the extended nature of technologies, they will still be ill-prepared for the workplace. They also recognize that more frequent use of computers, noted on a regular or always basis, will make them more prepared and more confident to perform with computers in the future. Furthermore, DECA students show that having an opportunity to assist others with computers makes them more confident in specific technology application. This kind of exposure provides them with the ability to apply more of what they know rather than just perform dictated classroom exercises.

Recommendations and Implications

The study reiterates the educators role in preparing employees for the workplace. Students that fail to provide profitable returns to an organization will be a part of high turnover while creating higher expenses for industry. The mere indifference shown by these DECA students suggests that business educators should continue to pursue actual education that enhances self-efficacy. Educators must be able to apply (home study and classroom use) the expanded capabilities of all the technologies noted in this study. If educators can provide more exposure for the students, students as future employees will be able to contribute worth to organizations. Also, educational institutions must continually monitor the classroom computer equipment for its capabilities so that up-to-date exposure is provided to the students. Educators should frequently, expose students to new applications and software used by industry, in a timely fashion, rather than remaining with the educational industry standard. Student performance confidence is increased when they are provided with an orientation to the unknown that, in turn, peaks student inspiration and curiosity for learning. Moreover, educators should allow other more proficient students to enhance their computer skills by helping others so that a complete learning environment exists.

Business educators and administrators should be increasingly required to establish collaborative arrangements in order to more closely mirror educational programs with actual work. Organizations and educators should partner to provide cutting edge, state-of-the-art, and vaguely unknown opportunities for learning which will promote continued challenge and vigor in workplace performance. Business educators and institutions that fail to provide the productivity required in the workplace will lose their support and reputations for developing human resources. This study supplies business administrators and educators with the means to initiate fruitful programs that parallel workforce needs and strengthen support for career education partnerships.

Further research should continue to explore computer technology advancements for their impact upon performance. More investigative comparisons should be made of performance confidence between students with and without adequate home use of computers. Educators should continue with more applied learning mechanisms (computer assistance) in the classroom and review their impact upon performance and what performance capabilities were increased or decreased.

REFERENCES

Bandura, A. (1977). Self-efficacy: Toward a unifying theory of behavioral change. *Psychological Review, 84,* 191-215.

Bandura, A. (1982). Self-efficacy mechanism in human accuracy. *American Psychologist, 37,* 122-147.

Bandura, A. (1988). Organisational applications of social cognitive theory. *Australian Journal of Management, 13,* 275-302.

Barling, J., & Beattie, R. (1983). Self-efficacy beliefs and sales performance. *Journal of Organizational Behavior Management, 5* (1), 41-51.

Compeau, D. R., & Higgins, C. A. (1995). Computer self-efficacy: Development of a measure and initial test. *MIS Quarterly* [On-line]. *19.* Available: http://www.ebscohost.com/cgi-bin/epwnorb/page.

Christoph, R. T., Schoenfeld, G. A., & Tansky, J. W. (1998). Overcoming barriers to training utilizing technology: The influence of self-efficacy factors on multimedia-based training receptiveness. *Human Resource Development Quarterly, 9* (1), 25-28.

Chowdhury, J. (1993). The motivational impact of sales quotas on effort. *Journal of Marketing Research, 30,* 28-41.

DECA: An association of marketing students. (1999). [On-line]. Available: http://www.deca.org/introduction/info.html

Decker, C. A., (1998). Training transfer: Perceptions of computer use self-efficacy among university employees. *Journal of Vocational Technical Education, 14,* 29-39.

Decker, C. A., (1999a). Technical education transfer: Perceptions of employee computer technology self-efficacy. *Computer in Human Behavior, 15,* 161-172.

Decker, C. A. (1999b). The performance dimensions of self-efficacy. *International Journal of Business Disciplines, 9,* 7-19.

Earley, P. C. (1994). Self or group? Cultural effects of training on self-efficacy and performance. *Administrative Science Quarterly, 39,* 89-117.

Ellen, P. S., Bearden, W. O., & Sharma, S., (1991). Resistance to technological innovations: An examination of the role of self-efficacy and performance satisfaction. *Journal of the Academy of Marketing Science, 19* (4), 297-307.

Eyring, J. D., Honson, D. S., & Francis, D. J. (1993). A cross-level units of analysis zpproach to individual differences in skill acquisition. *Journal of Applied Psychology, 78* (5), 805-814.

Garavan, T., & McCracken, C. (1993). Introducing end-user computing: The implications for training and development. *Industrial and Commercial Training, 25* (7), 8-14.

Gist, M. E. (1987). Self-efficacy: Implications for organizational behavior and human resource management. *Academy of Management Review, 12,* 472-485.

Gist, M. E. (1989). The influence of training method on self-efficacy and idea generation among managers. *Personnel Psychology, 42,* 787-805.

Gist, M. E., & Mitchell, T. R. (1992). Self-efficacy: A theoretical analysis: Of its determinants and malleability. *Academy of Management Review, 17* (2), 183-211.

Heilman, M. E., & Hornstein, H. A. (1982). *Managing human forces in organizations.* Homewood, IL: Irwin.

Henry, J. W., & Stone, R. W. (1994). A structural equation model of end-user satisfaction with a computer-based medical information system. *Information Resources Management Journal, 7* (3), 21-33.

Hill, T., Smith, N. D., & Mann, M. F. (1987). The role of efficacy expectations in predicting the decision to use advanced technologies: The case of computers. *Journal of Applied Psychology, 72,* 307-313.

Huck, S. W., Cormier, W. H., & Bounds, W. G. (1974). *Reading statistics and research.* New York: Harper & Row.

Jex, S. M., & Gudanowski, D. M. (1992). Efficacy beliefs and work stress: An exploratory study. *Journal of Organizational Behavior, 13* (5), 509-517.

Karsten, R. & Roth, R. M. (1998). The relationship of computer experience and computer self-efficacy to performance in stroductory computer literacy courses. *Journal of Research on Computing Education* [On-line]. *31.* Available: http://www.ebscohost. com/cgi-bin/epwnorb/page.

Kellenberger, D. W. (1996). Preservice teacher's perceived computer self-efficacy based on achievement and value beliefs within a motivation framework. *Journal of Research on Computing Education* [On-line]. *29.* Available: http://www.ebscohost. com/cgi-bin/epwnorb/page.

Kinzie, M. B., & Delcourt, M. A. B. (1991). *Computer technologies in teacher education: The measurement of attitudes and self-efficacy.* Chicago, IL: American Educational Research Association. (ERIC Document Reproduction Service No. ED 331 891)

Latham, G. P., & Frayne, C. A. (1989). Self-management training for increasing job attendance: A follow-up and a replication. *Journal of Applied Psychology, 74* (3), 411-416.

Lee, C., & Gillen, D. J. (1989). Relationship of type A behavior pattern: Self-efficacy perceptions on sales performance. *Journal of Organizational Behavior, 10* (1), 75-81.

Mathieu, J. E., Martineau, J. W., & Tannenbaum, S. (1993). Individual and situational influences on the development of self-efficacy: Implications for training effectiveness. *Personnel Psychology, 46* (1), 125-147.

Mitchell, T. R., Hopper, H., Daniels, D., George-Falvy, J., & James, L. R. (1994). Predicting self-efficacy and performance during skill acquisition. *Journal of Applied Psychology, 79* (4), 506-517.

Murphy, C. A., Coover, D., & Owen, S. V. (1988). *Assessment of computer self-efficacy: Instrument development and validation.* New Orleans, LA: National Council on Measurement in Education. (ERIC Document Reproduction Service No. ED 307 317)

Parker, L. E. (1993). When to fix it and when to leave: Relationships among perceived control, self-efficacy, dissent, and exit. *Journal of Applied Psychology, 78,* 949-959.

Phillips, J. C., & Russel, R. K. (1994). Research self-efficacy, the research training environment, and research productivity among graduate students in counseling psychology. *Counseling Psychologist, 22* (4), 628-641.

Popper, M., & Lipschiz, R. (1993). Putting leadership theory to work: A conceptual framework for theory based leadership development. *Leadership and Organization Development Journal, 14 (7),* 23-27.

Prieto, L. R., & Altmaier, E. M. (1994). The relationship of prior training and previous teaching experience to self-efficacy among graduate teaching assistants. *Research in Higher Education, 35* (4), 481-497.

Qutami, Y., & Abu-Jaber, M. (1997). Student's self-efficacy in computers skills as a function of gender and computer learning style at Sultan Oaboos University. *International Journal of Instruction Media* [On-line]. *24.* Available: http://www.ebscohost.com/cgi-bin/epwnorb/page.

Sing-Fai, T. (1996). Self-efficacy as a predictor of computer skills learning outcomes of individuals with physical disabilities. *Journal of Psychology Interdisciplinary & Applied* [On-line]. *130.* Available: http://www.ebscohost.com/cgi-bin/epwnorb/page.

Stevens, C. K., Bavette, A. G., & Gist, M. E. (1993). Gender differences in the acquisition of salary negotiation skills: The role of goals, self-efficacy, and perceived control. *Journal of Applied Psychology, 78* (5), 723-735.

Tabacknick, B. G., & Fidell, L. S. (1983). *Using multivariate statistics.* New York: Harper & Row.

Torkzadeh, G., & Koufteros, X. (1994). Factorial validity of a computer self-efficacy scale and the impact of computer training. *Educational and Psychological Measurement, 54* (3), 813-821.

Wolf, K. N. (1997). Predicting positive self-efficacy in group problem-solving. *Human Resource Development Quarterly, 8,* 155-169.

Zhang, Y. & Espinoza, S. (1998). Relationship among computer self-efficacy, attitudes toward comuters, and desirability of learning computing skills. *Journal of Research on Computing in Education* [On-line]. *30.* Available: http://www.ebscohost.com/cgi-bin/epwnorb/page.

Chapter 6

Marketing Success: Self-Concept and Individual Antecedents in Marketing Simulations

Carol A. Decker
Lincoln Memorial University, TN, USA

Abstract

Marketing program successes are only as good as those employees who are responsible for their implementation. With increasing emphasis being placed upon the accountability of marketing efforts, an examination of marketing personnel self-concepts and their characteristics is essential. The study examined simulation marketing student self-concept and individual antecedents for their influence upon marketing success. With such a probe of marketing students, a foundation is provided for organizational personnel to determine the potential success of marketing projects and indications of the most appropriate development initiatives for marketing decision makers. Age was significant in determining self-concept and negative correlations were found among self-concept and success. Conclusions and recommendations are provided for marketing success preparation.

Introduction

The plethora of techniques available for improving knowledge and skills without any behavioral understanding or measure of accountability is a major contributor to organizational marketing failures. As noted by Semler (1997), alignment of culture and facilitation of skills will be the catalyst of high achievement in the workplace. Thus, a study of self-concept from participation in a simulated marketing learning situation and participant characteristics is provided. Cantor defines self-concept (as cited by Corinne & Sigelman, 1998) as the difference between how an individual realistically defines themselves and their desired image. Measures of self-concept resulting from a simulated workplace and individual antecedents will provide developmental guidance for marketing personnel and determinants of marketing success.

Problem Statement

Emphasizing accountability, marketing programs and their successes are recognized as only as good as those employees who are responsible for their implementation. Thus, an examination of marketing personnel self-concepts and their characteristics is essential. For generalization, such an investigation should closely parallel the real-world marketing environment. Therefore, the purpose, here, is to explore individual self-concepts resulting from a simulated marketing environment and individual characteristics for their influence upon marketing success. While many studies suggest individual self-concept and characteristics to be determinants of performance, no studies exist that measure self-concept derived from a simulated marketing environment and individual antecedents for their impact on marketing success.

REVIEW OF LITERATURE

A previous advertising campaign by Heinz emphasizing "anticipation" is brought to life as organizations continue to search for ways to anticipate their success. It so happens, that the ketchup was produced in such a way to create anticipation. The approach to creating marketing success is no different. Marketing personnel must be produced (developed) so that successful marketing efforts are created. Kennedy, Lawton, and Walker (2001) consequently recognized that rapidly changing workplaces prompt new paradigms in business education. Anido, Llamas, and Fernandez (2001) and Dacko (2001) suggested internet-based simulations and educational technology innovations, respectively, as means of enhancing global workplace experience. Businesses such as Compliance Technologies have found that simulations were useful in teaching actual decision making tactics that are in accord with real business environments (Online training program fits bill for fair lending compliance effort, 2002). Celsi (2002) also recognized that technology application training and education resulted in extended knowledge, applied activities, and more current job knowledge capabilities. Further, Glynn and Rajendran (1993) suggested that program accountability could be measured through student perceptions. In the midst of simulation technology education investigations, Beier and Kanfer (2001) noted self-concept as a determinant of knowledge. Even so, self-concept can be distinguished beginning in early individual ages (Huang & Michael, 2002; Marsh, Ellis, & Craven, 2002). Self-concept measures can, also, be used to coordinate people and programs (Malhotra, 1981). Individual, group, and experience characteristics are noted as being related to student self-concept measures (Bray, 2001; Libby & Eibach, 2002; Oyserman, Coon, & Kennelmeier, 2002).

Thus far, such a review would exemplify the following theoretical framework:

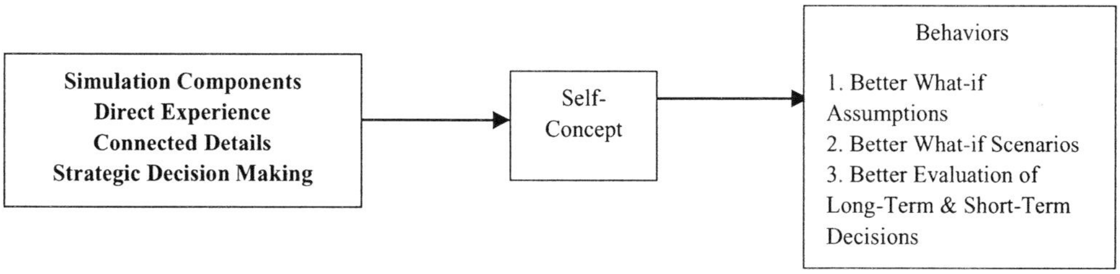

Figure 1. System Dynamics Simulation Model

Note. From "Simulation modeling for leadership development: A systems dynamics approach," by J.T. Eggers and R. Brown, 2002, *Corrections Today, 64*, p. 181-182. Adapted with permission.

Theoretical Framework

Despite these reinforcements, little research has manifested the connection of simulation education involvement, self-concept assessments and individual characteristics for determining, in particular, marketing success. However, many studies are prevalent that identify influences upon self-concept, success, and simulation influences on success and self-concept. In reinforcing the above theory, current studies suggest that simulation experiences, details, and decision making generally increase individual success (Andrews & Manrai, 1998; Swaak & de Jong, 2001; van der Veen, Maurits, & Browne 2001; Waks, 2001; Washbush & Gosen, 2001). In particular, some studies noted the importance of web-based simulations for facilitating successful learning and consequential work behavior (Maharg, 2001; Michau, Gentil, & Barrault 2001; Poustie 2001). Other studies posited the simulation impact on success in non-workplace related situations. Russomano (2000) determined individual disability impairments were improved through simulation experiences. Simulations were found by many researchers to improve success in engineering, manufacturing throughput and cycle time, cardiac arrest management efficiency, ball game decision making, patient anesthesia medical application, mathematical learning, medical practice, geological mapping, and legal techniques such as negotiation skills, draft preparation, and documentation searches (Chang, Krishnamurthy, Fowler, Men, & Kim 2001; Chung, Harmon, & Baker, 2001; Haider & Frensch, 2002; Insel, 1995; Mayer & Mautone 2002; Morgan, 2002; Poustie, 2001; Raab, 2002; Schwid 2001; Treadwell & Grobler 2001). Russomano (2000), Solomon (2001), and Gray (1992) discovered that simulations were important in improving self-concept. Further, Martinot and Monteil (2000), through the use of a matching strategy simulation, suggested that simulations improved self-concept.

While simulation participation, obviously, supports workplace success, many studies resolved that self-concept, too, influenced success in the workplace (Aaker & Lee, 2001; Chiungjung & Michael, 2002; Dolich & Shilling 2001; Grier, 2001; Grubb, 1967; Houston & Walker, 1996; Ireson 2002; Krishnamurthy, 2002; Reed, 2002; Russomano, 2000; Townsend, Moore, Tuck, & Wilter, 1998; Warshaw & Davi, 1984; Wiener & Goolsby, 1990; Wilson, 2002). In the only marketing related studies, Grier (2001), Heath and Scott (1998), Hogg and Cox (2000), Mehta (1999), and Rao (1972) noted self-concept as a determinant of increased product brand and image usage and ad effectiveness. Thereby, brand and brand images were more successful with increased individual self-concept. Bargozzi (1995) and Sirgy (1991) indicated that higher self-concept levels resulted in better consumer behavior and decision-making. Additionally, many studies noted individual characteristics such as age, gender, academic achievement or grade point average, occupational interests, grade or group level discipline or major (Ackerman, Bowen, Beier, & Kanfer, 2001; Alfeld-Liro, & Sigelman, 1998; Aries, Oliver, Blount, & Christaldi, 1998; Corinne & Sigelman, 1998; Charmaz, 2002; Grier, 2001; Hogg & Cox, 2000; Houston & Walker, 1996; Ireson, 2002; Martinot & Monteil, 2000; Mboya, 1998; McColl, 2002; Veenaman, 2002; Verkuyten, 2002) as supporting success. In a contrast, Adams, Prescher, Beehr, and Lepisto (2002) a positive self-concept resulting in organizational commitment had no connection to retirement intent. Veenaman (2002) concluded that achievement was needed for simulated learning to be appropriate. Also, Martinot and Monteil (2000) inferred that high achievement resulted in high self-concept.

Although numerous characteristics and self-concept are recognized as influencing success in the workplace, many studies reflect upon determinants of self-concept. Particularly, numerous studies suggested that self-concept is influenced by age, experience, development, grade point average, gender, group or classification, and occupational interest or class major (Abu-Hilal & Aal-Hussain, 1997; Carducci, 1998; Deathe, 2002; Gribbons, Tobey, & Michael, 1995; Haberstroh, 2002; Helitzer, Cunningham-Sabo, Vanleit, & Crowe, 2002; Hoegh & Martin, 2002; Hoza, Pelham Jr., Dobbs, Owens, Pillow, 2002; Ireson, 2002; Kling, Ryff, & Essex 1997; Laliberte-Rudman, 2002; Marsh, 1998; Marsh & Craven, 2002; McColl, 2002; Ren, Qian, Wang, & Gu, 2002; Stern, 1988; Verplanken & Holland, 2002). However, Coyne (2002) found that gender, classification, and curriculum choice had no affect on self-concept in middle school math students while achievement did have a main affect on self-concept.

Hypotheses

Considering prior studies and the original purpose of assessing self-concept from a simulation environment and individual antecedents for their influence upon marketing personnel success, the following hypotheses are appropriate for the study:

Ho_1: There is no significant difference in marketing success among marketing students, as measured by marketing class grade, with regard to self-concept.

Ho_2: There is no significant difference in marketing success among marketing students, as measured by marketing class grade, with regard to age, gender, development, classification, major, or grade point average.

Ho_3: There is no significant interaction among the characteristics of age, gender, development, classification, major or grade point average, as measured by marketing class grade, with regard to self-concept.

Ho_4: There is no significant difference in self-concept among marketing students, as measured by Malhotra's Self-Concept Scale, with regard to age, gender, development, classification, major or grade point average.

Ho_5: There is no relationship between marketing success among marketing students, as measured by marketing class grade, and self-concept among marketing students, as measured by Malhotra's Self-Concept Scale.

METHODOLOGY

Research Design

A descriptive ex post facto design was used to examine causal inferences through single instance implicit comparisons. According to Best (1977), descriptive research, as in this study, describes and interprets and is concerned with the present. This design was appropriate for its control of intrasession history through responses to a fixed instrument (Campbell and Stanley, 1963).

Population and Sample

The Malhotra Self-Concept Scale, distributed via e-mail, was employed to gather data from all undergraduate students participating in the Experiencing Marketing in the Marketplace and Business Management Marketplace Web-based Simulations in the 2001-2002 academic year (N = 50). Using a purposive sampling method, the prudency of the researcher, and the National Education Association research bulletin (1960), the entire population was surveyed to ensure a 95% confidence level. An overall response rate of 34% was obtained. Of the participants, 1 (6%) was 19 years of age, 3 (18%) were 20 years of age, 8 (47%) were 21 years of age, 2 (11.5%) were 24 years of age, 2 (11.5%) were 31 years of age, and 1 (6%) was 41 years of age. With regard to gender, 10 (59%) were male and 7 (41%) were female. Seventy-one percent of the respondents had only experienced one simulation, while 29% had participated in two simulation-learning experiences. In terms of classification, 10 (59%) were senior level status, 6 (35%) were juniors, and 1 (6%) was a sophomore. Concerning their occupational interests or major, 7 (41%) were management majors, 3 (18%) were marketing majors, 3 (18%) were general business majors, 2 (12%) were accounting majors, 1 (6%) was a management/marketing major, and 1 (6%) was a physical education major.

Instrumentation

Malhotra's Self-Concept Scale (Malhotra, 1981) was selected for its: (a) ability to measure self-concepts, person concepts, and product concepts, and (b) its applicability of coordinating organizational efforts with self-concept information. The Malhotra Self-Concept Scale consisted of a 15-item semantic differential scale ranging from rugged-delicate to modest-vain. The students were asked to select, on a scale of 1-5, their level of self-concept on each of the 15 items. For example, if students considered themselves most delicate, they would have rated themselves a level 5 on that item. The scale was validated using the test-re-test reliability of the scale. The overall reliability of the scale was assessed at the individual level for each measure of self-concept. The average test-re-test correlation for the ideal self-concept measure was .80 ranging to .98. A comparison with the results obtained in previous marketing studies shows that the scale items have an acceptable level of test-re-test reliability. Given the multidimensional nature of the scale, it is not very meaningful to estimate an overall measure of internal consistency. However, where appropriate, the coefficient alpha was computed for each subset of scale items that make up a particular factor. The coefficients range from .5 to .7 (Malhotra, 1981). An interval data score was totaled from the Malhotra Self-Concept Scale based upon each student's responses. Student data maintained by the school and instructor were also used to determine age, gender, development, classification, success, major area of study, and grade point average. The variable age was determined from birth date data at the time of participation in the simulation. The variable development, based upon prior research, was based upon whether the student had participated in either one or two simulation experiences. The variable classification had four levels, based upon the student's designated class level being either freshman = 1, sophomore = 2, junior = 3, or senior = 4. The student's success level was their final interval marketing course grade. The student's major area of study had six levels: 1 = Management, 2 = Marketing, 3 = General Business, 4 = Accounting, 5 = Management/Marketing, 6 = Physical Education.

Utilizing Dillman's (1978) Total Design Method, which is noted for systematic steps and high response rates, a cover letter detailing the purpose of the study and the Malhotra Self-Concept Scale was sent via e-mail to all students in the population. At their leisure, they were instructed to send the questionnaire back to the researcher via e-mail. Two weeks later, the same procedure was used in a follow-up attempt to collect the data. Yet, a third attempt was made to collect the data via e-mail.

Data Analysis

Data analysis involved frequencies and percentage to describe the study participants. Analysis of variance (ANOVA) was used to determine significant differences between self-concept, age, gender, development, classification, major, grade point average, and success. Where significant differences were found, the Tukey HSD post hoc test was used to address all pairwise comparisons among independent variable subscales. The Pearson's product moment correlation procedure was used to determine any relationship among self-concept and success.

FINDINGS

The prior stated hypotheses are the basis for presenting the findings for this study. The findings presented analyzed: (a) significant differences between self-concept and success, (b) significant differences between age, gender, development, classification, major, grade point average, and success, (c) age, gender, development, classification, major, grade point average, and self-concept interactions and success, (d) significant differences between age, gender, development, classification, major, grade point average, and self-concept, and (e) the relationship between self-concept and success.

Ho_1: There is no significant difference in marketing success among marketing students, as measured by marketing class grade, with regard to self-concept.

With self-concept being the independent variable and success being the dependent variable, the ANOVA procedure disclosed no significant difference in marketing success with regard to self-concept. Therefore, no follow-up tests were necessary and null hypothesis one was not rejected.

Ho_2: There is no significant difference in marketing success among marketing students, as measured by marketing class grade, with regard to age, gender, development, classification, major or grade point average.

Analyzing hypothesis two revealed no significant differences in marketing success as the dependent variable with regard to age, gender, development, classification, major, or grade point average. Again, no follow-up tests were necessary and null hypothesis two was not rejected.

Ho_3: There is no significant interaction among the characteristics of age, gender, development, classification, major, or grade point average, as measured by marketing class grade, with regard to self-concept.

The ANOVA procedures for interaction independent variables of age x self-concept, gender x self-concept, development x self-concept, classification x self-concept, major x self-concept, and grade point average x self-concept revealed no significant differences in marketing success. Thus, null hypothesis three was not rejected.

Ho_4: There is no significant difference in self-concept among marketing students, as measured by Malhotra's Self-Concept Scale, with regard to age, gender, development, classification, major, or grade point average.

ANOVA Table 1 demonstrates a significant difference in self-concept, being the dependent variable, and age, as the independent variable. At 5 degrees of freedom and an F-value of 7.185, age was significant at .003. The R-square also revealed that age represented .766 as a statistical model for self-concept. However, no significant differences were found in self-concept with regard to gender, development, classification, major, or grade point average.

Table 1. Analysis of Variance for Self-Concept Scores of Respondents Classified by Age

	df	SS	F	Pr>F
Self-Concept				
Age	5	245.08	7.185	.003*

* $p < .05$

a. R Squared = .766

The Tukey HSD post hoc test analysis was not necessary because, according to frequency data, less than two participants fell into certain age groups. Thus, no clear differences in age groups could be determined.

Ho_5: There is no relationship between marketing success among marketing students, as measured by marketing class grade, and self-concept among marketing students, as measured by Malhotra's Self-Concept Scale.

The Pearson's product moment correlation procedure, as shown in Table 2, indicates a negative correlation between marketing success and self-concept. With a correlation coefficient of -.239, hypothesis five can be rejected.

Table 2. Pearson Product Moment Correlation Matrix for Self-Concept and Success

	Success
Self-Concept	-.239

Conclusions and Recommendations

Using a simulated learning environment, self-concept and individual antecedent measures are a baseline for determining actual marketing work environment success. Study results illustrated necessary attributes to be considered when accounting for marketing success. Clearly, a person's self-concept (or their feeling about themselves) doesn't determine marketing success. This, herein, indicates that organizational programs implemented to enhance individual well-being are unnecessary for increasing work performance efficiency. Moreover, must speculation could be made as to other supplemental conditions to self-concept that may produce marketing success such as workplace relationships. In other words, a individual may not feel very good about themselves, but because of their connections or relationships with higher powers within the organization, they may be more successful or vice versa. Organizations can perhaps conclude marketing success through individual associations. From the study, it can also be said that marketing success if afforded to any gender and at any age. Further, individual marketing success can surface regardless of the level of marketing experience, level of education, area of interest, or grade point average. In parallel, marketing organizations, in need of marketing personnel, have no need to search for particular levels of simulation experience, education, major interests, or grade point averages. Coupled with the first conclusion, individual determination may be another attribute to consider, along with self-concept, in determining marketing success. With this in mind, individual marketing projects can be successful regardless of age, gender, experience, level of education, major interests, or grade point average.

Additional evidence suggests that age, gender, development (simulation experience), classification (level of education), occupational interests, or grade point average coupled with self-concept do not impact marketing success. Organizations that attempt to increase self-concept at any age, for any gender, at any level of experience, at any level of education, for any occupational interest, or any grade point average background, will not result in better marketing performance. Other results suggest that marketing organizations should consider age when accounting for marketing performance efficiency. This can be applied to the theory behind this study, in that, age can enhance self-concept and thus, increase performance success. Concerning marketing success, gender, development (simulation experience), level of education, occupational interest, or grade point average should not be considered as enhancers to self-concept for marketing success. While age explains a great part of a person's individual feelings about themselves, there is room for other explanatory antecedents.

Last of all, evidence shows that as an individual develops more positive feelings about themselves, marketing success may actually decline and vice versa. This further emphasizes the consideration of "the looking glass self". Individuals can classify themselves higher or lower that an outside party perceives them. Organizations, therefore, should be concerned with developing an individual's specific marketing skills which may infer or increase positive individual feelings that may produce a greater confidence.

Conclusively, this study provides guidance for marketing personnel development and significant concerns in producing successful marketing initiatives. However, it opens greater doors to further self-concept analysis. Therefore, the following recommendations should be considered:

1. Simulated marketing education should be studied with greater populations and perhaps diversity in individual locations.
2. Self-concept should be analyzed as to its impact upon success in specific responsibilities in marketing performance i. e. advertisement creation.
3. Self-concept should be studied for marketing success while considering simulation teaching techniques and the weight placed on simulations performance in course grade.
4. Ethnicity and race should be considered in further self-concept studies as the population permits.
5. Investigation should be made of antecedents that represent the remaining 24% variation in the self-concept model.
6. Further research should seek to combine self-concept and self-efficacy relationships in marketing success.

REFERENCES

Aaker, J. L., & Lee, A. Y. (2001). "I" seek pleasures and "we" avoid pains: The role of self-regulatory goals in information processing and persuasion. *Journal of Consumer Research, 28*(1), 33-49.

Abu-Hilal, M. M., Aal-Hussain, A. A. (1997).Dimensionality and hierarchy of the SDQ in a non-Western milieu: A test of self-concept invariance across gender. *Journal of Cross - Cultural Psychology, 28*(5), 535-553.

Ackerman, P. L., Bowen, K. R., Beier, M. E., & Kanfer, R. (2001). Determinants of individual differences and gender differences in knowledge. *Journal of Educational Psychology, 93*, 797-825.

Adams, G. A., Presher, J., Beehr, T. A., & Lepisto, L. (2002). Applying work-role attachment theory to retirement decision-making. *International Journal of Aging & Human Development, 54,* 125-137.

Alfeld-Liro, C., & Sigelman, C. K. (1998). Sex differences in self-concept and symptoms of depression during the transition to college. *Journal of Youth and Adolescence, 27*(2), 219-244.

Andrews, R. L., & Manrai, A. K. (1998). Simulation experiments in choice simplification: The effects of task and context on forecasting performance. *Journal of Marketing Research, 35*(2), 198-209.

Anido, L., Llamas, M., Fernandez, M. J. (2001). Internet-Based learning by doing. *IEEE Transactions on Education, 44,* 193.

Aries, E., Oliver, R. R., Blount, K., & Christaldi, K. (1998). Race and gender as components of the working self-concept. *The Journal of Social Psychology, 138,* 277-290.

Bagozzi, R. P. (1995). Reflections on relationship marketing in consumer markets. *Journal of the Academy of Marketing Science, 23,* 272-277.

Best, J. W. (1977). *Research in education.* Englewood Cliffs, NJ: Prentice Hall.

Bray, D. (2001). An evaluation of factors influencing the academic self-concept, self-esteem and academic stress for direct and re-entry students in higher education. *Educational Psychology, 21*, 455-472.

Campbell, D. T., & Stanley, J. C. (1963). *Experimental and quasi-experimental designs for research.* Chicago: Rand McNally.

Carducci, B. J. (1998). *The psychology of personality: Viewpoints, research, and applications.* Belmont, CA: Brooks/Cole.

Celsi, R. L. (2002). Discontinuous classroom innovation: Waves of change for marketing education. *Journal of Marketing Education, 24,* 64-72.

Chang, Y. S., Krishnamurthy, V., Fowler, J. W., Men, J., & Kim, R. (2001). Laser marking study proves benefits of simulation for backend tools. *Solid State Technology, 44,* 85-89.

Charmaz, K. (2002). The self as habit: The reconstruction of self in chronic illness. *Occupational Therapy Journal of Research, 22,* 31S-41S.

Chung, G. K. W. K., Harmon, T. C., & Baker, E. L. (2001). The impact of a simulation-based learning design project on student learning. *IEEE Transactions on Education, 44,* 390-399.

Coyne, E. P. (2002). Curricular choice and mathematical self-concept in middle school students: A pilot study. *Dissertation Abstracts International, 62,* 9-A. (UMI No. 2968)

Dacko, S. G. (2001). Narrowing skill development gaps in marketing and MBA programs: The role of innovative technologies for distance learning. *Journal of Marketing Education, 23*, 228-239.

Deathe, K. (2002). Ability grouping: year group differences in self-concept and attitudes of secondary school pupils. *Westminster Studies in Education, 25,* 7-18.

Dillman, D. (1978). *Mail and telephone surveys: The total design method.* New York: John Wiley & Sons.

Dolich, I. J. & Shilling, N. (2001). A critical evaluation of 'The Problem of Self-Concept in Store Image Studies'. *Journal of Marketing, 35*(1), 1-2.

Eggers, J. T., & Brown, R. (2002). Simulation modeling for leadership development: A systems dynamics approach. *Corrections Today, 64*(2), 181-182.

Glynn, K. A., & Rajendran, K. N. (1993). Perceptual-based student outcomes assessment process in the marketing curriculum. *Journal of Education for Business, 69*, 11-18.

Gray, C. (1992). Enterprise trainees' self-construals as entrepreneurs. *International Journal of Personal Construct Psychology, 53,* 307-322.

Gribbons, B. C., Tobey, P. E., Michael, W. B. (1995). Internal-consistency reliability and construct and criterion-related validity of an academic self-concept scale. *Educational and Psychological Measurement, 55,* 858.

Grier, S. A., & Deshpande, R. (2001). Social dimensions of consumer distinctiveness: The influence of social status on group identity and advertising persuasion. *Journal of Marketing Research, 38*(2), 216-224.

Grubb, E. L., & Grathwohl, H. L. (2001). Consumer self-concept, symbolism and marketing behavior: A theoretical approach. *Journal of Marketing, 31*(4), 1-2.

Haberstroh, S., Oyserman, D., Schwarz, N., Kuehnen, Ulrich, & Ji, L. (2002). Is the interdependent self more sensitive to question context than the independent self? Self-construal and the observation of conversational norms. *Journal of Experimental Social Psychology, 38,* 323-329.

Haider, H., & Frensch, P. A. (2002). Why aggregated learning follows the power law of practice when individual learning does not: Comment on Rickard (1997, 1999), Delaney, et al. (1998), and Palmieri (1999). *Journal of Experimental Psychology: Learning, Memory, & Cognition, 28*(2), 392-406.

Heath, A. P., & Scott, D. (1998). The self-concept and image congruence hypothesis: An empirical evaluation in the motor vehicle market. *European Journal of Marketing, 32*(11-12), 1110-1123.

Helitzer, D. L., Cunningham-Sabo, L. D., Vanleit, B., & Crowe, T. K. (2002). Perceived changes in self-image and coping strategies of mothers of children with disabilities. *Occupational Therapy Journal of Research, 22,* 5-33.

Hoegh, D. G., & Bourgeois, M. J. (2002). Prelude and postlude to the self: Correlates of achieved identity. *Youth & Society, 33*(4), 573-594.

Hogg, K. Cox, A. J., Keeling, K. (2000). The impact of self-monitoring on image congruence and product/brand evaluation. *European Journal of Marketing, 34,* 641-666.

Houston, M. B., & Walker, B. A. (1996). Self-relevance and purchase goals: Mapping a consumer decision. *Journal of the Academy of Marketing Science, 24*(3), 232-245.

Hoza, B. P., Jr., W. E., Dobbs, J., Owens, J. S., & Pillow, D. R. (2002). Do Boys With Attention-Deficit/Hyperactivity Disorder Have Positive Illusory Self-Concepts? *Journal of Abnormal Psychology, 111,* 268-279.

Huang, C., & Michael, W. B. (2002). Multitrait-meltimethod analyses of scores on a Chinese version of the dimensions of self-concept scale. *Educational and Psychological Measurement, 62*(2), 355-372.

Insel, P. M. (1995). Changing your self-concept. *Healthline, 14,* 6.

Ireson, J. (2002). Within-class ability grouping: placement of pupils in groups and self-concept. *British Educational Research Journal, 28,* 249-264.

Kennedy, E. J., Lawton, L., & Walker, E. (2001). The case for using live cases: Shifting the paradigm in marketing education. *Journal of Marketing Education, 23*, 145-151.

Kling, K. C., Ryff, C. D., & Essex, M. J. (1997). Adaptive changes in the self-concept during a life transition. *Personality and Social Psychology Bulletin, 23,* 981-990.

Krishnamurthy, P. & Kumar, P. (2002). Self-other discrepancies in waiting time decisions. *Organizational Behavior & Human Decision Processes, 87*(2), 207-226.

Laliberte-Rudman, D. (2002). Linking occupation and identity: Lessons learned through qualitative exploration. *Journal of Occupational Science, 9,* 12-19.

Libby, L. K., & Eibach, R. P. (2002). Looking back in time: Self-concept change affects visual perspective in autobiographical memory. *Journal of Personality and Social Psychology, 82*(2), 167-179.

Maharg, P. (2001). Negotiating the web: Legal skills learning in a virtual community. *International Review of Law, Computers, & Technology, 15,* 345-361.

Malhotra, N. K. (1981). A scale to measure self-concepts, person concepts, and product concepts. *Journal of Marketing Research, 18,* 456-464.

Marsh, H. W. (1997). Causal effexts of academic self-concept on academic achievement structure equation models of longitudinal data. *Journal of Educational Psychology, 89,* 41-54.

Marsh, H. W. (1998). Age and gender effects in physical self-concepts for adolescent elite athletes and nonathletes: A multicohort-multioccasion design. *Journal of Sport & Exercise Psychology, 20,* 237-259.

Marsh, H. W., Ellis, L. A., & Craven, R. G. (2002). How do preschool children feel about themselves? Unraveling measurement and multidimensional self-concept structure. *Developmental Psychology, 38,* 376-393.

Martinot, D. & Monteil, J. (2000). Use of the self-concept in forming preferences by French students of different levels of academic achievement. *Journal of Social Psychology, 140*(1), 119-132.

Mayer, R. E., Mautone, P., & Prothero, W. (2002). Pictorial aids for learning by doing in a multimedia geology simulation game. *Journal of Educational Psychology, 94,* 171-185.

Mboya, Mzobanxi, (1998). Self-concept of academic ability as a function of sex, age, and academic achievement among African adolescents. *Perceptual and Motor Skills, 87,* 154-161.

McColl, M. A. (2002). Occupation in stressful times. *American Journal of Occupational Therapy, 56,* 350-353.

Mehta, A. (1999). Using self-concept to assess advertising effectiveness. *Journal of Advertising Research, 39,* 81-89.

Michau, F., Gentil, S., & Barrault, M. (2001). Expected benefits of web-based learning for engineering education: examples in control engineering. *European Journal of Engineering Education, 26,* 151-169.

Morgan, P. J. (2002). Experiential learning in an anaesthesia simulation centre: analysis of students' comments. *Medical Teacher, 24,* 23-27.

National Education Association Research Division. (1960). Table for determining sample size from a given population. *Research Bulletin, 38,* 99.

Online training program fits bill for fair lending compliance effort. (2002, May 27). *National Mortgage News, 26,* 14.

Oyserman, D., Coon, H. M, & Kennelmeier, M. (2002). Rethinking individualism and collectivism: Evaluation of theoretical assumptions and meta-analyses. *Psychological Bulletin, 128,* 372.

Poustie, M. R. (2001). Engaging students and enhancing skills: Lessons from the development of a web-supported international environmental law conference simulation. *International Review of Law, Computers, & Technology, 15,* 331-345.

Raab, M. (2002). T-ECHO: Model of decision making to explain behaviour in experiments and simulations under time pressure. *Psychology of Sport & Exercise, 3,* 151-171.

Rao, V. R. (2001). Changes in explicit information and brand perceptions. *Journal of Marketing Research, 9,* 209-214.

Reed, A. (2002). Social identity as a useful perspective for self-concept-based consumer research. *Psychology & Marketing, 19,* 235-266.

Ren, G., Qian, M., Wang, Y., & Gu, B. (2002). Translated title: Family environment and the development of self-esteem, self-concept, and external-internal control in children with ADHD. *Chinese Mental Health Journal, 16,* 268-272.

Russomano, L. (2000). Achievement, locus of control, self-concept, social problem-solving training and the acquisition of prosocial skills in children. *Dissertation Abstracts International, 61,* 2-A. (UMI No. 485)

Schwid, H. A. (2001). Components of an effective medical simulation software solution. *Simulation & Gaming, 32,* 240-249.

Semler, S. W. (1997). Systematic agreement: A theory of organizational alignment. *Human Resource Development Quarterly, 8, 23-40.*

Sirgy, M., Johar, J. S., Samli, A. C., & Claiborne, C. B. (1991). Self-congruity versus functional congruity: Predictors of consumer behavior. *Journal of the Academy of Marketing Science, 19,* 363-375.

Solomon, M. R. (2001). The role of products as social stimuli: A symbolic interactionism perspective. *Journal of Consumer Research, 10,* 319-329.

Stern, B. B. (2001). Sex-role self-concept measures and marketing: A research note. *Psychology & Marketing, 5,* 85-101.

Swaak, J., & de Jong, T. (2001). Discovery simulations and the assessment of intuitive knowledge. *Journal of Computer Assisted Learning, 17*(3), 284-294.

Townsend, M. A. R., Moore, D. W., Tuck, B. F., & Wilton, K. M. (1998). Self-concept and anxiety in university students studying social science statistics within a co-operative learning structure. *Educational Psychology, 18,* 41-54.

Treadwell, I., & Grobler, S. (2001). Students' perceptions on skills training in simulation. *Medical Teacher, 23*(5), 476-483.

van der Veen, A. M., Lustick, I. S., & Miodownik, D. (2001). Studying performance and learning with ABIR: The effects of knowledge, mobilizing agents, and predictability. *Social Science Computer Review, 19,* 263-280.

Veenman, M. V. J., Prins, F. J., & Elshout, J. J. (2002). Initial inductive learning in a complex computer simulated environment: The role of metacognitive skills and intellectual ability. *Computers in Human Behavior, 18,* 327-341.

Verkuyten, M., & Brug, P. (2002). Ethnic identity achievement, self-esteem, and discrimination among Surinamese adolescents in the Netherlands. *Journal of Black Psychology, 28*(2), 122-141.

Verplanken, B., & Holland, R. W. (2002). Motivated decision making: Effects of activation and self-centrality of values on choices and behavior. *Journal of Personality & Social Psychology, 82*(3), 434-447.

Waks, L. T. (2001). Computer mediated experience and education. *Educational Theory, 51,* 415-433.

Warshaw, P. R., & Davis, F. D. (1984). Self-understanding and the accuracy of behavioral expectations. *Personality & social Psychology Bulletin, 10*(1), 111-118.

Washbush, J., & Gosen, J. (2001). An exploration of game-derived learning in total enterprise simulations. *Simulation & Gaming, 32*(3), 281-297.

Wiener, J. L., & Goolsby, J. R. (1990). Personal communication in marketing: An examination of self-interest contingency relationships. *Journal of Marketing Research, 27*(2), 227-231.

Wilson, C. (2002). Hypnosis in the rebuilding of self following marriage breakdown. *Australian Journal of Clinical & Experimental Hypnosis, 30,* 66-72.

Chapter 7

OPTIMIZING THE DURABILITY AND GENERALIZABILITY OF KNOWLEDGE AND SKILLS

Alice F. Healy, Carolyn J. Buck-Gengler, Immanuel Barshi, James T. Parker, Vivian I. Schneider, William D. Raymond, N. Noelle LaVoie and Anita R. Bowles

University of Colorado at Boulder, CO,USA

Paul Pauli

University of Tübingen, Germany

Julia A. Fisher and Lyle E. Bourne, Jr.

University of Colorado at Boulder,CO, USA

ABSTRACT

The work summarized here is based on our recent research into optimizing the durability and generalizability of knowledge and skills, which followed from our earlier work on optimizing the long-term retention of knowledge and skills and, subsequently, on improving the training of foreign languages. Unlike our research on foreign language training, the present research involved a broader and more general consideration of knowledge and skills. It examined performance in a variety of laboratory and naturalistic tasks, some involving language and some nonlinguistic. Extending our previous research, the main focus was on enhancing the *generalizability*, or transferability, of what was learned and retained. This new direction was prompted by our prior finding of training conditions that promote durable memory but constrain generalizability. The more specific issues that we addressed include the following: (a) variations in contextual interference as a means to promote durability and generalizability of knowledge and skills, (b) communication about navigating in space, (c) response execution in a spatial array, (d) the underlying causes of durability and specificity in digital data entry, (e) the relationship between the learning of examples and the learning of rules in conceptual tasks, (f) types of strategies and strategy transitions in category learning tasks, (g) the durability and use of general and specific knowledge in quantitative estimation, and (h) managing rapidly presented information.

INTRODUCTION

The work summarized in this chapter is based on our current research into optimizing the durability and generalizability of knowledge and skills, which follows from our earlier work on optimizing the long-term retention of knowledge and skills (Healy & Bourne, 1995) and, most recently, on improving the training of foreign languages (Healy & Bourne, 1998). Unlike our research on foreign language training, this research involves a broader and more general consideration of knowledge and skills. The research we are conducting examines training, durability, and generalizability in a variety of laboratory and naturalistic tasks, some involving language and some nonlinguistic. Extending our previous research, the main focus is on enhancing the *generalizability*, or transferability, of what is learned and retained. This new direction is prompted by two main outcomes of earlier studies. First, we found an amazing durability of knowledge and skills in certain domains which we were able to attribute to a phenomenon known as "procedural reinstatement" (Healy et al., 1992). Second, we found that typically the more durable the memory for a certain kind of knowledge or skill, the less generalizable it is beyond the circumstances in which training took place (Healy et al., 1993, 1995). Thus, we discovered training conditions that promote durable memory but constrain generalizability. In our current work, we are trying to discover ways of promoting the generalizability of those durable memories. This focus is particularly crucial to military tasks because training circumstances can rarely capture the full set of circumstances under which these tasks are encountered in the field. The Army has come to rely increasingly on the use of distributed interactive simulation and virtual reality as training mechanisms. Although these novel procedures seem likely to be effective, it is not yet clear how durable and generalizable the acquired knowledge and skills will be beyond the training environment. Therefore, some of our research examines principles underlying the retention and transfer of behaviors acquired under such new training technologies.

Our studies focus on three interrelated topics—*training*, *durability*, and *generalizability*—which correspond to the three fundamental processes associated with knowledge and skill utilization—*acquisition*, *retention*, and *transfer* (Healy, 1998). We view these three aspects as an interrelated set of concepts that cannot be examined or manipulated independently of one another. Indeed, in most of our studies the interaction of these three aspects is a major concern.

The more specific issues that we are addressing include the following: (a) variations in contextual interference as a means to promote durability and generalizability of knowledge and skills, (b) communication about navigating in space, (c) response execution in a spatial array, (d) the underlying causes of durability and specificity in digital data entry, (e) the relationship between the learning of examples and the learning of rules in conceptual tasks, (f) types of strategies and strategy transitions in category learning tasks, (g) the durability and use of general and specific knowledge in quantitative estimation, and (h) managing rapidly presented information.

Variations in Contextual Interference as a Means to Promote Durability and Generalizability of Knowledge and Skills

Our earlier work on optimizing training conditions included a project by Schneider, Healy, Ericsson, and Bourne (1995) on the acquisition and retention of logic rules. This project examined William Battig's (1972, 1979) contextual interference effect, defined as superior memory and greater transfer for materials that are presented under conditions of high interference. Such interference with the learning of a particular association is created by competition from the simultaneous occurrence of other stimuli or other associations to be learned. In our earlier study, we showed that varying the processing requirements of a task from trial to trial interferes with acquisition of the task skill but aids retention and transfer of that skill. Theoretically, items that have more contextual interference require more processing, and are thus learned more slowly. But if well learned initially, these items will be retained as well as, or better than, the low-interference items. This finding is of clear importance to the study of long-term retention because it implies that the methods used to optimize acquisition of new materials are not necessarily those that will optimize later retention.

In our earlier study (Schneider et al., 1995), we compared practice schedules in which different logical rules were either intermixed or blocked together over trials. We used a display meant to simulate an aircraft instrument monitor with four panels, only one of which was relevant (or operational) on any given trial. The subjects' task was to decide whether or not the stimulus configuration in the relevant panel signaled an emergency. Signals displayed in each panel involved a different logical rule for the decision about emergency status. Our experiments included one group of subjects given blocked practice (in which all trials within a block involved the same rule and panel, although the particular stimulus configuration or signal in the panel varied) and a second group given random practice (in which both the rule and the stimulus configuration varied from trial to trial). All subjects started with an acquisition phase followed by retention test blocks, with some tests occurring after a delay. The random group yielded longer response times and more errors during the acquisition phase than did the blocked group, in accord with theoretical expectations about contextual interference effects. In contrast, blocked practice led to longer latencies and more errors on both immediate and delayed retention tests than did random practice, again in accord with predictions based on contextual interference. In sum, our findings supported the principle that contextual interference produces slower learning but better retention.

In a subsequent study by Schneider, Healy, and Bourne (1998), we studied the effects of a similar manipulation of contextual interference on acquisition and retention of foreign language vocabulary. We performed two experiments on different methods for practicing vocabulary in a foreign language. In both of these experiments, non-French-speaking college students learned the association between 25 French words and their English equivalents. In the acquisition (i.e., training) phase, on each trial, all 25 French-English word pairs were presented to subjects on the computer screen in blocks of five. After a given block was shown for study, subjects were given the five French words alone one at a time and asked to type the English equivalent. For subjects in the blocked order acquisition phase, the five word pairs in each group were related; for example, one group of word pairs consisted of body parts: *back*, *mouth*, *face*, *finger*, *eyes*. For subjects in the mixed order acquisition phase, the five word pairs in each group were unrelated and consisted of one word from each of the five categories, for example, *back* (body part), *airplane* (vehicle), *plate* (tableware), *ham* (food), *shirt* (clothes). The subjects were next presented with another block of five pairs, and so on until all 25 word pairs had been presented and tested.

On the following training trial, all 25 word pairs were presented and tested again, using a new grouping of words in the mixed order acquisition condition but the same five-word groups in the blocked order acquisition condition, although the groups were presented in a new order and the five word pairs in a group were presented in a new order in the blocked order acquisition condition. After the subjects correctly responded on two consecutive trials to all 25 words, they were tested with all 25 French words one at a time and asked to type in the corresponding English word. Subjects in Experiment 1 were given training trials until they responded correctly to all 25 words two times in a row, whereas all subjects in Experiment 2 were given exactly three training trials. Following training, subjects were given an immediate retention test on all 25 French words one at a time and asked to type in the corresponding English word. Although there were no breaks within the test, the presentation of the French words at the test maintained the same type of groupings (blocked or mixed) as during training, but a new grouping of word pairs was used for the mixed acquisition condition and a new ordering of groups and of words within groups was used for the blocked order acquisition condition. One week later, subjects returned and took the retention test again. Subjects then repeated the acquisition phase to assess ease of relearning. During this second session, half of the subjects were in the same learning and testing condition (blocked or mixed) as they were in during initial acquisition, whereas the remaining subjects were in the other condition.

We expected that the blocked order initial acquisition group would be faster to learn the words but might be less accurate on the immediate test and especially on the retention test 1 week later. As we have found in the past, the method that leads to the fastest way to learn something is not necessarily the best method for retaining the learned information. We predicted that the subjects who were in the mixed order initial acquisition condition would be faster to relearn the words than would the subjects in the blocked order initial acquisition condition. Our results provided some, but not complete, support for these predictions. In particular, as shown in Fig. 1, we found a marginally significant interaction between initial acquisition condition (blocked order and mixed order) and session (learning and relearning) on the total time to respond during the first three trials of training in Experiment 1. Subjects in the blocked order acquisition condition were faster than those in the mixed order acquisition condition during learning, but the opposite was found during relearning. Furthermore, as shown in Fig. 2, we found a significant interaction between initial acquisition condition and trial number on the number of errors during the first three trials of training in Experiment 1. Specifically, subjects in the blocked order acquisition condition made fewer errors than those in the mixed order acquisition condition on the first trial, but the pattern reversed by the third trial, so that subjects in the mixed order acquisition condition later made fewer errors than those in the blocked order acquisition condition.

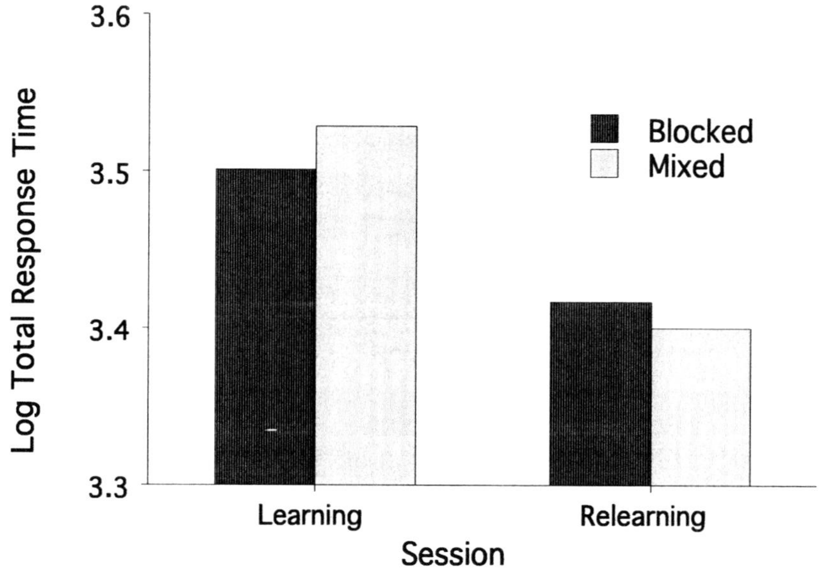

Figure 1. Log total response time as a function of session and acquisition condition (Schneider, Healy, & Bourne, 1998, Experiment 1).

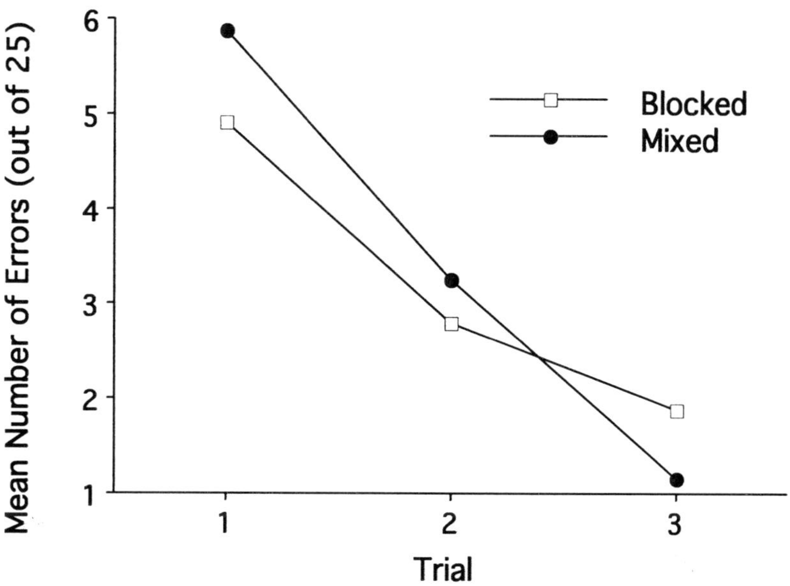

Figure 2. Mean number of errors as a function of trial number and acquisition condition (Schneider, Healy, & Bourne, 1998, Experiment 1).

Experiment 2 was just like Experiment 1 except that after the subjects were presented all 25 word pairs three times, they were administered the immediate test, whether or not they had achieved the criterion of correctly responding twice to all 25 words. As shown in Fig. 3, our analysis of the accuracy data yielded a significant interaction between initial acquisition condition (blocked order and mixed order) and session (learning and relearning) on the number of errors. Subjects in the blocked order acquisition condition made fewer errors than those in the mixed order acquisition condition during learning, but the opposite was found during relearning (i.e., subjects in the mixed order acquisition condition made slightly fewer errors than those in the blocked order acquisition condition).

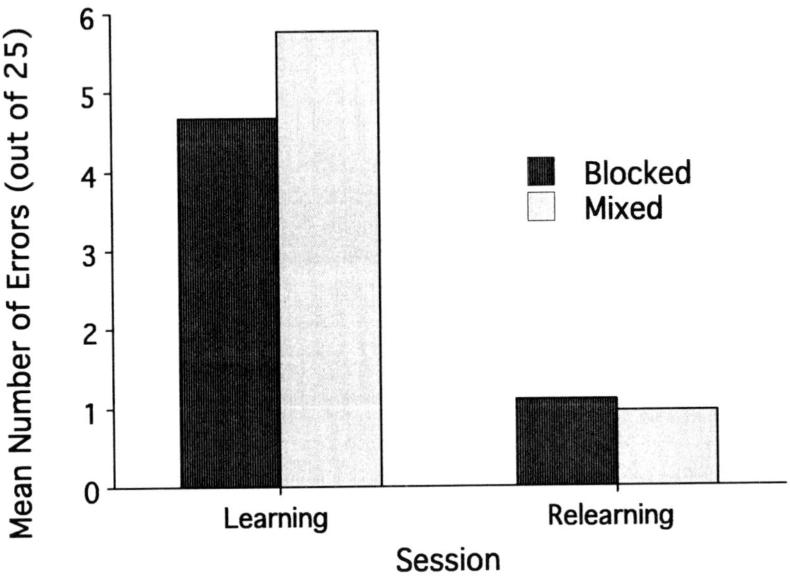

Figure 3. Mean number of errors as a function of session and acquisition condition (Schneider, Healy, & Bourne, 1998, Experiment 2).

Across the two experiments, we found that subjects in the blocked order acquisition condition were faster and more accurate than those in the mixed order acquisition condition during learning, but were slower and less accurate during relearning. These findings are in accordance with predictions based on contextual interference. They imply that blocking vocabulary by category, which is frequently done in foreign language courses, may aid initial acquisition but may not yield optimal retention.

In addition, as shown in Fig. 4, we found a significant interaction for the tests indicating that subjects did better on the retest, a week later, if the retest matched the type of training they received. That is, subjects who were trained in the blocked condition and given a blocked retest and subjects trained in a mixed condition and given a mixed retest made fewer errors on the retest than did subjects trained in one condition and given the opposite type of retest. Note that both blocked and mixed tests involved the same word pairs; only the grouping of word pairs varied. This finding provides, therefore, strong support for our procedural reinstatement principle (e.g., Healy et al., 1992), which specifies that retention is optimal when the procedures (or cognitive operations) used during learning are reinstated (or duplicated) during the subsequent test of retention. Our finding enables us to extend this principle to new materials, namely vocabulary pairs, and to a new set of procedures, namely those involving grouping of the materials.

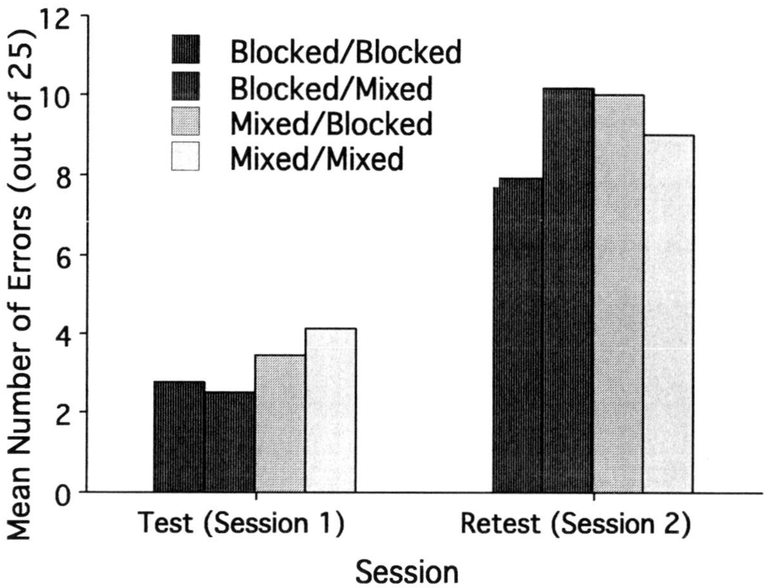

Figure 4. Mean number of errors as a function of session and acquisition condition in each session (Schneider, Healy, & Bourne, 1998, Experiment 2).

In a more recent follow-up experiment (Schneider, Healy, & Bourne, in press, Experiment 1), we sought to determine whether contextual interference improves transfer as well as retention of foreign vocabulary items. Vocabulary learning is fundamentally bidirectional and requires two complementary sets of translation processes, one from the foreign language to the native language, and the other in the opposite direction. In our earlier experiments just reviewed (Schneider et al., 1998), we examined only one of these processes because subjects were always cued with the French word and responded with the English word. In this follow-up experiment, we examined both sets of processes as well as transfer from one set to the other. Specifically, half of the subjects were trained and given an immediate retention test with French words as cues and English words as responses, as in our earlier experiments, and the remaining half of the subjects were trained and given an immediate retention test instead with English words as cues and French words as responses. As in Experiment 2 of our earlier study, subjects were all given exactly three training trials before the immediate retention test. In the second session 1 week after training, subjects were given a delayed retention test and then retrained either with the same set of cues and responses as in the first session or with the opposite set. All four combinations of first set and second set were examined, so that four different groups of subjects were employed—French cues in both sessions, English cues in both sessions, French cues in the first session and English cues in the second session, or English cues in the first session and French cues in the second session. There were 16 subjects in each of these groups, with each group divided into four subgroups of 4 subjects and with the subgroups varying in terms of the combination of practice schedules employed in the first and second sessions (either blocked in both sessions, mixed in both sessions, blocked in the first session and mixed in the second session, or mixed in the first session and blocked in the second session).

Although we did include the standard contextual interference manipulation in this experiment, comparing blocked and mixed practice schedules, that is a relatively weak manipulation of task difficulty. In contrast, the translation direction manipulation in the present experiment is a very strong manipulation of task difficulty. Hence, the manipulation of translation direction provides a much more powerful test of the effects of task difficulty on foreign vocabulary acquisition than does variation in practice schedules.

In fact, we found that the manipulation of translation direction created opposite effects on training and testing, similar to earlier demonstrations of contextual interference. As shown in Fig. 5, accuracy was much higher during learning when subjects were given French cues (and thus responded with familiar English words) than when they were given English cues (and thus responded with foreign French words). However, importantly, those subjects given the more difficult task during learning involving English cues showed somewhat (but not significantly) higher accuracy during relearning than did subjects given the easier task during learning involving French cues. Likewise, as shown in Fig. 6, those subjects given the more difficult task during learning involving English cues showed lower accuracy on the immediate retention test but higher accuracy on the delayed retention test than did subjects given the easier task during learning involving French cues, and the difference between the two groups on the delayed retention test was statistically significant. Because all subjects were given the same number of training trials, this advantage for the subjects given more difficult training cannot be attributed to a greater exposure to the vocabulary items, which would only exaggerate this effect in a natural learning situation outside the laboratory.

The question arises whether the advantage we found on the delayed retention test for the more difficult training with English cues applies to both easy and difficult retention tests. As shown in Fig. 7, the same advantage is found for both types of delayed retention tests although it is somewhat larger for the more difficult retention test involving English cues; the interaction of training cues and delayed retention test cues was marginally significant. It is interesting to note that subjects given the more difficult training showed higher accuracy on the delayed retention test than did subjects given the easier training even on the easier delayed retention test. That is, subjects trained with English cues did better on the delayed retention test than did subjects trained with French cues even when the delayed retention test involved French cues.

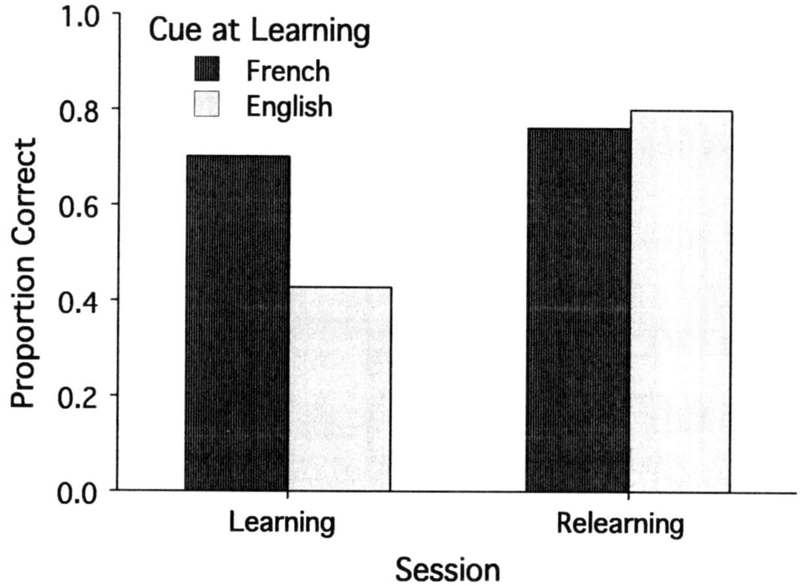

Figure 5. Proportion correct as a function of session and cue at learning (Schneider, Healy, & Bourne, in press, Experiment 1).

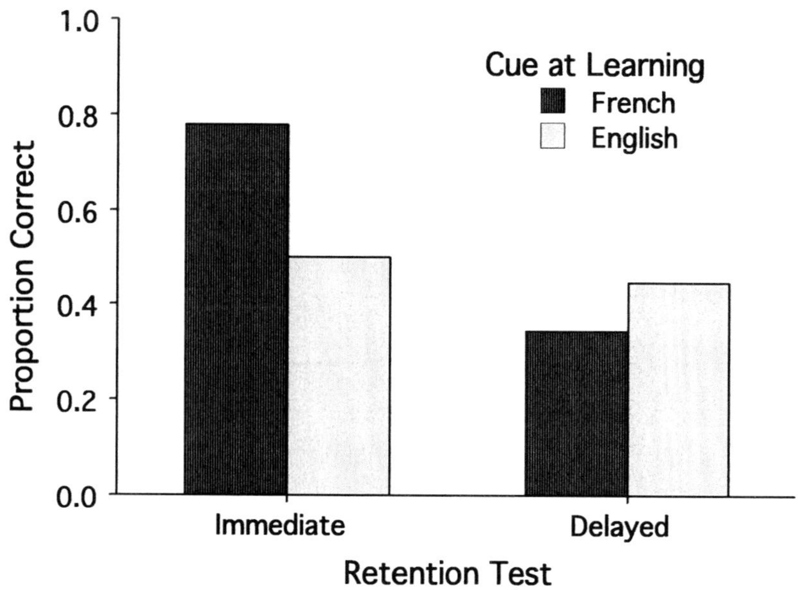

Figure 6. Proportion correct as a function of retention test type and cue at learning (Schneider, Healy, & Bourne, in press, Experiment 1).

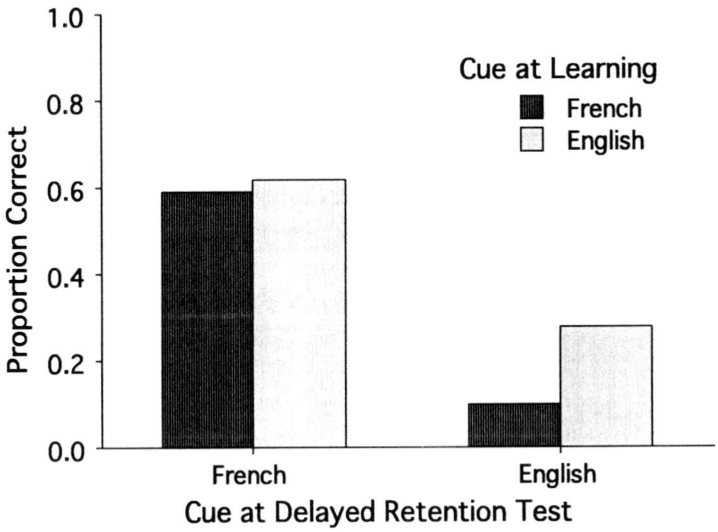

Figure 7. Proportion correct in delayed retention test as a function of cue at delayed retention test and cue at learning (Schneider, Healy, & Bourne, in press, Experiment 1).

In our earlier studies involving contextual interference (Schneider et al., 1998, 1995), we examined retention but not transfer. The present findings demonstrate that task difficulty during study aids transfer to a new task as well as long-term retention of the original task. Further, in our earlier studies (Schneider et al., 1998, 1995), contextual interference was caused by presenting stimuli in a random or mixed order rather than in a blocked order. The present findings demonstrate a similar pattern of results for a manipulation of task difficulty and suggest the hypothesis that any method for enhancing task difficulty during training may lead to improvements in task retention or transfer during subsequent tests. In our future research, we plan to test this hypothesis by exploring the effects on retention and transfer of various methods for enhancing task difficulty during training. An alternative hypothesis is that our manipulation of translation direction simply produces other forms of contextual interference. That is, in the English-to-French direction, learners are required to learn the French words as responses in addition to learning the association between English and French words. This response learning requirement is not present in the French-to-English direction and may provide a source of contextual interference through competing process requirements. By the same token, the French-to-English direction requires the subjects to learn the French words as stimuli. This greater stimulus learning also could provide competition with learning the French-English associations. If this analysis is correct, then response learning must compete more strongly with associative learning than does stimulus learning. The possibility that there is competition from the response and stimulus learning processes with associative learning is consistent with the Battig (1972, 1979) contextual interference principle. Thus, one implication of this alternative hypothesis is the need for a series of experiments to examine the separate and interactive contributions of stimulus, response, and associative learning to foreign language vocabulary acquisition.

COMMUNICATION ABOUT NAVIGATING IN SPACE

We have completed a series of five new experiments involving communication about navigation in space. This work focuses on effective communication in a specific concrete domain in which it has life-or-death consequences, namely communication between air crews and air traffic controllers. We chose that domain in part because of the success of our earlier research in this area, in which we developed an experimental paradigm that is analogous to the natural flight situation (see Barshi, 1997; Barshi & Healy, 1998, for summaries of our earlier studies on this topic).

To begin, we conducted a detailed analysis of an actual half-hour recording of air-ground communication at an international airport (see Barshi, 1997). This analysis revealed two main findings. First, flight crews were only likely to misunderstand the air-traffic controllers' instructions when these instructions included four or more propositions. Second, among these long instructions, more misunderstanding occurred when the speech rate was fast than when it was slow. The following excerpt from this recording is an example of a partial and erroneous read-back by the pilot of Expo Flight 92. The controller said, "Expo nine two you're niner miles from Laker, turn left heading three one zero, maintain three thousand till established on the localizer, cleared ILS runway two eight right approach, maintain one seven zero knots until Laker." The pilot responded, "That's two one, uh zero on the heading and, uh- one seventy till Laker." In this example, the pilot read back the wrong compass heading (he said "*two* one zero" instead of "*three* one zero"), though he was correct on the speed. He also neglected to acknowledge the altitude restriction and the approach clearance and omitted the aircraft identification. This omission prevented the controller from ascertaining that the intended crew received the correct clearance. Such a partial and erroneous read-back is hazardous and could lead to a fatal accident.

In prior laboratory investigations (Barshi, 1997; Barshi & Healy, 1998), we independently manipulated the number of propositions contained in each message and either the number of words in each proposition or the speech rate at which the propositions were uttered. Specifically, in our task, subjects heard instructions like those given by air traffic controllers; repeated the instructions aloud, as pilots are expected to do; and then followed the instructions, navigating in the space displayed on the computer. The repetition of the instructions provided us with a measure of immediate memory, whereas the implementation of the instructions given provided us with a strict test of both message memory and comprehension. The instructions described movements in a grid of four 4 X 4 matrices stacked one on top of another representing a three-dimensional space on a computer screen, as illustrated in Fig. 8. A sample instruction including three units was: "Turn left two squares; climb down one level; move forward one step." The subject immediately repeated this instruction. Next, to demonstrate comprehension, the subject used the computer mouse to follow the instructions, by clicking each appropriate square on the grid in the order specified, as shown in Fig. 8.

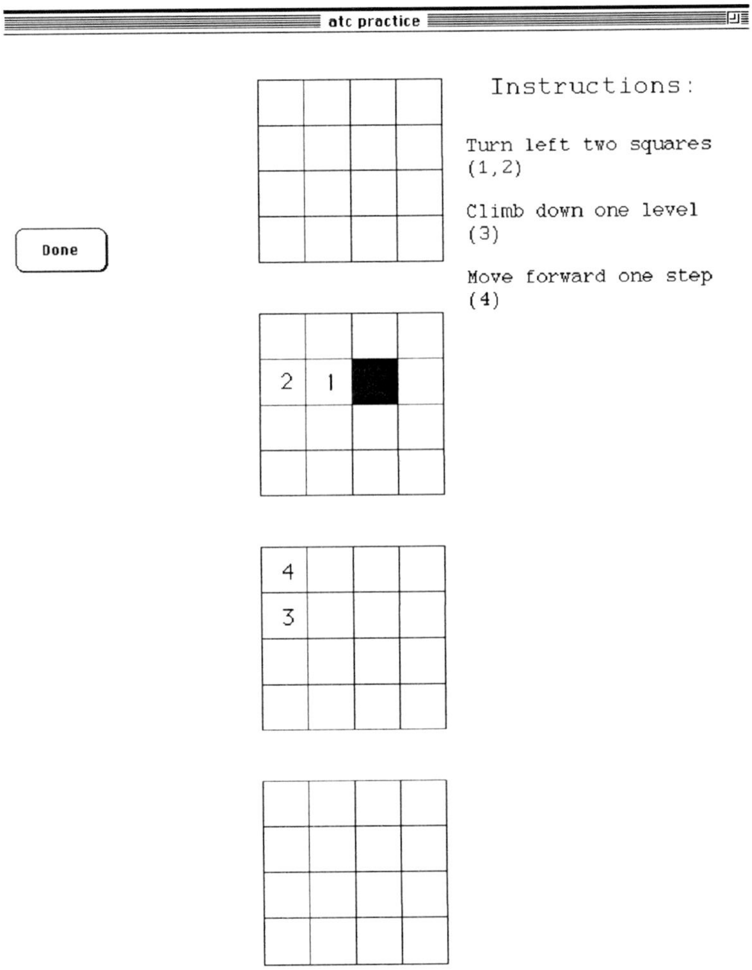

Figure 8. This figure (and Figs. 10a and b) illustrates the screen seen by the subjects in the experiments. Neither the numbers on the grids nor the instructions actually appeared on the screen. The numbers in parentheses indicate the corresponding squares on the grids that subjects should enter when following the given instructions.

We found substantial improvements in message comprehension with training and large effects of the number of propositions in the message but essentially no effects of either the number of words in a proposition or the speech rate, whether we varied rate by manipulating the pause duration between words or the duration of the words themselves.

In our previous studies (Barshi, 1997; Barshi & Healy, 1998), subjects heard instructions for movement in the whole space and on all three dimensions. We have found time and again that performance in the experimental task breaks down when the instructional string contains more than three units of instruction. Because the experiments simulate a three-dimensional environment, and because the order of the instructions is fixed (i.e., subjects move in three different dimensions in a fixed sequence and after moving through all three, a new set starts), we wanted to verify that the breakdown in performance after one cycle of three dimensions is indeed a cognitive limit rather than an artifact of the experimental design. In Experiment 1

(Barshi, 1998; Barshi & Healy, 1997, 2001), in a between-subjects design, we compared movement along all three dimensions with movement along only two of the three dimensions of the same space. In the two-dimension condition, rather than having two cycles of three dimensions in the longest instructional strings of six units (e.g., turn left two squares, climb down one level, move forward one step; turn right one square, climb up two levels, move back three steps), we had three cycles of two dimensions for the same string length of six units (e.g., turn left two squares, climb down one level; turn right one square, climb up two levels; turn left one square, climb down one level). The three-dimension condition is a replication of our previous work.

If performance breaks down in the same place in both the three-dimension and the two-dimension conditions (i.e., between the third and the fourth instructions), then indeed the performance limit is a cognitive one, and not an artifact of the experimental design. In fact, as shown in Fig. 9, we found no reliable difference between the two conditions. Furthermore, as also evident in Fig. 9, in both conditions the largest drop in performance is between messages of three units and messages of four units. Hence, we have verified that the breakdown in performance as a function of message length is indeed a cognitive limit rather than an artifact of our earlier experimental design.

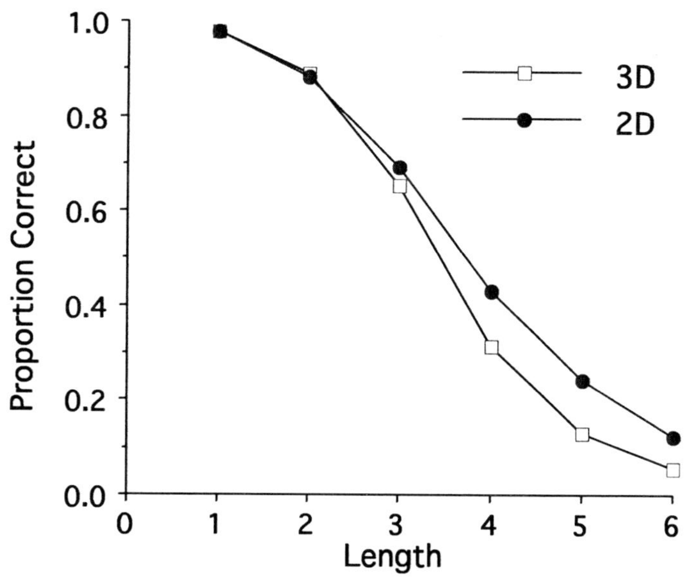

Figure 9. Proportion correct as a function of message length and dimension condition in Experiment 1 (Barshi & Healy, 2001).

In Experiment 2 (Barshi, 1998; Barshi & Healy, 1997, 2001), we compared memorizing and following directions for movements in only two of the three dimensions in two matrix conditions, one (like that used in Experiment 1) requiring a mental representation or visualization of a three-dimensional space and the other requiring a mental representation of only a two-dimensional space.

Specifically, in both conditions subjects were shown the complete space including all four matrices. The instructions were to move left or right and up or down, but subjects were never told to move forward or back. Importantly, the words heard were the same in the two matrix conditions, but as shown in Fig. 10a, the instructions led the subjects to move from one matrix to another in the multi-matrix condition (which was like the two-dimension condition used in Experiment 1), and, as shown in Fig. 10b, the same instructions kept the subjects in one matrix in the single-matrix condition. Thus, in the single-matrix condition movement occurred along the picture plane (i.e., along the plane shown as matrices on the computer screen), whereas in the multi-matrix condition movement occurred outside the picture plane (i.e., along a plane that is not shown directly on the computer screen because it is perpendicular to the matrices). For example, when the subject in the multi-matrix condition

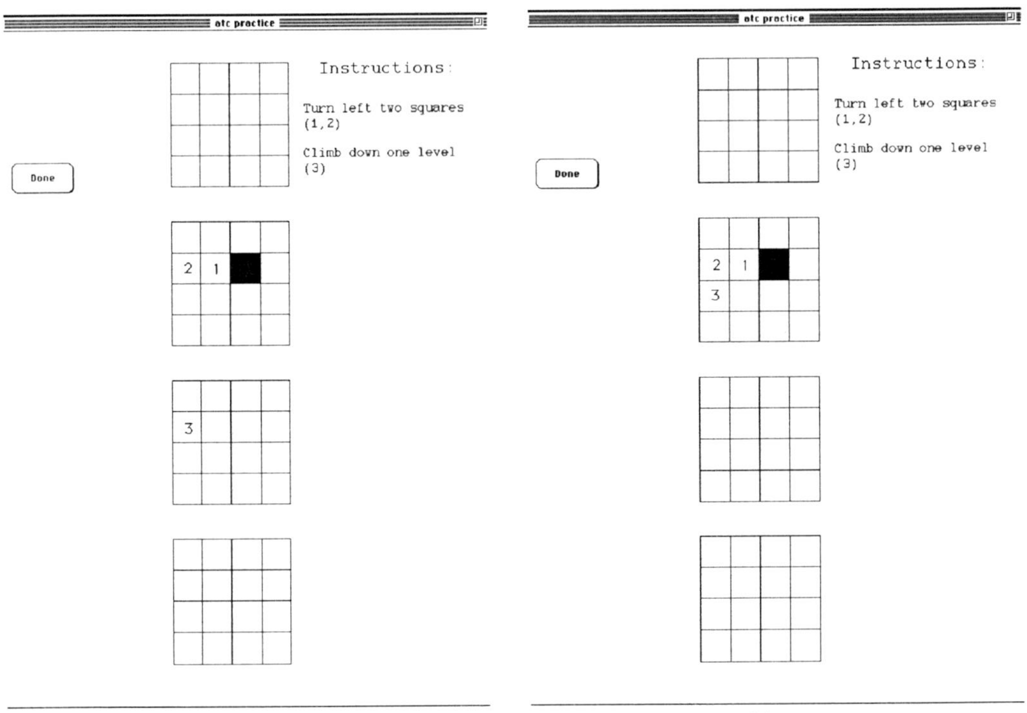

(a) (b)

Figure 10. This figure depicts the squares entered by subjects in the multi-matrix (Panel a) or the single-matrix (Panel b) condition when following the given instructions.

is told to "turn left two squares" (i.e., to go from the highlighted square to Step 1 and then to Step 2) in the sample shown in Fig. 10a, to do so the subject simply moves within the picture plane shown on the computer screen as a matrix. In contrast, when the same subject is then told to "climb down one level" (i.e., to go from Step 2 to Step 3) in the sample shown in Fig. 10a, to do so the subject must in effect move along a plane perpendicular to the matrices and thus not shown directly on the computer screen. Our hypothesis was that subjects would find it easier to represent a two-dimensional, as opposed to a three-dimensional, space because of the reduction in complexity. On that basis, we expected that subjects' memory for and execution of the instructions would be better in the single-matrix condition than in the multi-matrix condition. Note that the up-down directions seem somewhat less natural for the single-matrix condition than for the multi-matrix condition, but that difference should work against finding any advantage for the single-matrix condition.

As shown in Fig. 11, we found that our predictions were supported. We found that in both conditions subjects showed the same effect of message length as we had seen in all our previous studies. In addition, importantly, subjects' memory for and execution of the instructions were reliably better in the single-matrix condition when they had to represent mentally a two-dimensional space than in the multi-matrix condition when they had to represent mentally a three-dimensional space.

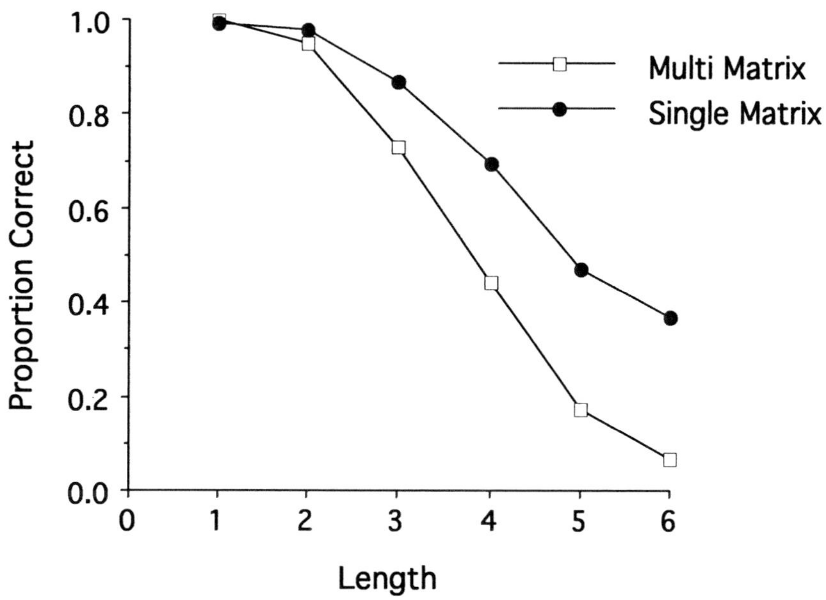

Figure 11. Proportion correct for manual movement responses as a function of message length and matrix condition in Experiment 2 (Barshi & Healy, 2001).

It is interesting to compare these findings from Experiment 2 to the findings from Experiment 1. In Experiment 1, there was no reliable difference between the two-dimension and three-dimension conditions, presumably because both required the subjects to form a three-dimensional mental representation and to move outside of the picture plane, although they differed in the number of dimensions along which subjects were instructed to move. In contrast, in Experiment 2, there was a large, significant difference between the single-matrix and multi-matrix conditions. Although both matrix conditions involved movement along only two dimensions, only the multi-matrix condition required the subjects to move outside of the picture plane and, presumably, only the multi-matrix condition required the subjects to form a three-dimensional mental representation.

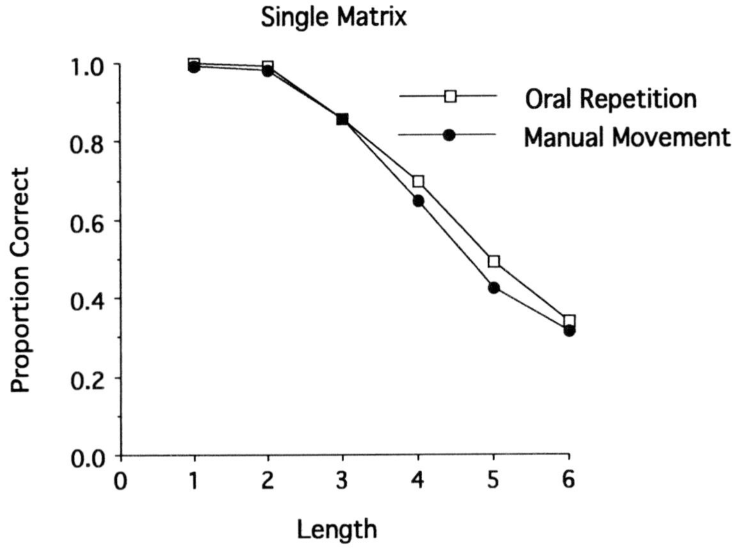

(a)

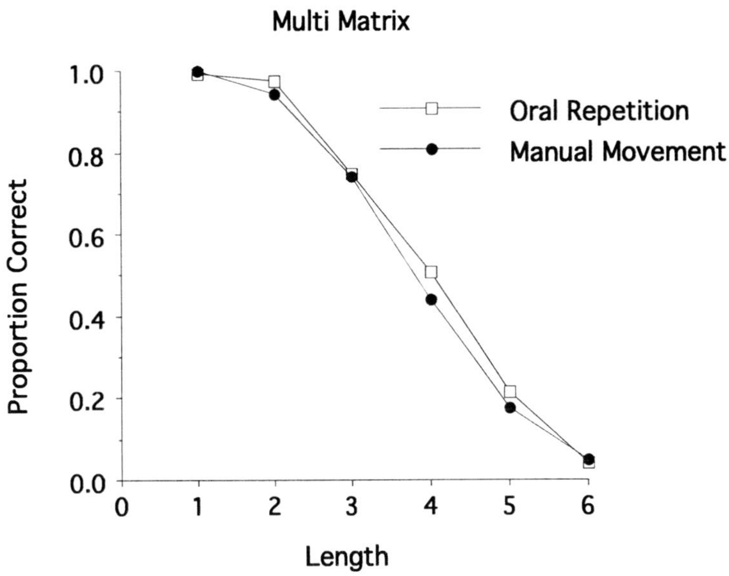

(b)

Figure 12. Proportion correct as a function of message length and response type in the single-matrix (a) and multi-matrix (b) conditions in Experiment 2 (Barshi & Healy, 2001).

As part of our experiments, subjects had to repeat aloud the instructions that they heard prior to following them, and we recorded their responses on a tape. If we were to find a difference between the two conditions of Experiment 2 in subjects' oral repetition responses, it would suggest that subjects' ability simply to repeat what they hear depends on the mental representation of the context to which the language input is relevant. We transcribed repetition responses from the majority of our subjects in that experiment so that we could analyze that part of the data. As shown in Fig. 12, we found in both conditions a very close correspondence between the accuracy of subjects' oral repetition responses and the accuracy of their manual movement responses. Importantly, as shown in Fig. 13, there was a large, significant difference between the two matrix conditions for the accuracy of the oral repetition responses. Subjects in the single-matrix condition were much more accurate than subjects in the multi-matrix condition even though the words they heard and were to repeat immediately were identical in the two conditions. These results provide strong evidence that subjects' immediate memory for a set of instructions is greatly affected by their mental representation of the space to which the instructions refer.

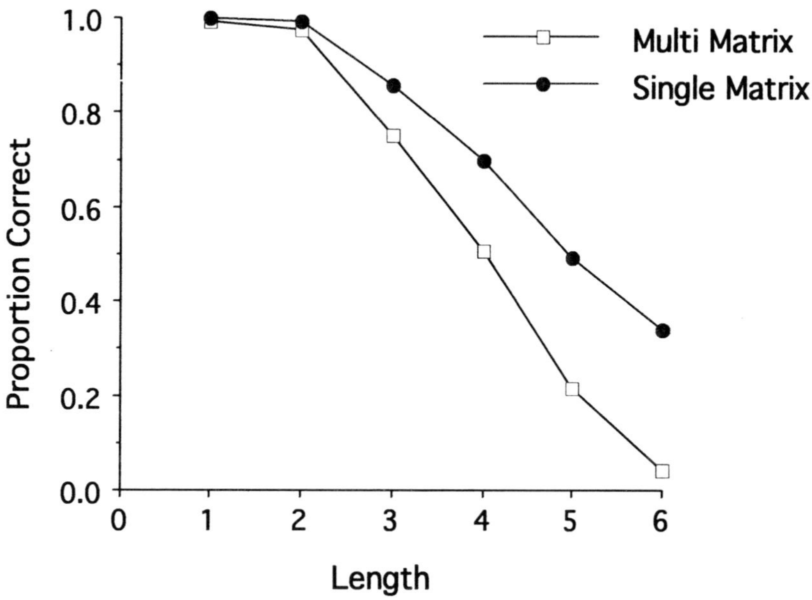

Figure 13. Proportion correct for oral repetition responses as a function of message length and matrix condition in Experiment 2 (Barshi & Healy, 2001).

The difficulty subjects had with the multi-matrix condition in this experiment could be attributed to difficulty in translating the two-dimensional visual display to a three-dimensional mental representation, rather than to the complexity of the mental representation itself. To test this hypothesis, we conducted the next two experiments (Barshi, 1998; Barshi & Healy, 2001). In Experiment 3, we compared two different visual displays, flat and perspective. The flat display was equivalent to the one used in our previous experiments and shown in Figs. 8 and 10. As shown in Fig. 14, the perspective display was drawn towards a vanishing point to provide additional visual cues as to the three-dimensional nature of the space. Each of the two displays was used in both the single-matrix and the multi-matrix conditions. There were 12 subjects in each of four groups, flat display single-matrix condition, flat display multi-matrix condition, perspective display single-matrix condition, and perspective display multi-matrix condition. All subjects heard the exact same instructions on 72 trials, broken down into six blocks of 12 trials each. Each block included two messages at each length from one to six instructional units, one message containing a redundant four words per unit (e.g., "turn left two squares") and the other message containing a minimal two words per unit (e.g., "left two").

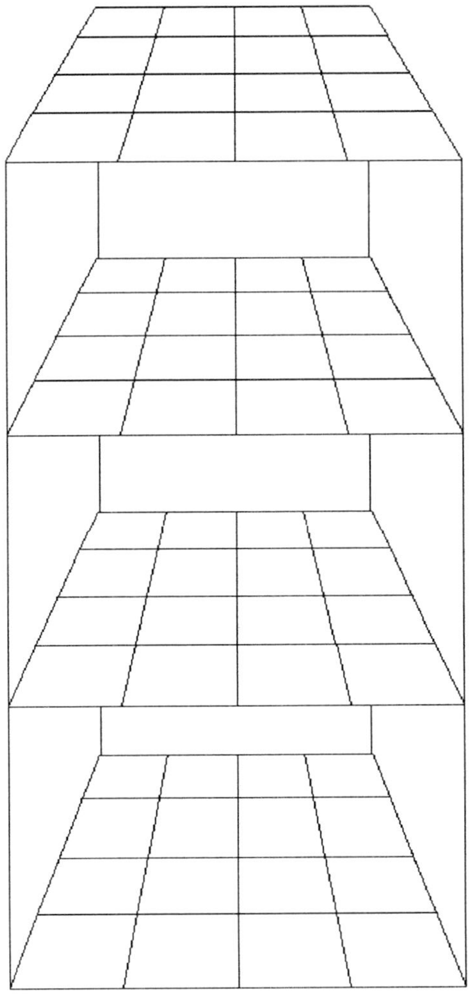

Figure 14. Perspective display used in Experiment 3 (Barshi & Healy, 2001).

Our results are shown in Fig. 15 in terms of the proportion of correct manual movement responses for each display as a function of matrix condition and message length. Once again we found a striking disadvantage for the multi-matrix condition relative to the single-matrix condition, and this disadvantage occurred for both displays. Although the difference between the two matrix conditions was somewhat smaller for the longer messages with the perspective display than for the same messages with the flat display, the effect of matrix condition was significant even for the perspective display. The three-way interaction of display, matrix condition, and message length was significant. For the longer messages, the perspective display helped in the multi-matrix condition but hurt in the single-matrix condition.

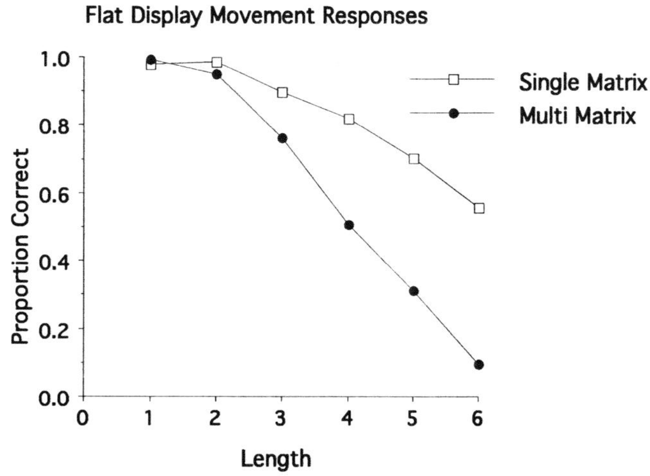

(a)

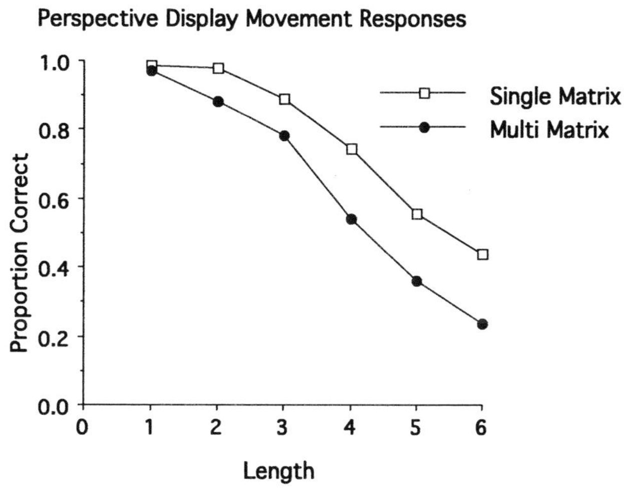

(b)

Figure 15. Proportion correct for manual movement responses as a function of message length and matrix condition for the flat display (a) and perspective display (b) in Experiment 3 (Barshi & Healy, 2001).

The pattern of results for the subjects' oral repetition responses was similar to that found for the movement response, as shown in Fig. 16. Once again, there is a significant advantage for the single-matrix condition relative to the multi-matrix condition with both the perspective and flat displays.

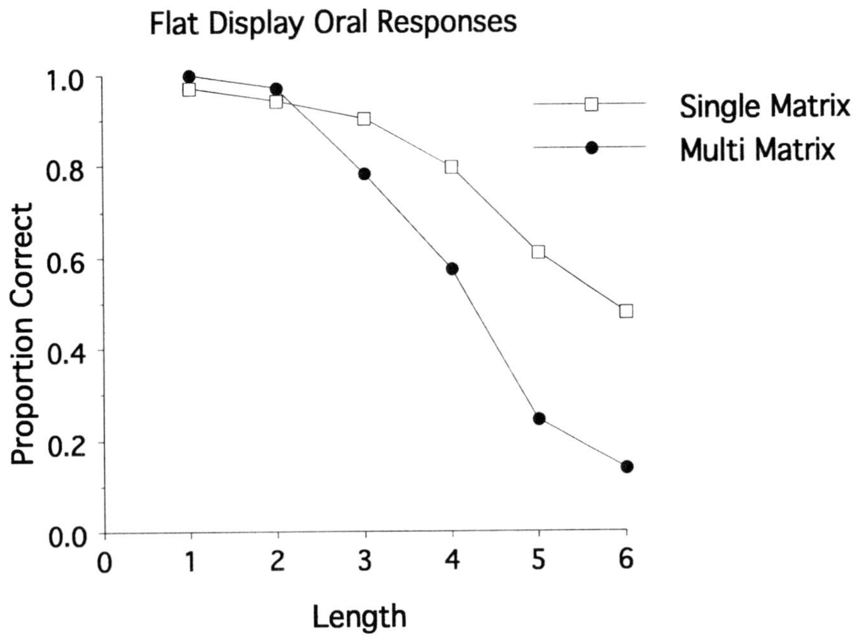

(a)

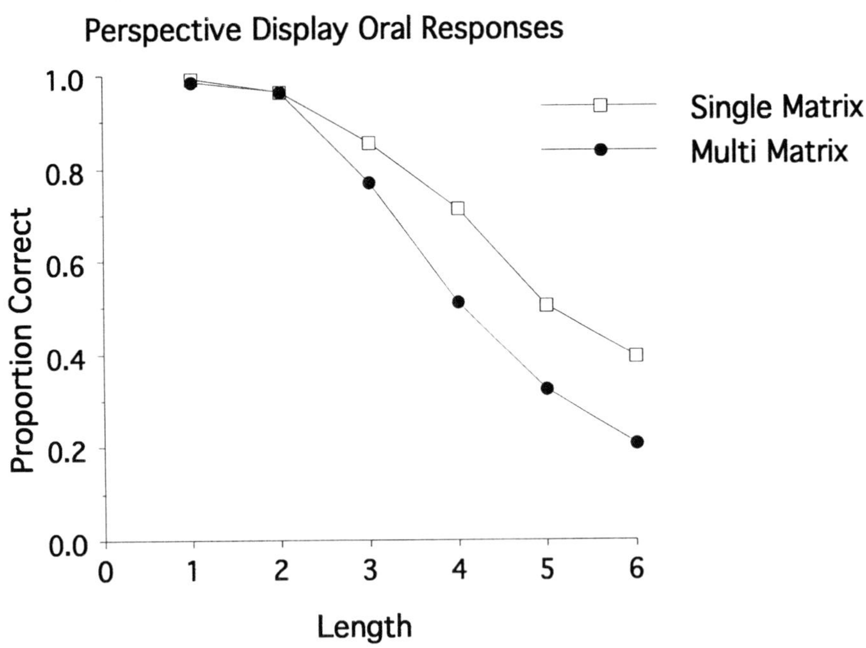

(b)

Figure 16. Proportion correct for oral repetition responses as a function of message length and matrix condition for the flat display (a) and perspective display (b) in Experiment 3 (Barshi & Healy, 2001).

The results of this experiment suggest that the difference between matrix conditions is due to the complexity of the required mental model independent of display characteristics. Nevertheless, because the perspective display was shown on a two-dimensional computer screen, a translation process is still needed between the visual display and the three-dimensional mental representation. Experiment 4 was designed to test for the importance of such a translation process.

In this experiment, we compared six groups of subjects, with 12 subjects in each group. We repeated the manipulation of matrix condition (multi-matrix vs. single-matrix), but instead of comparing two different computer displays, we compared three different physical displays in which subjects had to touch their finger to the appropriate squares on the matrices as they followed the given instructions. In one case, subjects touched a flat sheet of paper showing the "flat" computer display used in the previous study. In a second case, subjects touched a flat sheet of paper showing the "perspective" display used in the previous study. In the third case, as shown in Fig. 17, subjects touched a three-dimensional "model" of the space. We videotaped the subjects' hand movements and audio taped their oral repetition responses. Because we found previously essentially the same pattern of results for redundant and minimal messages and for oral repetition and manual movement responses, we included only the redundant messages in the present experiment and analyzed only the oral repetition responses. Further, because we found previously that subjects' performance was on the ceiling for messages containing a single instructional unit and was near the floor for messages containing six instructional units, in the present experiment, subjects heard only messages varying in length from two to five instructional units.

Figure 17. Three dimensional model of the navigation space used in Experiment 4 (Barshi & Healy, 2001).

We found no reliable difference in proportions of correct responses between the three different physical displays in the subjects' oral repetition responses (flat = .733, perspective =

.740, model = .731). Moreover, there were no reliable interactions involving the display. In contrast, as shown in Fig. 18, we did find a clear difference between matrix conditions as well as a strong effect of message length and a significant interaction of matrix condition and message length. All three displays yielded a disadvantage for the multi-matrix condition relative to the single-matrix condition on the longer messages. In other words, with all three physical displays, we replicated our previous results with the two computer displays, showing a clear performance advantage with longer messages for the single-matrix condition relative to the multi-matrix condition. We also established that the visual stimuli or computer display used are far less important in determining performance than are the instructions given and the nature of the task.

These results suggested that the disadvantage for the multi-matrix relative to the single-matrix condition is because the multi-matrix but not the single-matrix condition requires subjects to use a three-dimensional mental representation of the space. This suggestion is consistent with "array" theories of mental imagery, including Kosslyn's (1980), according to which mental imagery is done in two dimensions and information about the third dimension is coded only indirectly and must be derived from the two-dimensional representation.

Although we interpreted the difference between the single-matrix and multi-matrix tasks as reflecting the cost of forming a three-dimensional mental representation of the space, the difference may reflect instead the fact that only the multi-matrix condition required the subjects to move outside of the picture plane. To determine whether the use of a three-dimensional representation is crucial, we conducted a fifth experiment in which we compared two different sets of instructions (Healy, Parker, & Barshi, 2001). In one set of instructions (model), we showed the subjects a three-dimensional model of the space, and we made it clear to the subjects that the computer display represents that model, as in our previous experiments. In contrast, in the other set of instructions (board), we showed the subjects a two-dimensional piece of paper containing a copy of the computer display, and we told the subjects that the display represents four checkerboards. Thus, the board instructions do not include a three-dimensional representation of the space, and the notion of a checkerboard provides a good cover story for why the subjects would have to move into the same position on a different matrix, without using the three-dimensional model. This experiment included the single-matrix and multi-matrix manipulation between subjects, and it included six message lengths, all of which involved minimal messages that did not refer to three-dimensional movement (e.g., "left two, down one"). As in our previous experiments, we recorded subjects' oral repetition responses as well as their manual movement responses in this experiment.

If the difference between the two matrix conditions depends on the three-dimensional representation of the space, we should find this difference only for the model set of instructions, not for the board set because only the model involves three dimensions. If, on the other hand, the difference between the two matrix conditions depends on the fact that only in the multi-matrix condition do subjects move outside of the picture plane, we should find a difference for the board set of instructions as well as for the model set of instructions because in both cases the subjects must move along a dimension not directly shown in the picture plane.

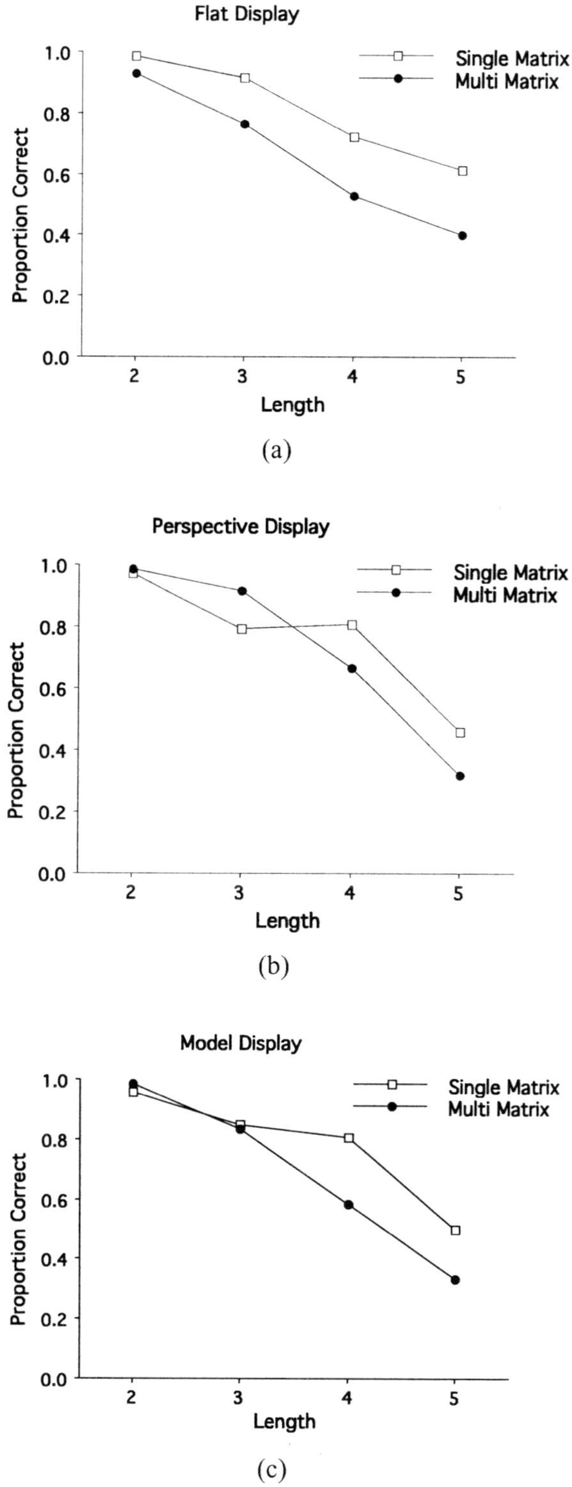

Figure 18. Proportion correct as a function of message length and matrix condition for the flat display (a), perspective display (b), and model display (c) in Experiment 4 (Barshi & Healy, 2001).

The results for the manual movement responses are summarized in Fig. 19 in terms of proportion of correct responses as a function of message length and matrix condition for both sets of instructions. It is clear that performance decreased significantly across increases in message length. Also, performance was significantly worse in the multi-matrix condition than in the single-matrix condition, and there was a significant interaction between matrix condition and message length because the effect of matrix condition was not evident for the shortest message lengths where performance approached the ceiling. Importantly, this same pattern was found for both the board and model sets of instructions; neither the interaction between matrix condition and instruction set nor the three-way interaction between matrix condition, instruction set, and length was significant. These findings provide strong evidence that the difference between the two matrix conditions does not depend on a three-dimensional representation of the space but rather on the fact that only in the multi-matrix condition do subjects move outside the picture plane.

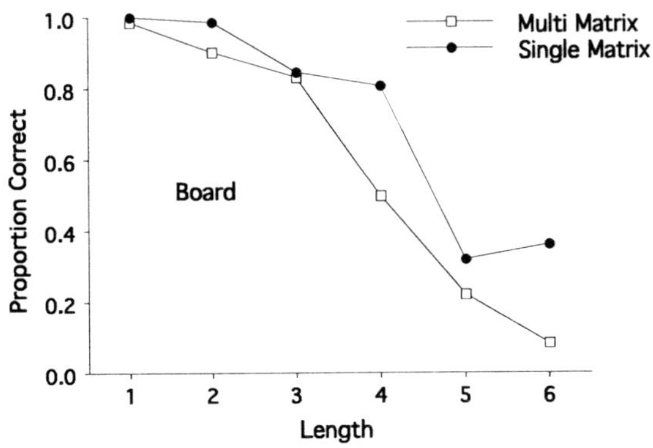

(a)

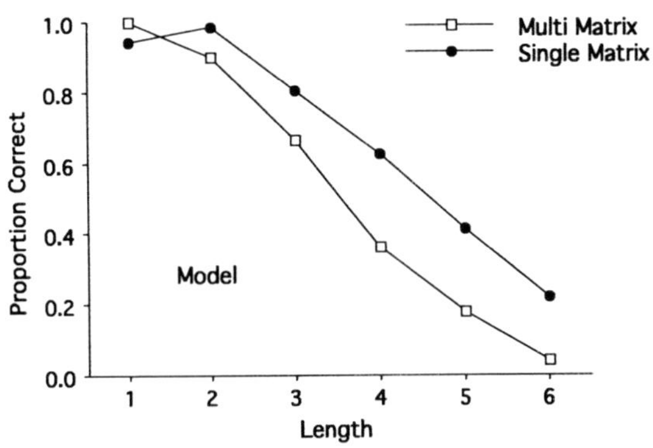

(b)

Figure 19. Proportion correct for manual movement responses as a function of message length and matrix condition for the board (a) and model (b) instructions in Experiment 5 (Healy, Parker, & Barshi, 2001).

Although the two instruction sets showed the same effects of matrix condition, there was an important difference between them. As shown in Fig. 20, performance was worse with the model set of instructions than with the board set of instructions, although this disadvantage for the model instructions was not evident at the shortest lengths (where performance was close to the ceiling) or at Message Length 5; the interaction between instruction set and length was significant and the main effect of instruction set was marginally significant. The disadvantage for the model set of instructions relative to the board set does point to a cost in the requirement to form a three-dimensional mental representation.

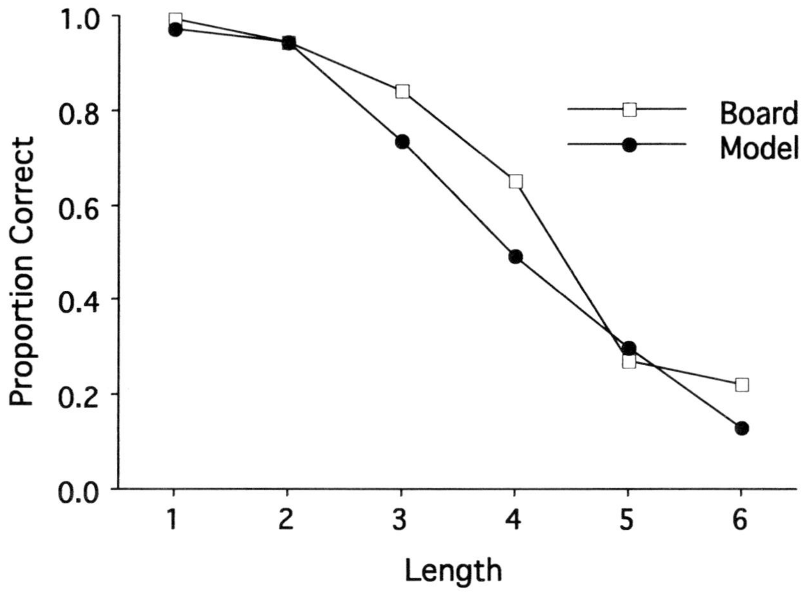

Figure 20. Proportion correct for manual movement responses as a function of message length and instructions in Experiment 5 (Healy, Parker, & Barshi, 2001).

Of greatest interest are the results for the oral repetition responses. These results are summarized in Fig. 21 in terms of proportion of correct responses as a function of message length and matrix condition for both the board and model sets of instructions. As with the manual movement responses, performance decreased significantly across increases in message length. Also, performance was significantly worse in the multi-matrix condition than in the single-matrix condition, and there was a significant interaction between matrix condition and message length because the effect of matrix condition was not evident for the shortest message lengths. Importantly, this same pattern was found for both the board and model sets of instructions; neither the interaction between matrix condition and instruction set nor the three-way interaction between matrix condition, instruction set, and length was significant. These findings provide additional support that the difference between the two matrix conditions depends on the fact that only in the multi-matrix condition do subjects move outside of the picture plane. More importantly, these findings document the fact that subjects' ability to remember and repeat back immediately the words in a set of instructions that they hear depends on the meaning and requirements of those instructions.

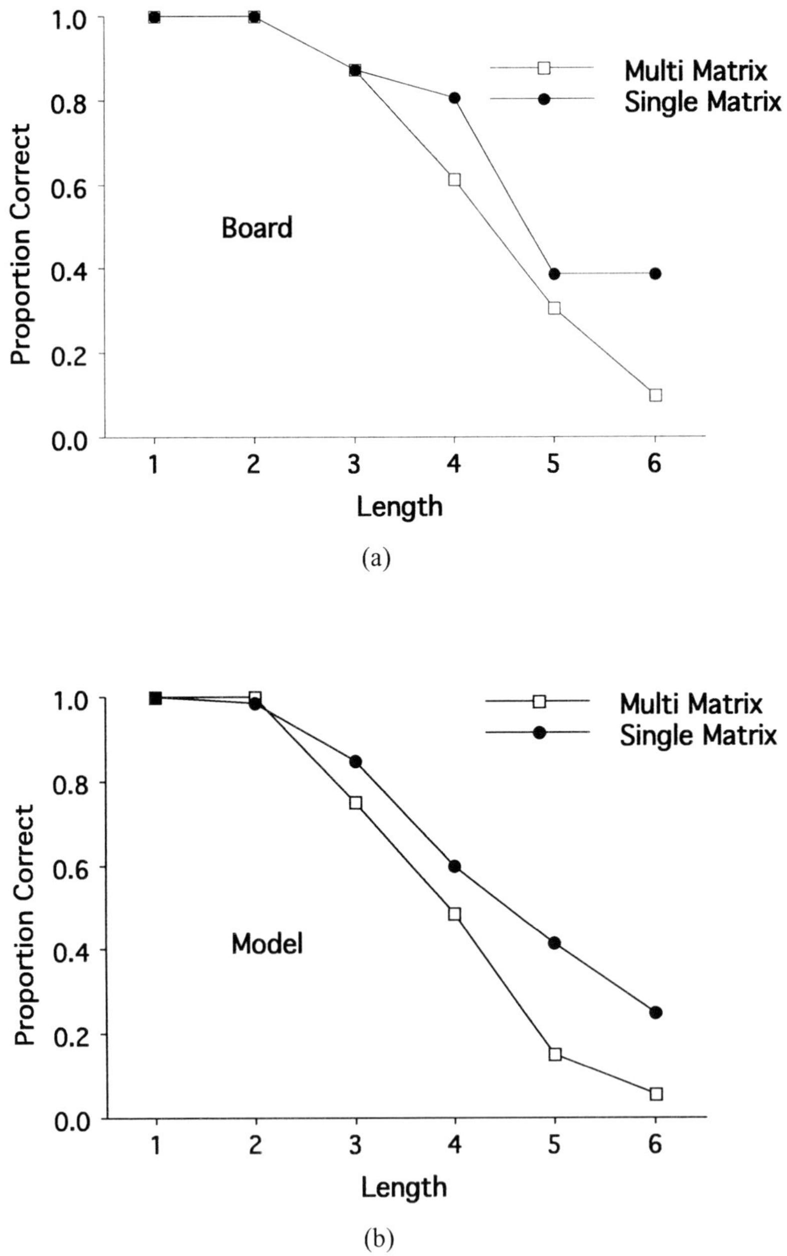

Figure 21. Proportion correct for oral repetition responses as a function of message length and matrix condition for the board (a) and model (b) instructions in Experiment 5 (Healy, Parker, & Barshi, 2001).

As with the manual movement responses, there was an important difference between the two instruction sets in the oral repetition responses. Fig. 22 shows that performance was worse overall on the model set of instructions than with the board set of instructions, except at the shortest lengths (where performance was close to the ceiling). For this measure, the main effect of instruction set was significant and the interaction between instruction set and message length was marginally significant. Again, the disadvantage for the model set of instructions relative to the board set points to a cost in the requirement to form a three-dimensional mental representation even on repeating back the words in the instructions, in accord with array theories of mental imagery.

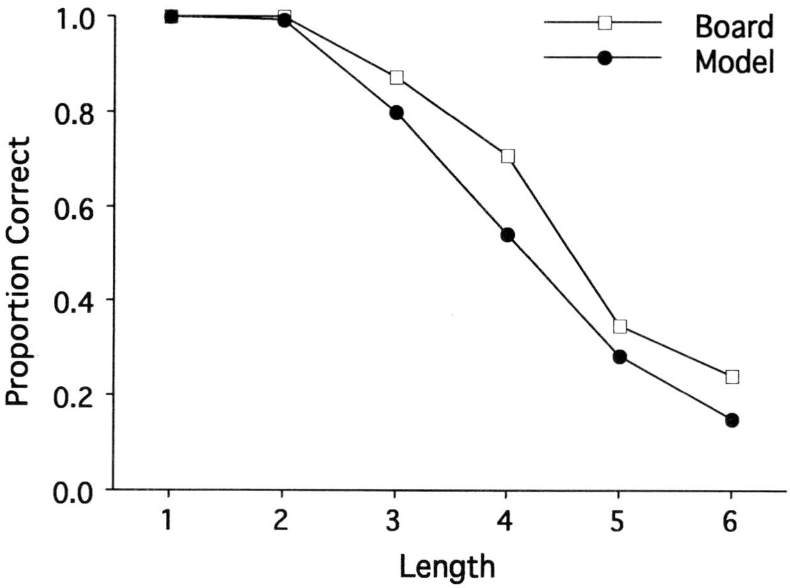

Figure 22. Proportion correct for oral repetition responses as a function of message length and instructions in Experiment 5 (Healy, Parker, & Barshi, 2001).

In summary, this experiment showed that subjects' ability to repeat back navigation instructions that they hear is affected by the length of the message and two aspects of the subjects' mental representation of the space through which they are navigating. First, there was a decrement in performance when the subjects had to move outside the picture plane. Second, there was a decrement in performance when the space was described as involving three dimensions. The first finding is reflected in the disadvantage for the multi-matrix condition relative to the single-matrix condition for both the board and model sets of instructions. The second finding is reflected in the overall disadvantage for the model set of instructions relative to the board set, especially with the longer messages.

More generally, these results provide powerful evidence that individuals' ability simply to repeat immediately what they hear strongly depends on their mental representation of the situation to which the language input is relevant.

Because we found that memory varied as a function of the number of dimensions that the subjects were required to represent, we are able to recommend that additional training be given whenever a task to be learned requires performance in three dimensions. Our results imply that three-dimensional displays, such as those found in virtual reality environments, might not be sufficient to eliminate the need for this additional training.

RESPONSE EXECUTION IN A SPATIAL ARRAY

Every laboratory and military task requires a variety of skills, some of which are perceptual, some cognitive, and some motoric. Many of the tasks that we have employed in our research tend to emphasize the cognitive component, but none of them is completely free from perceptual or motoric skills. In the present study, we focus on perceptual and motoric components, with a concentration on spatial aspects for the perceptual component. The primary goal is to develop general training principles relevant to the integration and coordination of skills for optimal performance in military tasks. These principles should extend to the refinement of training techniques based on virtual reality (VR), which depends heavily on new integrations and coordinations of perceptual, cognitive, and motoric skills. This study does not use VR per se, but rather examines analytically the spatial and motoric aspects of performance that might be woven into a VR system. As in our other studies, we are primarily interested in issues related to the durability and generalizability of these skills.

In order to facilitate the acquisition of new integrations and coordinations of various skills, it is important to understand and isolate the separate components or stages entailed in these tasks. Such tasks typically include three major processing stages: *perception* of an event or a stimulus, *decision* about how to respond (including memory retrieval and/or calculation), and *execution* of the selected response. These three stages are assumed to occur in a canonical order—perception, decision, execution—whether serial or cascading. The response execution stage of a cognitive task itself can be described as including two substages: *finding* the proper response among available alternatives and *moving* to (or selecting) that response alternative.

The finding and moving substages of response execution, if they are separable and measurable, are likely to be influenced by different variables and by the same variables in different ways. For example, the spatial location of the correct alternative might or might not influence finding the alternative, but it might be more difficult to move to some locations or in some directions in a spatial array than others.

We created a task environment in which the role of perception and decision is minimized or at least is held constant and independent of execution. We do this in a clock face format, patterned after some earlier work by Pauli et al. (1994). As shown in Fig. 23, we used four different clock faces, each with the numerals, 0 through 9, appearing in serial order on the circumference of the clock face, equidistant from a center start position. The specific clock faces used in this experiment differed along three dimensions of transformation or rotation. For the standard clock face (SC), the digits 0 through 9 are placed at equal intervals around the circumference with 0 at due North, 1 immediately to its right and 9 immediately to its left. One transformation (UDC) is to rotate the face clockwise so that 0 lays at due South and other digits maintained their relative position (with 1 to the left and 9 to the right). Two other transformations either flip the standard clock face right to left (SCC), such that 0 remains in the North position but 1 appears to the left and 9 to the right, or flip the standard clock face top to bottom (UDCC), such that 0 takes the South position and 1 appears to its right and 9 to its left.

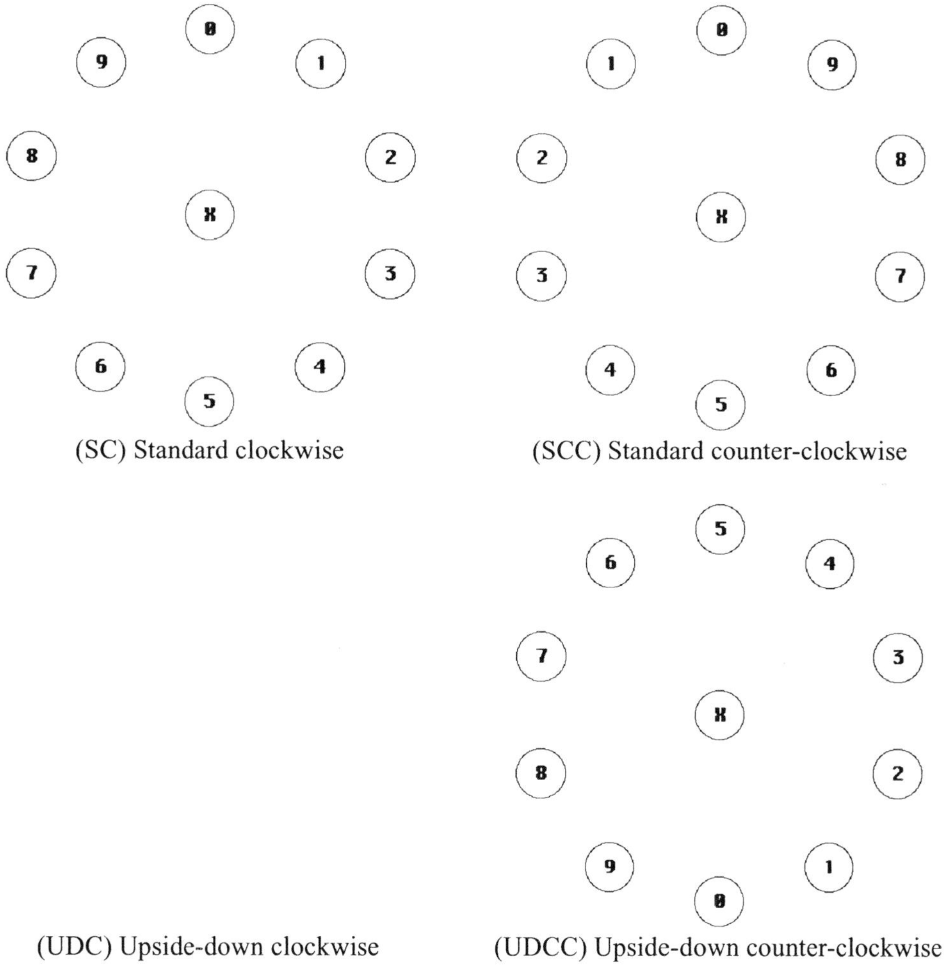

Figure 23. Clockfaces used in experiments
(Bourne, Healy, Pauli, Parker, & Birbaumer, 2001).

On a given trial, the subject places a cursor in a center start position using a mouse. A target numeral then appears above the start position, and the subject is instructed to find that numeral on the clock face and then move the cursor from the start position to that numeral's position on the clock face. We assume that time and accuracy of perceiving the target numeral and making the general decision to respond to it do not differ over numerals. However, finding the numeral's position on the clock face and moving to that position, the execution substages, might be affected by clock face arrangement, by numeral position, by numeral magnitude (or meaning), and possibly by other variables.

In a pair of new experiments (Bourne, Healy, Pauli, Parker, & Birbaumer, 2001), we varied three aspects of the response alternative system: (a) the structure of the set of alternatives, by comparing a standard clock face arrangement to the various transformed clock faces; (b) the spatial location of each alternative, by comparing responses to the upper versus the lower half and right versus left side on the clock face, independent of the overall structure of the clock face; and (c) the magnitude (or meaning) of alternatives, by analyzing responses to the different numerals.

Specifically, in our first experiment, subjects were trained on one of the four clock faces for 500 trials, arranged into 25 blocks of 20 trials, two with each of the numbers 0 to 9 as the target. After a short break of approximately 5 minutes, they were given the transfer test consisting of another 500 trials arranged in the same way. Transfer was either to the same clock face or to one of the other three clock faces, with training and transfer clock faces varied orthogonally. During training and transfer phases, subjects could not proceed to the next trial until the correct target digit was entered. For half of the subjects, there was no delay between when the subject placed the cursor in the start position and the onset of the target, whereas for the other half of the subjects, there was a variable delay of from .5 to 1.5 seconds. The dependent measures were time to leave the center fixation point (response initiation time) and time from response initiation to correct target identification (response movement time).

We found that the two response measures were affected differently by whether or not there was a variable delay between when the subject placed the cursor in the start position and the onset of the target. As shown in Fig. 24, there was only a slight, nonsignificant effect of this variable in training and transfer on initiation time, but in both phases there was a large and significant slowing of movement time by the variable delay. If we assume that the subject is in a state of readiness to respond at the outset of each trial, then a variable delay is more likely to disrupt movement time than initiation time because the subject cannot begin to find the target until the target actually appears. But subjects seem to be able to prepare to move even if they do not yet know where they will move.

As summarized in Fig. 25, we found that for both training and transfer phases, over all clock faces, initiation time shows a bow-shaped (inverted U) relationship to target number, the smallest digits generally producing the fastest finding responses. Movement time is essentially unaffected by target digit. This bow-shaped relationship between target number and initiation time depended significantly on clock faces, suggesting that digit location as well as digit magnitude played a role in initiation time.

Although we used four different serial clock face orientations in this experiment, it was not the case that every target number occurred in every clock face location. There is an unavoidable partial confounding of target number and spatial location. However, as shown in Fig. 23, four target numbers, 1, 4, 6, and 9, did occur in each of four different locations (top right, bottom right, bottom left, and top left) across the four different clock faces. These locations can be viewed as the corners of a rectangle. Separate analyses, limited to these four numbers and locations, were conducted to determine whether either digit magnitude or digit location had an independent influence on performance. Indeed, both variables were shown to influence at least one of the response time measures. As shown in Fig. 26, target number averaged across all four locations had a significant bowed effect overall. Furthermore, target number interacted with measure because the bowed effect was quite distinct for initiation time but essentially nonexistent for movement time. Further, as shown in Fig. 27, for both measures averaged across all four target digits, movements upward were considerably faster than movements downward but movements to the left and right did not differ.

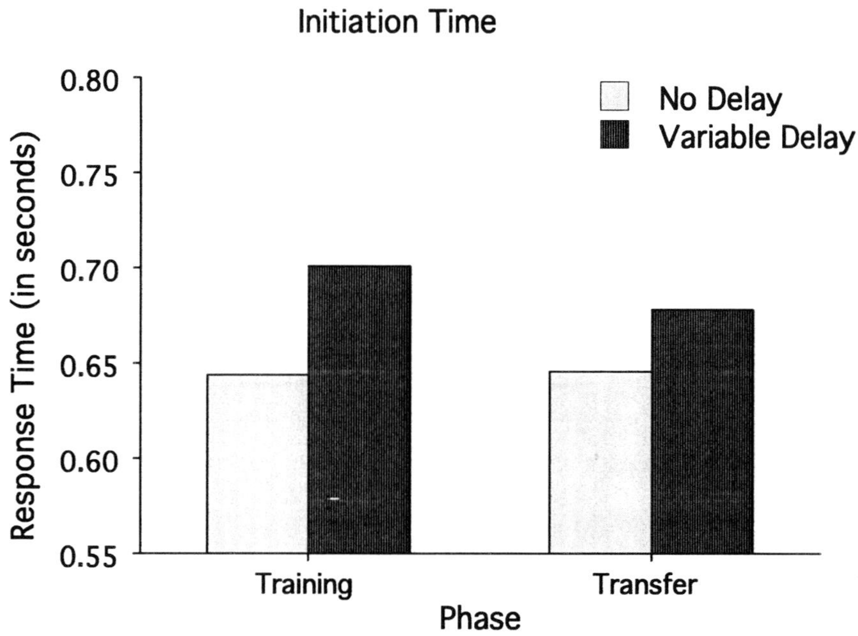

(a)

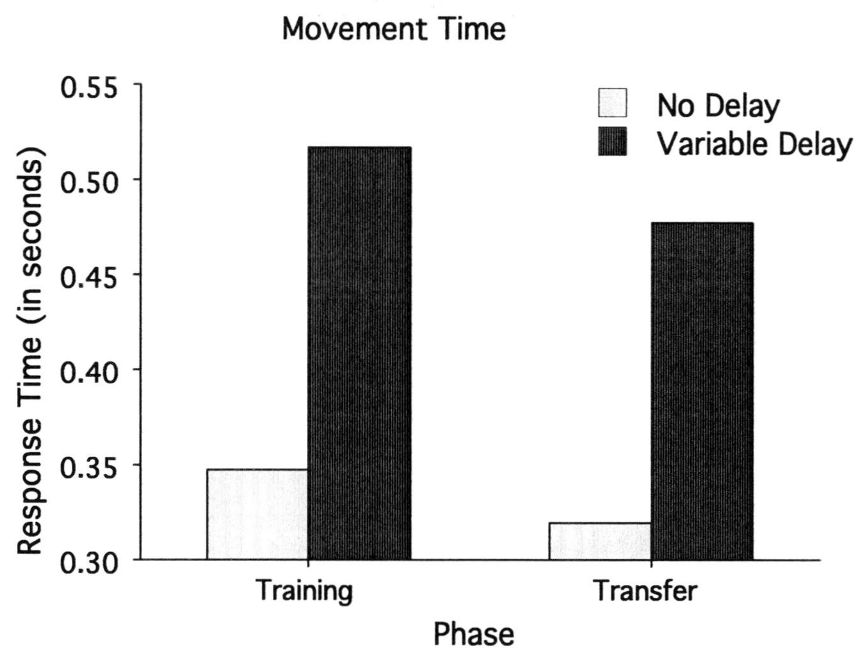

(b)

Figure 24. Response time as a function of phase and delay type for initiation time (a) and movement time (b) in Experiment 1 (Bourne, Healy, Pauli, Parker, & Birbaumer, 2001).

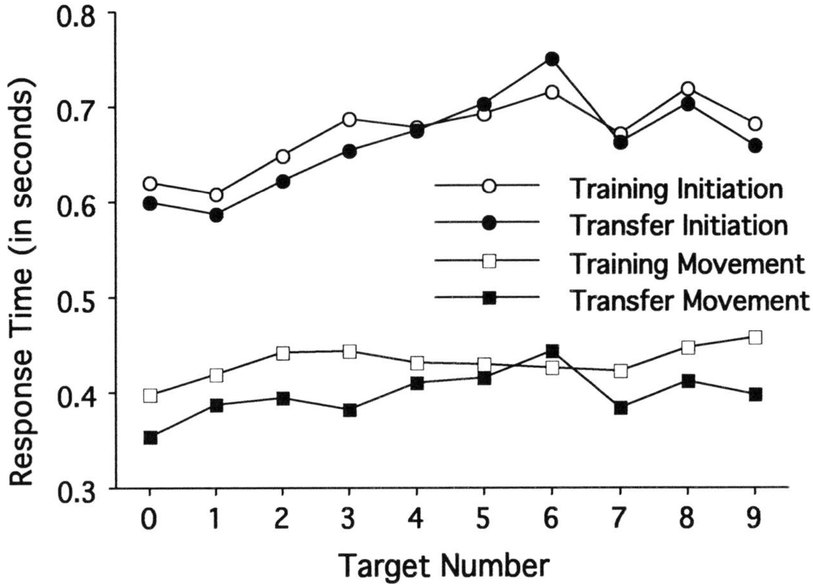

Figure 25. Response time as a function of target number, phase, and measure in Experiment 1 (Bourne, Healy, Pauli, Parker, & Birbaumer, 2001).

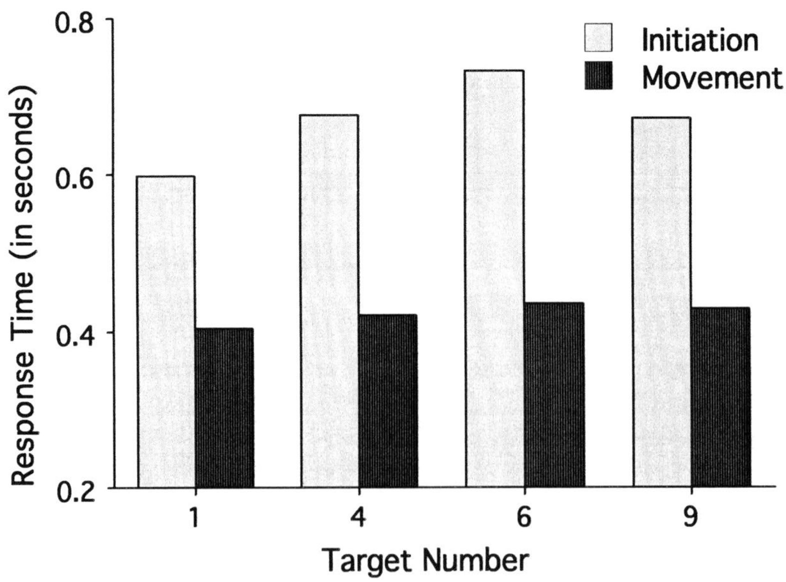

Figure 26. Response time as a function of target number and measure in Experiment 1 (Bourne, Healy, Pauli, Parker, & Birbaumer, 2001).

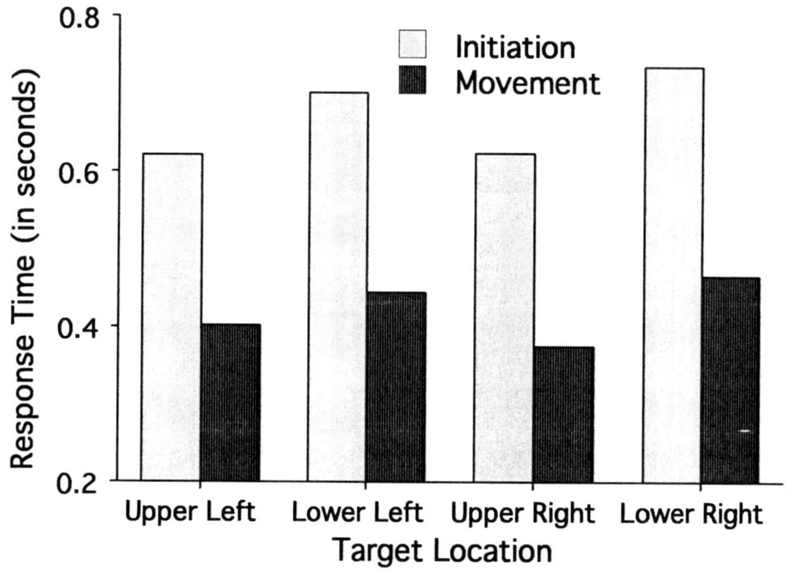

Figure 27. Response time as a function of target location and measure in Experiment 1 (Bourne, Healy, Pauli, Parker, & Birbaumer, 2001).

As shown in Fig. 28, skill acquisition was reflected in decreases in initiation and movement response times in both the training and the transfer phases, with one exception. Movement times reached an asymptote during training and, unlike initiation time, did not decrease further during transfer.

In both phases of the task used in Experiment 1, subjects were given no explicit feedback about the speed of their responses. It is possible that the availability of explicit feedback might affect either the asymptotic response time achieved by subjects or the rate at which they approached asymptote. We examined potential feedback effects in Experiment 2, using the same clock face speeded response task as in Experiment 1. Subjects were trained either with continuous trial-by-trial feedback on overall response times (the sum of initiation and movement times) or with periodic cumulative feedback which signaled average overall response time every 20 trials during training. In a transfer task, with either the same or a different clock face, subjects in all conditions were given no feedback (as in Experiment 1).

The design of Experiment 2 was like that of Experiment 1, with four exceptions. First and most importantly, during training subjects were given either trial-by-trial or periodic feedback on how quickly they entered target digits (overall response time). Second, only the standard clockwise and upside-down clockwise clock face displays were used (clock faces SC and UDC in Fig. 23). Third, because additional time was necessary to provide feedback, only 400 trials were included in each phase, so as to allow completion of the experiment within 1 hour. Finally, we used only the condition with a variable delay between when the subject placed the cursor in the start position and the onset of the target.

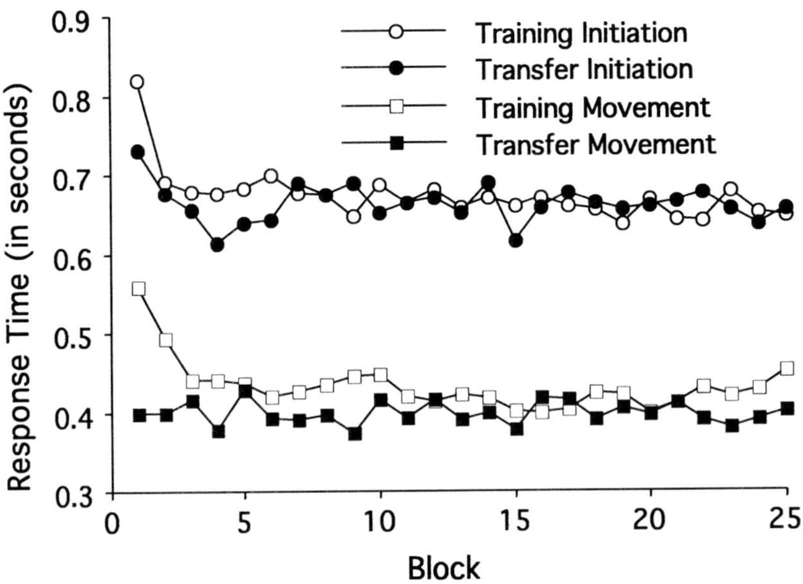

Figure 28. Response time as a function of block, phase, and measure in Experiment 1 (Bourne, Healy, Pauli, Parker, & Birbaumer, 2001).

The results of this experiment were largely consistent with those from Experiment 1. Specifically, as in Experiment 1 and as shown in Fig. 29, for both training and transfer, over both clock faces, initiation time shows a bow-shaped (inverted U) relationship to target number, the smallest digits generally producing the fastest finding responses, whereas movement time is essentially unaffected by target digit.

Further, as in Experiment 1 and as shown in Fig. 30, both initiation and movement response times decreased during training, but movement times reached an asymptote during training and, unlike initiation time, did not decrease further during transfer trials.

The variable of greatest interest in Experiment 2 was feedback. There were significant feedback effects, but they occurred entirely during training and did not carry over into transfer performance. Among these effects, the most important was a significant interaction of feedback and measure. As shown in Fig. 31, mean initiation time during training was somewhat faster in the trial-by-trial feedback condition than in the periodic feedback condition, and mean movement time was slightly slower in the trial-by-trial condition. We are inclined tentatively to attribute this pattern of results to the fact that trial-by-trial feedback might enhance a subject's state of readiness to begin a response without influencing the speed with which the response itself is completed. When feedback is no longer given, during transfer trials, the difference between measures is missing, as would be expected if the influence of feedback is local and applies only to trials immediately following its occurrence.

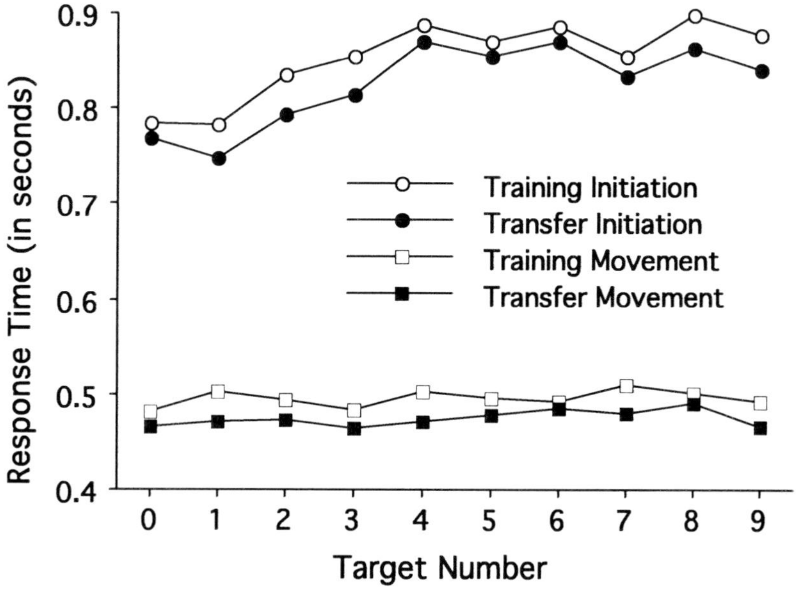

Figure 29. Response time as a function of target number, phase, and measure in Experiment 2 (Bourne, Healy, Pauli, Parker, & Birbaumer, 2001).

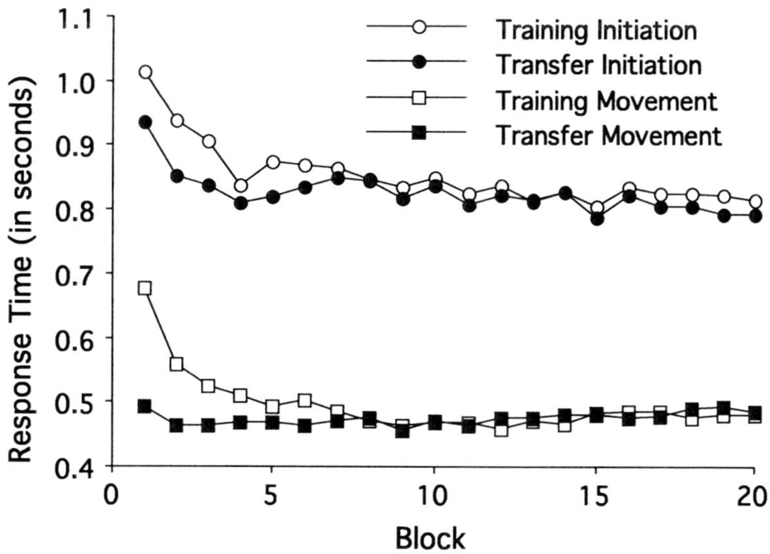

Figure 30. Response time as a function of block, phase, and measure in Experiment 2 (Bourne, Healy, Pauli, Parker, & Birbaumer, 2001).

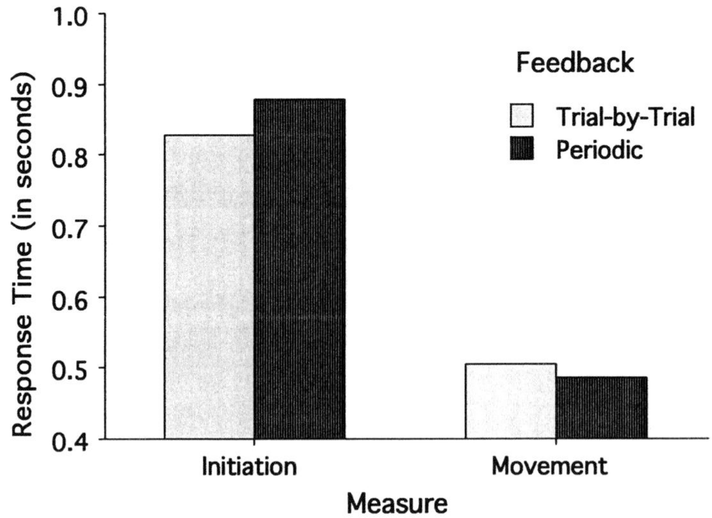

Figure 31. Response time as a function of measure and feedback type in Experiment 2 (Bourne, Healy, Pauli, Parker, & Birbaumer, 2001).

In summary, we found that the simple task of response execution improved considerably with training and that both the identity and the location of the target affected speed of response. Most interesting was our finding that the two component response time measures—initiation and movement time—showed quite different patterns of results. As mentioned earlier, the effects of feedback during training were different for the two measures. Further, initiation time was more sensitive than was movement time to some of the effects we found, including the effect of target number (but not the effect of target location) and the effect of practice during transfer (but not the effect of practice during training). However, there was one important case where movement time was more sensitive than initiation time, namely the effect in Experiment 1 of whether or not there was a variable delay between when the subject placed the cursor in the start position and the onset of the target. An explanation of these differences between the two measures must be based on the assumption that they reflect two distinct processing stages of simple response execution. Hence, it seems crucial in future studies to be alert to the potential component structure of response execution.

The Underlying Causes of Durability and Specificity in Digital Data Entry

Our previous work on training digit data entry skills (Fendrich, Gesi, Healy, & Bourne, 1995; Fendrich, Healy, & Bourne, 1991) demonstrated that subjects are significantly faster at typing numbers that have been typed previously during training compared to numbers that, in the context of the laboratory experiment, are "new" or unfamiliar. This advantage for typing "old" numbers is known as the repetition priming effect, and it is based largely on implicit or nondeliberate memory. We found that both perceptual and motoric processes contribute to the repetition effect. However, this earlier work did not differentiate between the surface percept and the abstract meaning of that percept. We conducted a new series of experiments (Buck-Gengler & Healy, 2001) to compare the contributions of these two types of perceptual (i.e., pre-response) processes. The first experiment in this series allowed us to determine whether both the surface percept and the abstract meaning of that percept are durably represented in memory, thus contributing to the repetition effect.

Specifically, the surface percept was manipulated by varying the presentation format of the numbers. One format involved presenting the numbers as numerals ("5 7 2 1") (which correspond to the labels on the data entry keys), whereas the other format involved presenting the numbers symbolically as words ("five seven two one").

In the experiment, subjects were trained in a digit data entry task over a single session and then tested in a second session after a 1-week delay. In both sessions, four-digit numbers were presented one at a time on a computer display screen until the subject responded. The subjects responded by entering the number into a computer terminal keyboard. Total response times for entering all four digits in a number, times to initiate the entry of the first digit of a number, and the response times of the individual keystrokes, as well as the accuracy of the entries were recorded on every trial. The presentation and subsequent entry of a four-digit number constituted one trial. During the training session, a set of 64 four-digit numbers was used as the learning set. The learning set was repeated for entry by the subject five times across five blocks of 64 trials during the training session, with a different order of numbers for each block.

The procedure used in the testing session was the same as in the training session except that there were only two blocks, rather than five. Sixty-four "new" four-digit numbers not previously entered by the subjects were intermixed with the 64 "old" numbers of the learning set.

One group of subjects trained with the numeric keypad to the right of the letter keys on the standard keyboard and was tested with the number row above the letters on the keyboard. A second group of subjects trained with the number row and was tested with the keypad. Thus, for both groups, the key configuration used in training did not match the key configuration used in testing to ensure that any repetition effect could not be attributed to the retention of the motoric component of the task. The training and testing presentation formats were orthogonally varied for all subjects. The two formats used were numeral and word. Half of the old numbers were in numeral format during training, and half were in word format, with the same format used on each one of the five repetitions of a given number. For half of the numbers in each training presentation format, the test format was the same as the training format, and for the other half the test format was different from the training format. Likewise, for the new numbers at test, half were in word format and half in numeral format.

In some of our earlier experiments (Fendrich et al., 1995, 1991), we gave subjects an explicit recognition test concerning a given number immediately before or after they entered the number. However, in this experiment, we included no such explicit test, and subjects were not explicitly instructed to remember the numbers or told that the numbers shown during testing were related to those shown during training. Hence, this experiment provided a relatively pure assessment of implicit memory.

Based on our previous research, which demonstrated a perceptual component of the long-term retention of the data entry skill as well as a motoric component, we predicted faster response times for the old than for the new numbers, even though we eliminated the motoric component by switching the response mode from training to testing. Of greatest interest was whether we would find an effect of the presentation format on the repetition effect. If old numbers were entered more quickly during testing than new numbers only when there was a match between the presentation formats during training and testing, then there would be evidence for a contribution of the surface percept to the repetition effect. Alternatively, if old numbers were entered more quickly at test than new numbers even when the presentation formats during training and testing did not match, then there would be evidence for a contribution of the abstract meaning of the percept to the repetition effect.

Our results provide clear-cut support for the second of these alternatives. Numbers presented in numeral format (*M* = 3.050 s) were typed much faster than numbers presented in word format (*M* = 3.604 s). In addition, old numbers (*M* = 3.301 s) were typed significantly faster than new numbers (*M* = 3.353 s), reflecting a repetition priming effect. As shown in Fig. 32, there was no significant difference between old numbers presented in the same format at test as at training and old numbers presented in a different format. However, there was an interesting significant interaction of format at test and format continuity. This interaction reflects the fact that for both test presentation formats, numbers presented as words in training had an advantage over numbers presented as numerals in training, suggesting that having to encode the number from words in training led to deeper or more complex processing than encoding them as numerals.

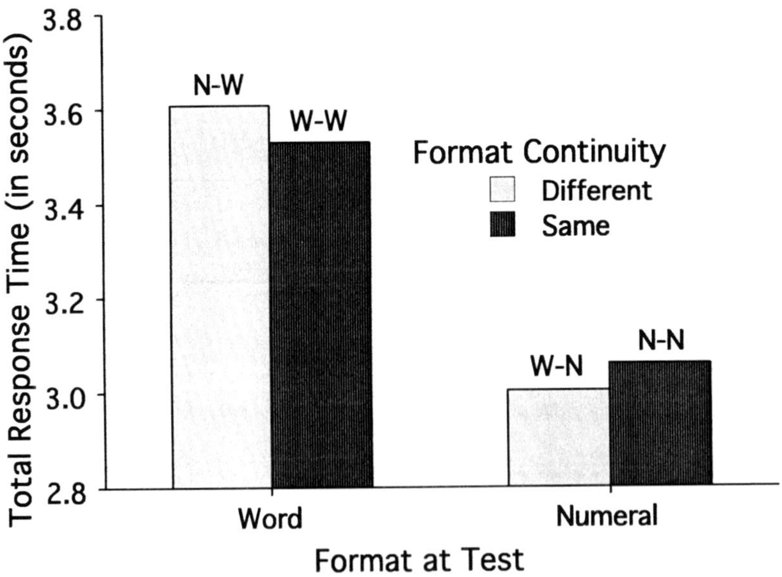

Figure 32. Total response time as a function of format at test and format continuity in Experiment 1 (Buck-Gengler & Healy, 2001). N-N = numerals at training and test; N-W = numerals at training and words at test; W-N = words at training and numerals at test; W-W = words at training and test.

As with the total response time, the time to initiate was faster for numbers in numeral format (*M* = 1.520 s) than in word format (*M* = 1.936 s). However, unlike for the total response time, there was no significant difference between initiating old and new numbers. Further, unlike for total response time, subjects responded faster to old numbers given in the same format in each session (M = 1.709 s) than to those given in different formats in each session (M = 1.745 s). This finding provides some support for the hypothesis that the repetition priming effect is sensitive to perceptual factors, at least for the time to initiate.

The issues for this experiment were two-fold: First was the hypothesis that, as in many other experiments giving evidence for a repetition priming effect, numbers learned in the training session (old numbers) would be entered faster than those seen for the first time in the test session (new numbers). Second was the question of whether there would be an influence of the presentation format on the magnitude of the repetition effect; that is, whether old numbers would be entered more quickly during testing than new numbers only when there was a match in formats between the training session and the test session, or whether the old numbers would be entered faster regardless of whether there was a match in format between the two sessions.

The results replicated the repetition priming effect found in our previous studies (Fendrich et al., 1995, 1991), in that older numbers were entered more quickly than newer numbers even though there was a 1-week delay between training and testing and there were no explicit memory instructions. This effect seems to come from the time spent actually typing the digits, as there was a difference in total response time but no difference in time to initiate. The results also extended the earlier findings in several interesting ways.

In one of our earlier studies (Fendrich et al., 1991), both motoric and perceptual contributions were implicated. The current experiment extended this work by making it clear that motoric repetition is not necessary to achieve the repetition effect, and enabled us to examine the perceptual contribution by itself. Because there was no overall difference in total response time (although there was in the time to initiate) between old numbers typed in the same format at test and at training, and those typed in different formats, it appears that the advantage for old numbers must come largely from the underlying abstract numerical concept, and not simply the surface percept. Even more interesting, however, is the fact that it appears that numbers coded as words at training had an advantage over those coded as numerals at training. This finding implies that the amount of processing necessary to reach the abstract underlying meaning given the particular format is an additional factor influencing the repetition effect. By this explanation, the numeral format is closer to the abstract meaning of the number than is the word format; thus greater processing is needed for numbers in the word format than for those in the numeral format.

In the first experiment, all numbers were typed as digits. Hence, there are two possible alternative explanations for the superiority of training with words. First, processing the word format takes longer than processing the numeral format perhaps because of the relative number of characters or word frequency. Second, the numeral format of the stimuli, but not the word format of the stimuli, matches the format of the digit responses. These two possible explanations cannot be eliminated in that experiment because of the use of only one response format, namely digits. Thus, in the second experiment of this series, a word-format-related response was added, to complement the numeral-format-related digit response.

This new response required translating in the opposite direction, from numeric input to alphabetic output. Letters were chosen as the word-related response, to make the response be a single character just as in the digit response. Specifically, subjects typed the initial letters of the digits. For example, they typed the letter *T* for the digit *2*. Typing digits versus letters was manipulated between subjects to allow each subject to become thoroughly trained in one response, rather than be confused by having to switch between responses.

The method used for Experiment 2 was, thus, the same as that used in Experiment 1, with five exceptions. First, we used a total of 64 subjects, whereas we used only 32 subjects in Experiment 1. Second, the stimulus numbers were made up of just six digits 1, 2, 4, 6, 8, and 9. (The digits 3, 5, and 7 were not used because they start with the same letter as 2, 4, and 6, respectively.) Third, the words were typed with the initial letters capitalized to make them more salient (e.g., “Two Four Eight One”). Fourth, half of the subjects typed digits, as in Experiment 1, and the remaining half typed the initial letters of the digits (e.g., T F E O). Fifth, there was a condition-specific key layout, with the keyboard marked with stickers to indicate the keys that were to be typed, as shown in Fig. 33. Only the six digits and corresponding six letters were used as possible responses. For the letter conditions, the stickers with the six letters were placed in alphabetical order over the number keys on the number row or keypad; for the digit conditions, the stickers with the six digits were placed in numerical order over the normal number keys. The same keys thus were used for both conditions; however, what a given key represented depended on the condition.

E	F	N	O	S	T

(a)

E	F
N	O
S	T

(b)

1	2	4	6	8	9

(c)

1	2
4	6
8	9

(d)

Figure 33. Keyboard layouts for each response type and typing location combination of Experiment 2: (a) letter-row, (b) letter-keypad, (c) digit-row, (d) digit-keypad (Buck-Gengler & Healy, 2001).

The results are summarized in Fig. 34 in terms of total response times for correct responses on the test as a function of presentation format at test and response format. Digit responses were faster overall than letter responses, and numbers presented as numerals were entered faster overall than those presented as words. More importantly, these two factors interacted so that when the presentation format of a particular number matched its response format (i.e., word presentation for letter responses and numeral presentation for digit responses), subjects typed the number faster than when the presentation format of the number did not match its response format. In fact, the subjects responding with letters actually typed numbers presented as words somewhat faster than numbers presented as numerals.

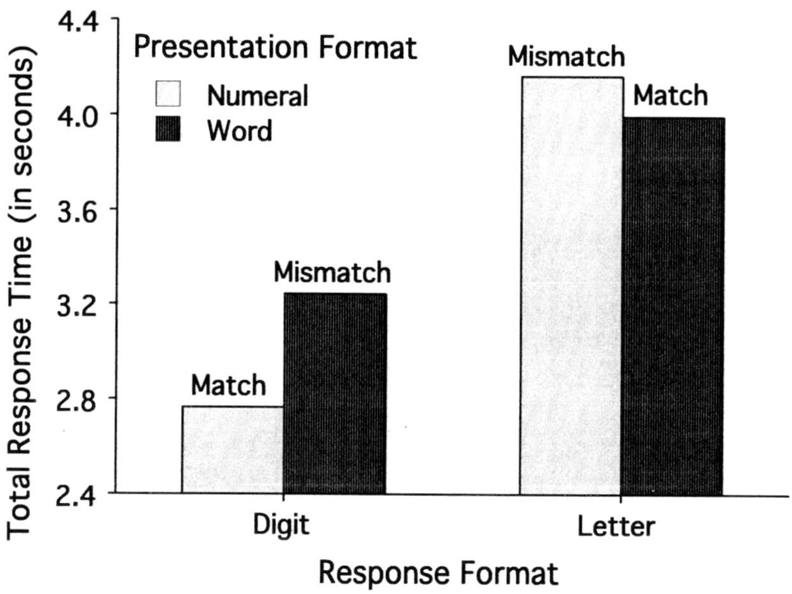

Figure 34. Total response time as a function of response format and presentation format at test in Experiment 2 (Buck-Gengler & Healy, 2001).

Also importantly, old numbers (M = 3.520 s) were typed significantly faster than new numbers (M = 3.561 s), indicating a repetition priming effect. Most importantly, as shown in Fig. 35, for the old numbers, format continuity interacted with test presentation format, as in Experiment 1. This interaction reflects the fact that within each presentation type at test, numbers in the word format at training were entered faster at test than those in the numeral format at training. And this interaction did not depend on response format; the three-way interaction was not significant.

In summary, Experiment 2 provided further evidence for the repetition priming effect even though, as in Experiment 1, there was a 1-week delay between training and testing and the key configuration manipulation ensured that the repetition effect could not be attributed to the retention of the sequence of motor movements. In addition, presentation format and response format interacted so that within each response format, numbers were entered faster with the matching presentation format. It is clear from this result that the word format can be faster than the numeral format. Thus, it is not time spent processing the word format that accounts for the previous interaction between presentation format at test and format continuity, which was still found in this experiment. Within each test presentation format, numbers presented in word format at training were entered more quickly at test than numbers presented in numeral format at training.

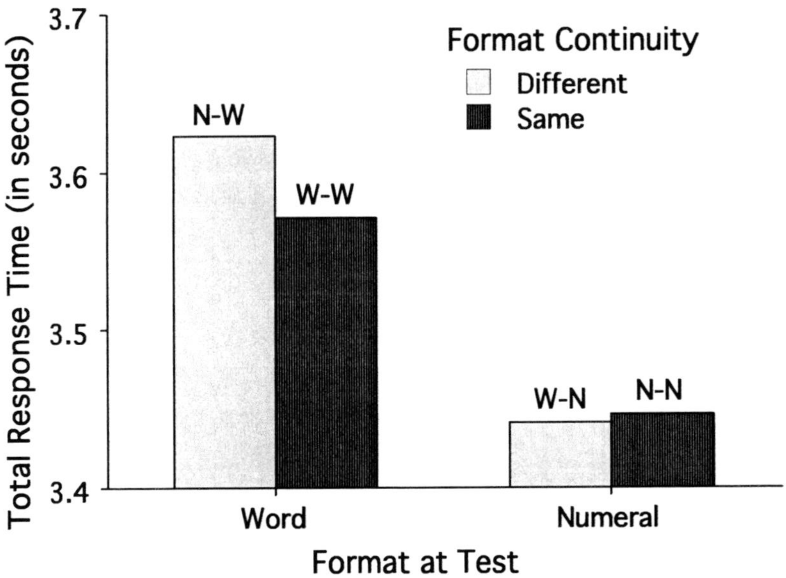

Figure 35. Total response time as a function of format at test and format continuity in Experiment 2 (Buck-Gengler & Healy, 2001). N-N = numerals at training and test; N-W = numerals at training and words at test; W-N = words at training and numerals at test; W-W = words at training and test.

We conclude that the superiority for training with words is due to the fact that word presentation at training leads to greater processing than does numeral presentation, presumably because the word format is farther from the abstract meaning of the number than is the numeral format, thus leading to facilitation of entry of those numbers at test.

The Relationship between the Learning of Examples and the Learning of Rules in Conceptual Tasks

The question arises in many contexts whether to teach an abstract rule or instead present a series of individual examples. To investigate this question, we examined a natural situation that people perform routinely in which their responses might be based either on a general rule or on their familiarity with specific examples covered by the rule (Raymond, Fisher, & Healy, in press; Raymond, Fisher, Healy, & Bourne, 1999).

In English, the indefinite article *a* is used preceding nouns and adjectives beginning with a consonant sound, whereas *an* is used preceding words beginning with a vowel sound. It is less commonly known that in most varieties of English (according to standard English dictionaries), a similar distinction pertains to the pronunciation of the definite article *the*. When *the* precedes a word that begins with a consonant sound, it is pronounced with the schwa phoneme /´/, as /D´/ (pronounced "thuh"). When it precedes a vowel sound (and also when it is given emphasis), it is pronounced with the phoneme /i/, as /Di/ (pronounced "thee"). In the present study, we examine the extent to which native adult speakers of English make use of that pronunciation distinction.

The indefinite and definite article pronunciation distinctions together define a division of words into two categories based on initial phonology, words coupled with *a* or /D´/ and words coupled with *an* or /Di/. Underlying this view is the assumption that such a categorical division is governed by a high-level, abstract construct or rule. This view was questioned earlier by Serwatka and Healy (1998), who referred to it as the "rule-and-category" paradigm and contrasted it to a "rote-and-analogy" paradigm, according to which such knowledge is organized not in terms of high-level generalizations, but rather in terms of low-level, or local, information. According to the rote-and-analogy paradigm, knowledge is particular to a single unit, such as a word or combination of words, rather than to the whole scope of the abstract construct. In the case of the article pronunciation distinctions, it is not clear which is the relevant unit according to the rote-and-analogy paradigm. Speakers might consider a combination of an article and an adjective or a noun (e.g., *the umbrella* /Di ´m'brEl´/), as they presumably do, for example, when learning the gender of nouns in languages like French, Spanish, or German. Alternatively, speakers might consider a combination of an article and an initial syllable (e.g., *the um* /Di ´m/), or a combination of an article and an initial phoneme (e.g., *the u* /Di ´/). That is, a syllable or a phoneme, rather than an entire word, could act as a cue for the article pronunciation. In other words, the question is at what level or levels of abstraction (e.g., vowel or consonant feature, specific phoneme, specific syllable, specific word) does the speakers' behavior become systematic or rule-like.

In earlier work, Healy and Sherrod (1994; Healy et al., 1998) tested whether the rote-and-analogy paradigm can explain the behavior of adult speakers of English with respect to the definite article pronunciation distinction. Subjects saw a list of nouns or adjectives presented one at a time. Half of the words began with a consonant and half with a vowel. In some conditions, for each word subjects were to choose between /D´/ and /Di/. In other conditions, subjects had to make a choice instead between *a* and *an*.

We found a very low error rate in the indefinite article conditions, indicating that subjects behaved in accordance with the abstract rule for that distinction. In contrast, the overall error rate in the definite article conditions was well below the chance level but high enough to preclude abstract rule usage.

We also examined error rates in definite and indefinite article conditions as a function of word frequency, whether there was a mismatch between the initial letter and the initial phoneme of the word, and the stress of the initial syllable. The results of these examinations provided additional evidence in favor of the rote-and-analogy paradigm and also gave insight into the nature of the local information subjects used. They indicated that subjects made decisions based on the word itself, on the sound of the initial letter, on the initial letter itself, and on the stress of the initial syllable.

Although our results were clear-cut, they suffer from an interpretive problem. The findings may at least in part reflect reading processes rather than processes used in speaking. For example, the influence of the initial letter (as opposed to the initial sound) of the word might be an artifact of subjects' seeing the words. In a new study including two experiments, we employed two new procedures involving only the auditory modality. In Experiment 1, 24 subjects heard a list of test words, all nouns and adjectives. They were presented one word at a time, each preceded by the possessive pronoun *his*. The subjects' task was simply to repeat each word after hearing it. In one condition, they were to replace the word *his* with the definite article (/D´/ or /Di/) before each word, and in another condition, they were to say the indefinite article (*a* or *an*) before each word.

One methodological difficulty with the procedure used in Experiment 1, however, is that the experimenter must determine whether the subject said /D´/ or /Di/ for each word. This determination is not always easy, even by trained linguists, because of vowel reduction and vowel assimilation to the following phoneme. To avoid this difficulty, in Experiment 2, we used a procedure based on word perception rather than word production. Specifically, 24 subjects again heard the list of words, but in this case, the words were heard twice in succession, once with each variant of the article. Again, there were two conditions, one involving a choice between /D´/ and /Di/ and the other involving a choice between *a* and *an*. The subjects' task was simply to choose the variant that sounded best.

In both of these experiments, we examined subjects' choice of articles with either regular or special words as a function of the initial sound category, word frequency, mismatch between the initial letter and initial phoneme, and the stress of the initial syllable.

For the 72 regular words, the error rate for definite articles was about 28% and for indefinite articles was less than 4% in both experiments, indicating that subjects responded in accordance with the abstract rule for the indefinite article (*a*/*an*) distinction but not for the equivalent definite article (*the*) pronunciation distinction. There are at least two reasons why performance was so much better for the *a*/*an* distinction. First, children are generally taught the abstract rule for *a* versus *an* but not for /D´/ versus /Di/. Second, in written text, the spelling difference between *a* and *an* provides visual cues and reinforcement for that distinction.

Along with regular words, we gave as stimuli two sets of nine special words beginning with the letter *h* and the phoneme /h/. One set had stress on the first syllable, and the other set had stress on the second syllable. According to the dictionary, such words with the second syllable stressed can be preceded by either article choice. For example, it is acceptable to say either "a historic" or "an historic." In contrast, such words with the first syllable stressed do not have such flexibility. For example, it would be acceptable to say "a helmet" but not "an helmet." The error proportions in both conditions of the two experiments on the special words beginning with the letter *h* and the phoneme /h/ are shown in Fig. 36. To allow us to examine the effects of syllable stress, all responses of *an* or /Di/ were considered errors for the purpose of this analysis. Subjects in the more objectively scored Experiment 2 showed a significant effect of syllable stress and those in Experiment 1 showed a marginally significant effect, indicating that all subjects, even those choosing between /D´/ and /Di/, were sensitive to this subtle cue, which occurs at the level of the syllable.

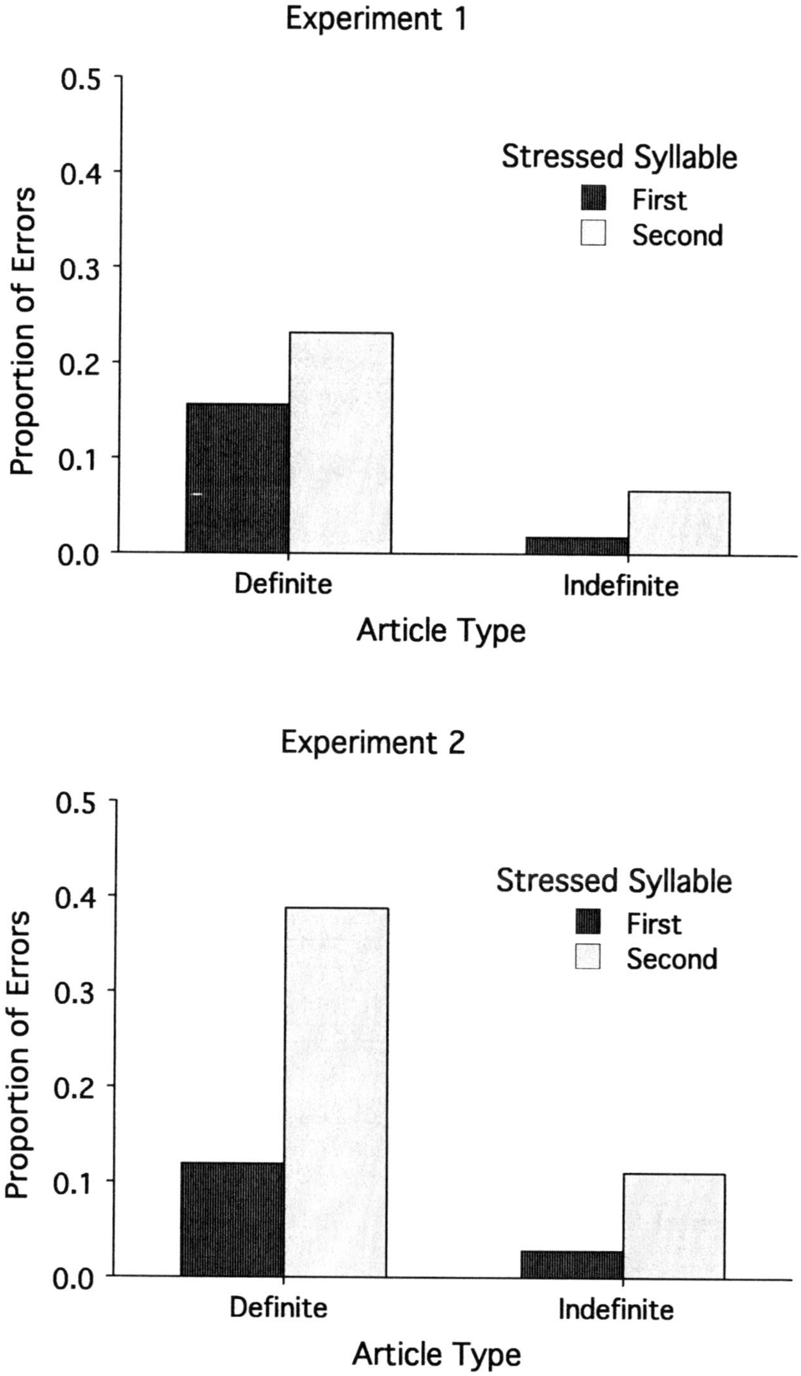

Figure 36. Proportion of errors in Experiments 1 and 2 as a function of article type and stressed syllable (Raymond, Fisher, & Healy, in press).

Another cue which may affect subjects' choices is the actual initial letter of the word, although the abstract distinction is based on the initial sound rather than the initial letter. In the earlier work by Healy and Sherrod (1994; Healy et al., 1998), we found evidence that subjects were sensitive to the initial letter of the word, but in that case subjects were making their choices based on written stimuli. In contrast, in the present experiments, the subjects saw no written material; instead the stimuli were heard. To provide a strict test of the importance of the initial letter on article choice, we made two types of comparisons. In the first type, we examined two sets of nine special words, all of which began with the same initial consonant sound /j/ (the initial sound in *use* and *youth*). One set of words began with the consonant letter *y* (e.g., *youth*), and the other set began with the vowel letter *u* (e.g., *use*). As shown in Fig. 37, we found in both experiments significantly more errors on words that began with the initial letter *u* than on words that began with the initial letter *y*, presumably because of the conflict between the spelling and pronunciation of the initial letter for the words beginning with the vowel letter *u* but the consonant sound /j/. These results clearly indicate that subjects who heard the stimulus words but did not see them or write them were nevertheless sensitive to the letters in the words, not just the sounds. This finding may not be so surprising for the indefinite article distinction because, as mentioned previously, subjects have learned that distinction in part through visual reinforcement from the letters seen in written text. However, this finding seems quite surprising for the definite article pronunciation distinction because no visual reinforcement from written text is provided when learning that distinction.

For the comparison of words beginning with the vowel letter *u* and the consonant letter *y*, the initial sound was a consonant sound /j/ in all cases and the correct answer was either *a* or /D´/. We made a similar analysis involving words in which the initial sound in all cases was a vowel sound and the correct answer was either *an* or /Di/. This comparison involves a set of 9 special words beginning with the silent consonant letter *h* (such as *honor* and *hourly*) and 36 regular words beginning with a vowel letter (such as *olive* and *auburn*). As shown in Fig. 38, in Experiment 2 we found significantly more errors on words that began with the initial letter *h* than on words that began with initial vowel letters, presumably again because of the conflict between the spelling and pronunciation of the initial letter for the words beginning with the consonant letter *h* but vowel sound. The results for the less objectively scored Experiment 1 are not so clear and there was neither a significant main effect nor a significant interaction involving initial letter in that case.

Our results as a whole provide strong evidence that subjects are sensitive to letters as well as to sounds when making article selections even when the stimuli are heard but not seen. Further, our results provide strong support for the rote-and-analogy paradigm and also confirm our previous findings concerning the nature of the local information subjects use. They indicate that subjects make decisions based on the sound of the initial letter, on the initial letter itself, and on the stress of the initial syllable. More generally, our findings indicate that individuals' knowledge of the article distinctions does not conform exactly to the simple, general abstract rule included in the dictionary, but rather is more complex and involves information at various levels of abstraction. Thus, individuals learn complex rules involving different levels of abstraction when given sufficient specific examples. The superior performance for the indefinite article distinction relative to the definite article distinction further indicates that individuals do benefit from explicit visual cues, reinforcement, and instruction concerning abstract, general rules.

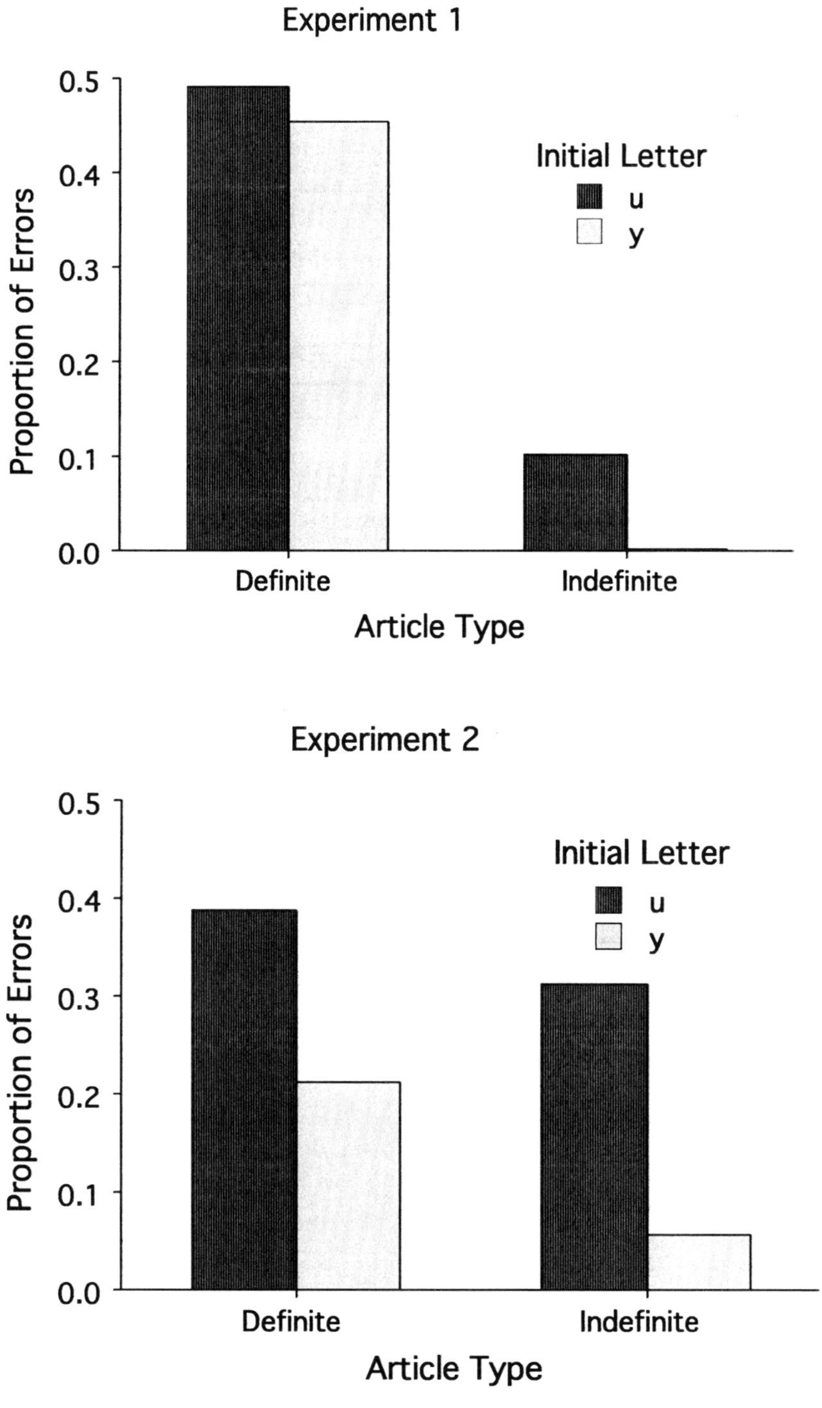

Figure 37. Proportion of errors in Experiments 1 and 2 as a function of article type and initial letter (*u* versus *y*) (Raymond, Fisher, & Healy, in press).

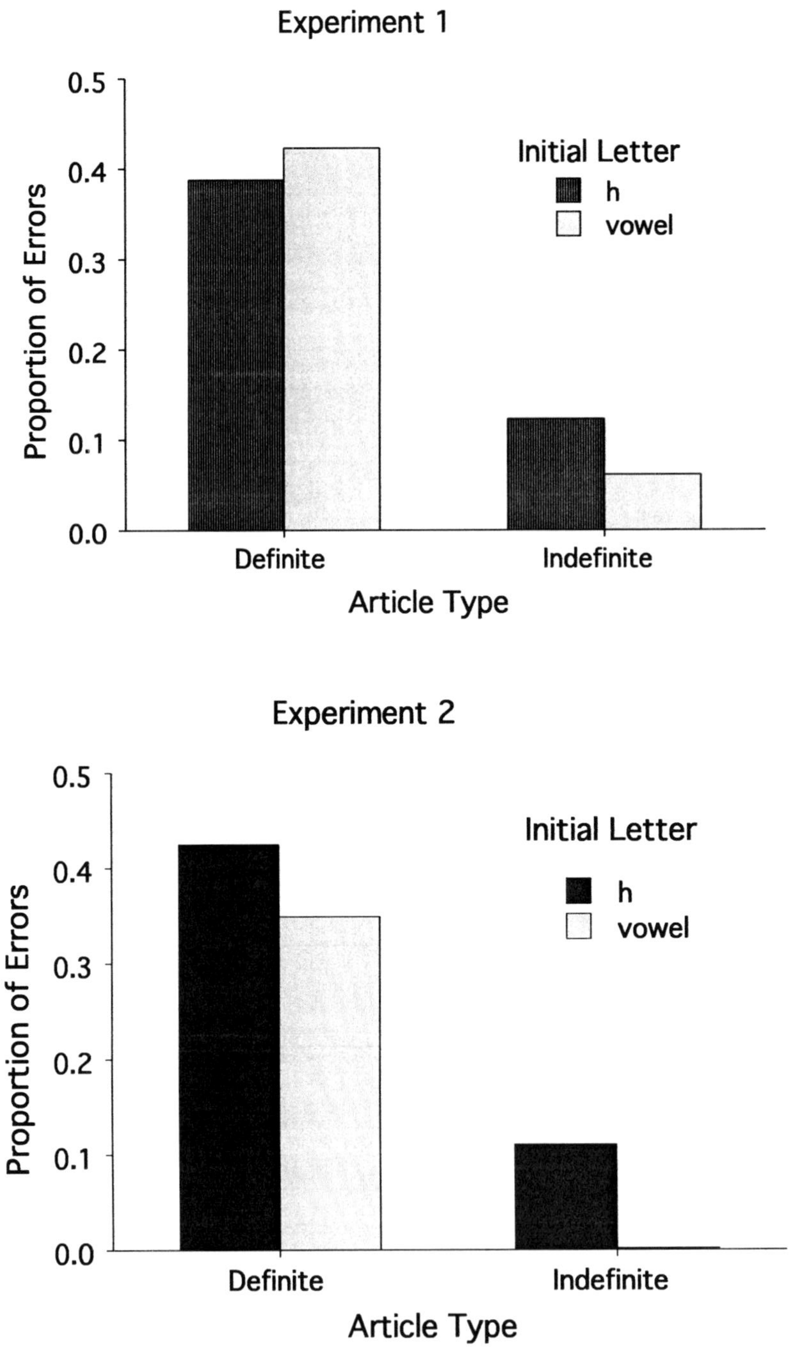

Figure 38. Proportion of errors in Experiments 1 and 2 as a function of article type and initial letter (*h* versus vowel) (Raymond, Fisher, & Healy, in press).

TYPES OF STRATEGIES AND STRATEGY TRANSITIONS IN CATEGORY LEARNING TASKS

Although many investigators of category learning tasks have relied exclusively on accuracy and response time measures, we have learned that these indices alone do not provide an accurate and complete picture of performance. An examination of subject strategies is helpful in understanding the cognitive processes underlying performance.

Cognitive psychologists have demonstrated two general types of strategy transitions in a variety of learning contexts. First, in concept learning experiments, subjects often adopt an initial strategy of learning arbitrary associations between items and responses by rote, but later make a transition to rule-based performance for all items once a rule is discovered (e.g., Homa, Sterling, & Trepel, 1981). A second type of strategy transition, from rule- (or algorithm-) based performance to instance- (or memory-) based performance, has been the focus of our more recent research on mental arithmetic (see, e.g., Rickard, 1997). The existence of these two opposing types of transition raises some important questions about strategy use during learning: Under what conditions does practice (i.e., experience with a domain) result in the abstraction of rules that are then used to guide performance? Under what conditions does rule-based processing eventually give way to instance-based processing? Are there learning contexts in which rule discovery and rule-based performance are important during the first stages of learning, only to be replaced by instance-based performance as skill improves? Are there learning contexts in which the opposite transition is likely to occur? We have developed an experimental framework for exploring the transitions between rule-based and instance-based performance that occur with practice which should help to answer some of these questions.

Initially, we conducted two preliminary experiments on these issues (Bourne, Healy, Parker, & Rickard, 1999; Healy et al., 1998). In the first, subjects were presented strings of three letters (such as LMV) one at a time, and were asked to indicate whether each string is "grammatical" or "ungrammatical," with feedback provided after every response. Grammaticality was determined by a rule unknown to the subject at the onset of the experiment. By the rule, a string is grammatical only if it can be rearranged to correspond to a sequence of adjacent letters in the alphabet. For example, LMV is ungrammatical, but RQP is grammatical because it can be rearranged as PQR. A single set of 12 strings (half grammatical, half ungrammatical) were presented to subjects repeatedly over multiple blocks of practice. Subjects were informed that an underlying rule might be useful if discovered but that their primary goal was to learn to respond as fast and as accurately as possible to each item. Subject strategies were probed after every trial to determine whether they guessed, used a rule, remembered the answer from a previous trial, or did something else. Analyses of the strategy data show that all subjects initially guessed at the answer to each stimulus, that most subjects subsequently discovered the operative rule, and that all subjects showed a transition to instance-based performance by the end of practice. Transitions like these, from guessing to rule-based performance and from rule-based to instance-based performance, had not been demonstrated previously within a single task.

A second preliminary study (Bourne et al., 1999; Healy et al., 1998) explored further the conditions under which rules, once discovered, continue to be used indefinitely, and the conditions under which a transition back to instance-based performance occurs. The effects of practice on frequency of use of rule- and instance-based strategies for pronunciation of the definite article *the* (/D´/ vs. /Di/, as described earlier) was the focus of this study. Grammatically correct performance in this case depends on whether the first sound of the word is a vowel, in which the correct pronunciation is /Di/, or a consonant, in which case the correct pronunciation is /D´/. For example, the correct pronunciation is "/Di/ answer" and "/D´/ story." In this task, rule use increased and instance-based performance decreased with practice, possibly because the rule for this task is simple enough to support very rapid responses.

The results of these two preliminary experiments differed in which of two strategies became dominant with practice. In the first, which involved artificial strings of letters, instance-based strategies became dominant, whereas in the second, which involved a naturalistic linguistic rule, rule-based strategies dominated. The tasks used in these two experiments differed in three potentially important ways which could have contributed to the strategy use and transitions observed. First, rule difficulty varied, with the rule for the string task being more difficult than the rule for the linguistic task. Second, the tasks varied in how natural they were, with the rule for the string task being arbitrary and artificial and the rule for the linguistic task being a part of natural language skill. Third, the rule in the string task was unknown to any subject in advance of the experiment, whereas the rule in the linguistic task was at least partially known to all subjects in advance. We examined the first two of these three differences in the following experiment to determine whether either of them is crucial in determining the strategy that eventually dominates (see Bourne et al., 1999, for a more complete description).

Specifically, we orthogonally varied rule difficulty and task naturalness. We compared two string rules, the one used previously (a string is grammatical only if it can be rearranged to correspond to a sequence of adjacent letters in the alphabet) and a second one that is simpler (a string is grammatical only if it corresponds *without rearrangement* to a sequence of adjacent letters in the alphabet). We also compared two linguistic rules, one like the simple pronunciation rule used previously that involves a choice between *thee* (/Di/) and *thuh* /D´/ and a second, more difficult one based on the count/mass distinction that involves a choice between *many* and *much* (Serwatka & Healy, 1998). In both easy and difficult rules and for both artificial and natural tasks, a single set of 12 stimuli (6 of each type) and 2 response choices were presented individually and repeatedly to subjects with feedback on their responses over 30 blocks of practice. On every second and third block, one novel stimulus was presented (with 10 of each type across blocks). As in our preliminary experiments, subjects were told that there was an underlying rule for categorizing the stimuli which might be useful if discovered but that they were primarily to learn to respond as fast and as accurately as possible to each item. After every trial, subject strategies were probed to determine whether they guessed, used a rule, remembered the answer from a previous trial, or used another strategy.

We found that the proportion of correct responses increased systematically over blocks of trials in all conditions, as shown in Fig. 39. There was an overall difference in accuracy between the artificial letter string task and the more natural linguistic tasks, with performance in the natural tasks starting (in Block 1) and remaining more accurate than in the artificial task throughout practice. Easy tasks were performed somewhat more accurately than hard tasks, but the difference was not statistically reliable.

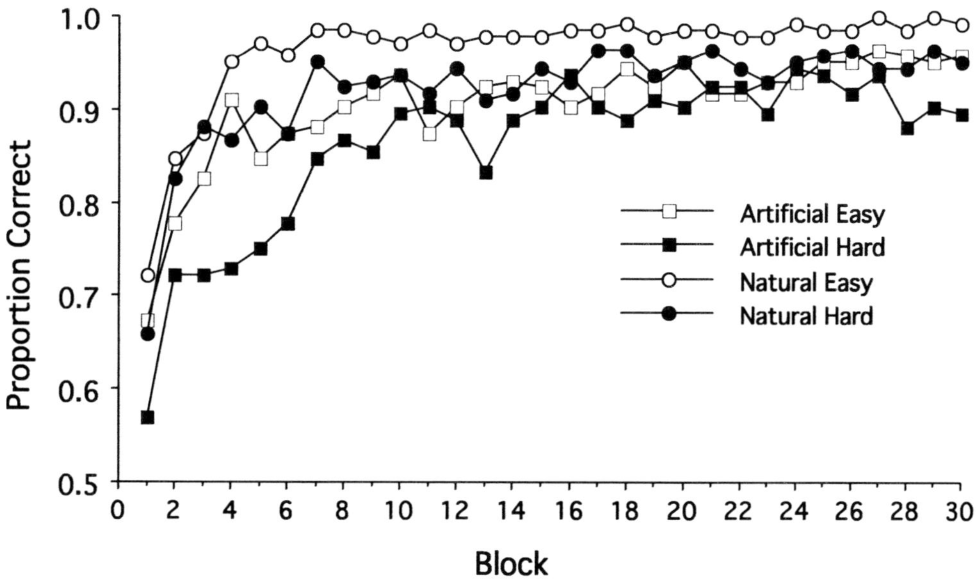

Figure 39. Proportion correct as a function of block, rule difficulty, and task naturalness (Bourne, Healy, Parker, & Rickard, 1999).

As shown in Fig. 40, the pattern of strategy use over blocks was much the same for three of the four task conditions, artificial easy, artificial hard, and natural hard. In each of these conditions, use of the instance strategy increased monotonically and was clearly the dominant strategy by the end of practice. Rule use first increased and then decreased, as instances became dominant. Guesses were frequent at the outset of practice but quickly dropped to insignificance. Other strategies were rarely reported. The picture was qualitatively and quantitatively different in the natural easy task, which was based on the *thee/thuh* pronunciation distinction. As in our previous experiment (see Bourne et al., 1999; Healy et al., 1998), rule use increased from the beginning to the end of practice and was clearly the dominant strategy after about 15 blocks of trials. Instance use increased and then decreased in frequency as the rule became dominant. Guesses and other strategies followed essentially the same pattern over trials as they did in the other three conditions.

As shown in Fig. 41, response time speed up over blocks of trials was reasonably well-fit by power functions. The response time analysis shows a clear and significant difference between the instance and the rule strategies in the artificial tasks, with performance being reliably slower when the rule strategy was used. This outcome is consistent with our preliminary experiment (see Bourne et al., 1999; Healy et al., 1998), in which only the more difficult artificial string task was used. Differences in response times between strategies were more complex in the natural tasks.

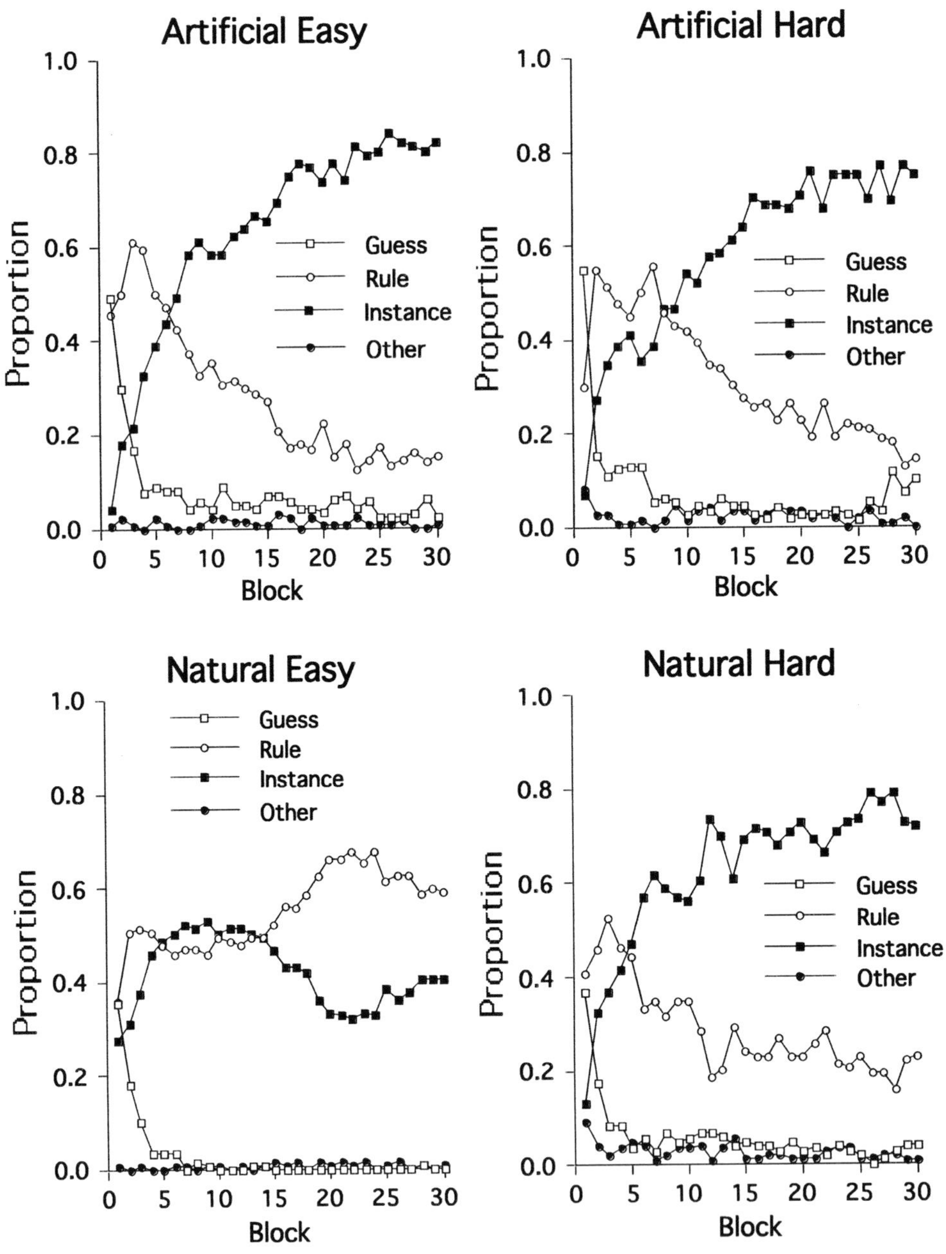

Figure 40. Proportion of strategy choices as a function of block and strategy, for each rule difficulty and task naturalness combination (Bourne, Healy, Parker, & Rickard, 1999).

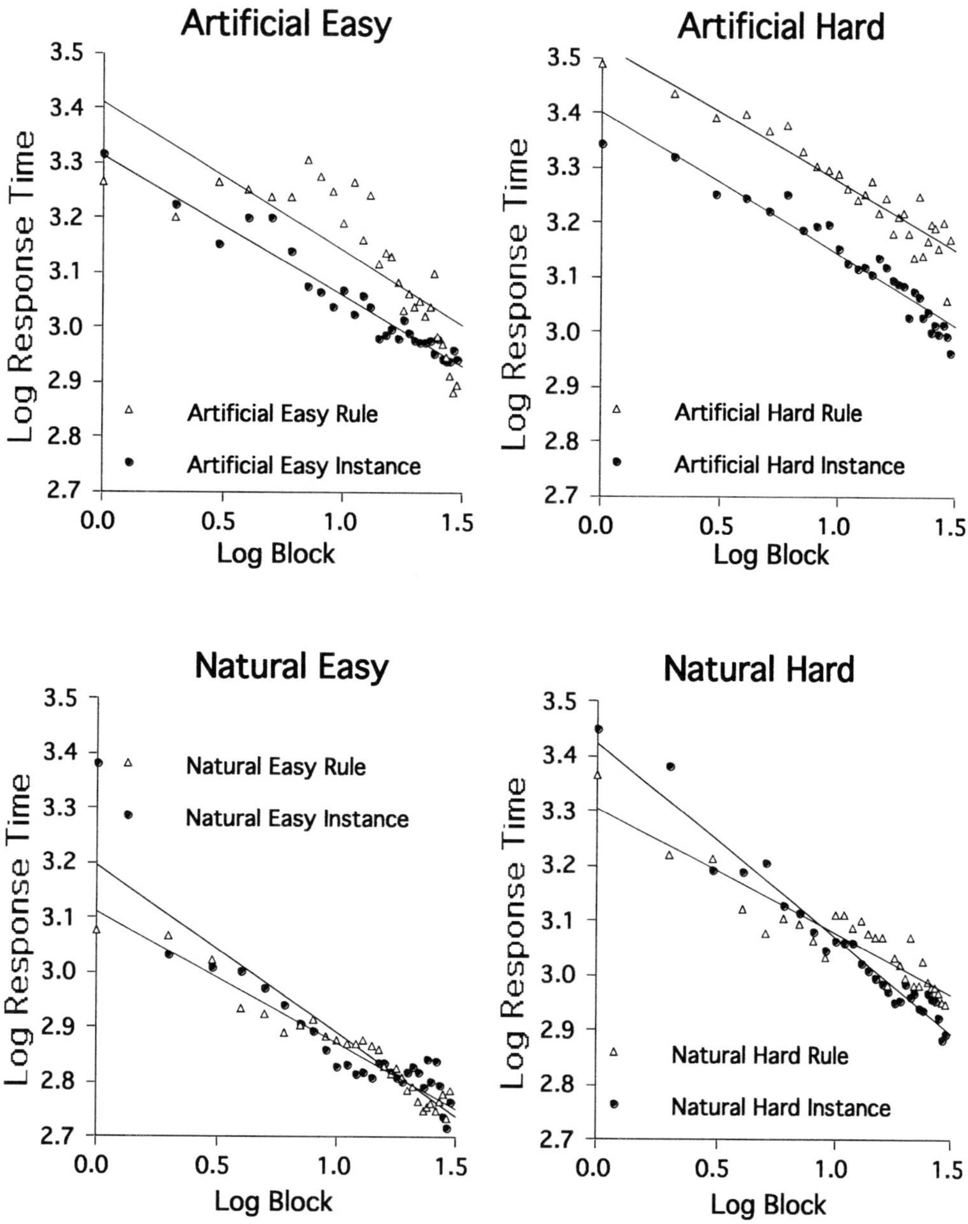

Figure 41. Log response time as a function of log block and strategy, for each rule difficulty and task naturalness combination (Bourne, Healy, Parker, & Rickard, 1999). Straight lines represent best-fitting power functions.

In the artificial tasks, response time speed up functions were nearly parallel and responses made under the instance strategy were faster both at the beginning and at the end of practice. Functions in the natural tasks converged when the discrimination was based on *thee/thuh*, just as they did in our preliminary experiment, with the instance strategy producing slower responses at the outset (and overall). In the difficult natural task, instance-based responses were slower at the outset, but rule and instance speed-up functions intersect and rule-based responses were slower by the end of practice (and overall). The intersecting functions for the natural tasks are due to the fact that the response time speed-up functions have a particularly steep slope for the instance strategy in those cases. This steep slope may occur because the instances in both natural tasks are common words, which are readily learned and retrieved from memory, unlike the letter strings used in the artificial tasks.

Response times were related to the dominant strategy at the end of practice. In artificial tasks, the dominant strategy was instance-based in both the easy and the difficult versions. In the natural linguistic tasks, the rule was dominant by the end of practice in the easy version, and rule-based responses were faster overall (although the functions converge). In the difficult version, although instance based responding was slower at the outset of practice, it was faster at the end when the instance strategy was dominant. In all tasks then, the dominant strategy determines which basis of responding is faster.

The proportion of trials on which subjects reported guess, rule-based, and instance-based strategies during training and on novel instances are shown in Table 1. The pattern of results was nearly the same for both easy and difficult artificial tasks. During training, guesses were relatively infrequent. As found in our earlier string experiment (Bourne et al., 1999; Healy et al., 1998), however, guesses were reported frequently and more often than instance strategies on novel stimuli. Instance-based responses were most common on the training stimuli, but rule-based responses were most common on the novel stimuli, as would be expected if subjects recognized that novel stimuli had not been seen previously. Strategy choices differed between the easy and difficult natural tasks. For the easier, *thee/thuh* task, guesses were uncommon and rule usage was the most frequently reported strategy on novel as well as on training items. There was a strong mixture of both rule-based and instance-based strategies for training items, but rule use predominated on novel items. These results support the claim that the rule was used or discovered by most subjects during practice, and that use of this rule generalized readily to new items. In the difficult natural task, the pattern of strategy selections on novel and training items was similar to that on the artificial tasks. Instance-based performance was typical of training whereas rule-based performance dominated on novel stimuli.

In conclusion, it is less a matter of task familiarity or task difficulty that affects strategy use than it is the cognitive demands of the strategy itself. That is, we found the same pattern of strategy use in artificial and natural tasks and in easy and difficult tasks. The only case in which strategy use deviated from a pattern in which instance-based responding became dominant after an early period of rule acquisition and rule use was in the *thee/thuh* pronunciation task, the natural easy task, in which rule use persisted and was the dominant basis of responding at the end of training. Why should this condition be unique? We suspect it is because this is the one case in which the rule was easier and quicker to employ on each trial than instance memory. This is not to say that subjects necessarily consciously applied the formal rule of pronunciation to each word. It is possible that the *thee/thuh* distinction, which is required often in natural discourse, was known and used implicitly by some subjects.

Table 1. Proportion of Strategy Reports during Training and on Novel Instances for Each Rule Difficulty and Task Naturalness Combination (Bourne, Healy, Parker, & Rickard, 1999)

	Strategy		
	Guess	Rule	Instance
Artificial Easy			
Training	.08	.29	.63
Novel	.23	.58	.15
Artificial Difficult			
Training	.08	.32	.58
Novel	.23	.57	.17
Natural Easy			
Training	.03	.55	.42
Novel	.18	.63	.19
Natural Difficult			
Training	.06	.29	.63
Novel	.13	.60	.16

More generally, we conclude that monotonic changes in overall measures of performance, such as accuracy and speed of response, during practice in categorization tasks might mask and certainly do not always reveal the complexity of changes in the learner's basis of response. The strategy learners developed or the response basis they reported in these experiments did not typically move in parallel over trials with overall performance measures. Overall measures changed systematically toward an asymptote, whereas strategy measures showed increases and decreases in the frequency of various strategies and transitions from one strategy to another at certain points in practice.

The Durability and Use of General and Specific Knowledge in Quantitative Estimation

In previous work (Healy et al., 1992), we demonstrated that in some tasks, learned performance is very durable over time whereas in other tasks, rapid decrements in retention are evident. We have interpreted this difference in terms of a principle of procedural reinstatement. Whenever we have been able to reinstate the original contextual and procedural conditions of training at the time of retention, retention test performance has been equivalent to performance at the end of training, with essentially no loss in speed or accuracy. In contrast, whenever the original contextual and procedural conditions of training could not be duplicated at the retention test, performance was markedly poorer at the retention test relative to the end of training. These findings imply that training effectiveness is optimized under conditions of procedural reinstatement. However, we have also found an undesirable side effect of the durable retention produced by procedural reinstatement, namely that it is highly specific to the trained task elements and does not readily generalize to variations on that task.

Implicit in the procedural reinstatement principle is the distinction between general and specific memories. The procedural reinstatement principle depends on the availability of specific memories, and, when specific memories are available, performance on a retention test is optimal. But even when specific memories are unavailable, the full benefits of training may not be totally lost because a learner can sometimes rely on general knowledge and skills relevant to the task. Performance based only on general memories, however, is never, nor is it expected to be, at the same level as that based on specific memories.

This point was addressed by Brown and Siegler (1996) in a study of quantitative estimation skills. The goal of their study was to determine the durability and generalizability of training on a limited set of examples, or "seed" facts. Specifically, subjects were first asked to estimate the population of each of a set of 99 countries. Following that pretest, in a study-test procedure, subjects were trained on the populations of a sample of 24 of the countries. They were then tested again on the full set of countries, and this test was repeated after a 6-month delay. At the immediate test, subjects were better on the trained countries than on the countries not trained, although they were better on the untrained countries than they were at pretest. Six months later, performance was no different on the trained and untrained countries, though again they were better than on the pretest. This pattern of results implies that subjects had specific knowledge of the populations of the trained countries immediately after training, which generalized to untrained countries, but that this specific knowledge was lost during the delay interval.

These findings are important to us for several reasons. First, they identify a training procedure that produces generalizable knowledge. Second, that procedure promises great economy of training because only a subset of the to-be-learned material needs to be provided during training with a subsequent benefit to all of the material. Third, within that procedure, both durable general and transient specific knowledge can be measured.

We have initiated a series of experiments to follow up on Brown and Siegler's (1996) findings. Brown and Siegler outlined two possible explanations for training effects of seeding in their population estimation task. First, there is the possibility that each trained country's population is stored in memory and then later used as an anchor for comparison when new populations are to be estimated. Second, individuals may form a framework for estimating populations by extracting general, or abstract, information from the population facts that they have been taught, essentially creating a set of mental descriptive statistics for reference, such as the average population and range of population values. Brown and Siegler concluded that abstraction, and not anchoring, was responsible for the seeding effect, because after a delay their subjects were not significantly more accurate when estimating the trained countries than the untrained ones.

In our initial experiment (LaVoie, Bourne, & Healy, 1999, in press), we corrected a flaw in the experimental design of the Brown and Siegler (1996) experiment. Whereas they pretested and posttested subjects on all country populations and used only a single subset of the countries as training seeds, we divided a set of 96 countries into four matched subsets of 24, using one subset as a pretest, a second as seeds, a third as an immediate posttest, and a fourth as a retention test after a 2-week delay, with the assignment of subsets to test lists counterbalanced across the 24 subjects we tested. On the immediate posttest, subjects were given the subset of countries assigned to that test followed by a retest of the pretest and seeding subsets. Likewise, on the delayed retention test, subjects were given the subset of countries assigned to that test followed by retests of the pretest, seeding, and immediate posttest subsets. The dependent measure we examined was the order of magnitude error (OME), which is the absolute value of the log to the base 10 ratio of the estimated country population value divided by the actual country population value.

The results are summarized in Fig. 42, in terms of the OME on each of four tests (pretest, end of seeding, immediate posttest, and delayed retention test), with the value for each test from the subset of countries assigned to that test. We replicated the seeding effect found by Brown and Siegler (1996), because performance on the untrained countries on the immediate posttest and the delayed retention test was superior to that on the pretest. However, as shown in Fig. 43, unlike the results of Brown and Siegler, our subjects showed superior performance on the trained countries relative to the untrained countries (those assigned to the immediate posttest) at both the immediate posttest and the delayed retention test, thereby demonstrating reliable long-term memory for the specific seeding information. Our retention interval was only 2 weeks, whereas that of Brown and Siegler was 6 months. To rule out the possibility that their longer retention interval was responsible for the discrepancies between our results and theirs, we retested 8 of our 24 subjects after 2 months. The pattern of results for these subjects on the 2-month retention test (shown in Fig. 44) was similar to that on the 2-week retention test, with an advantage for the trained countries, or specific seeding information, still evident, although it seems that subjects were losing specific memories to some extent without losing general memories.

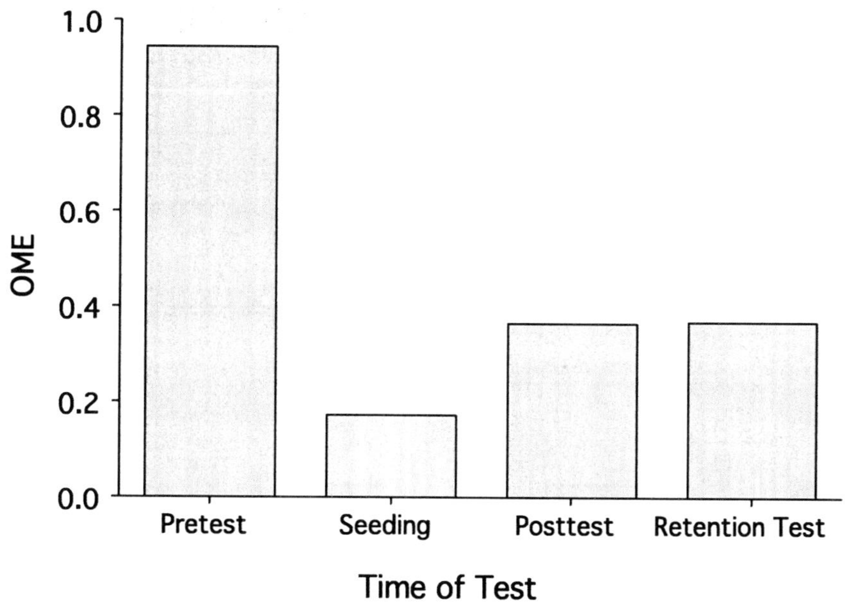

Figure 42. OME as a function of time of test in Experiment 1 (LaVoie et al., in press).

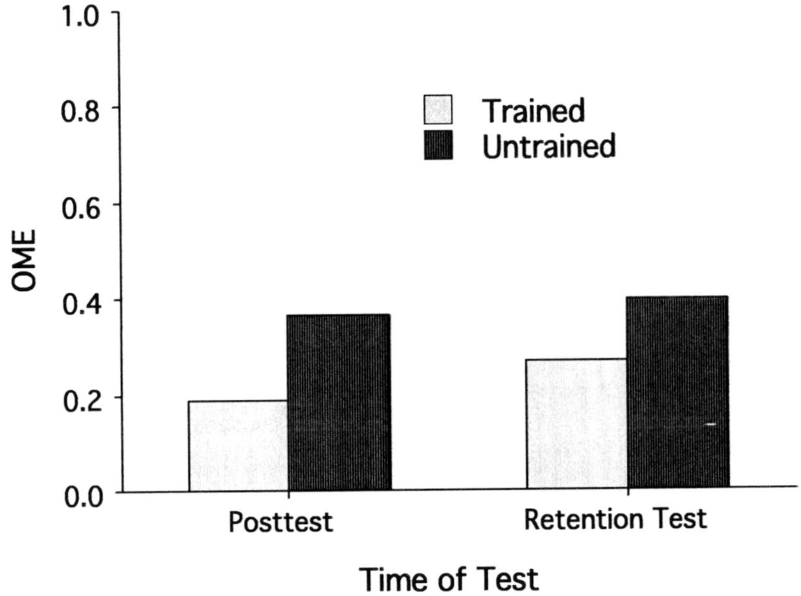

Figure 43. OME as a function of time of test and training status in Experiment 1 (LaVoie et al., in press).

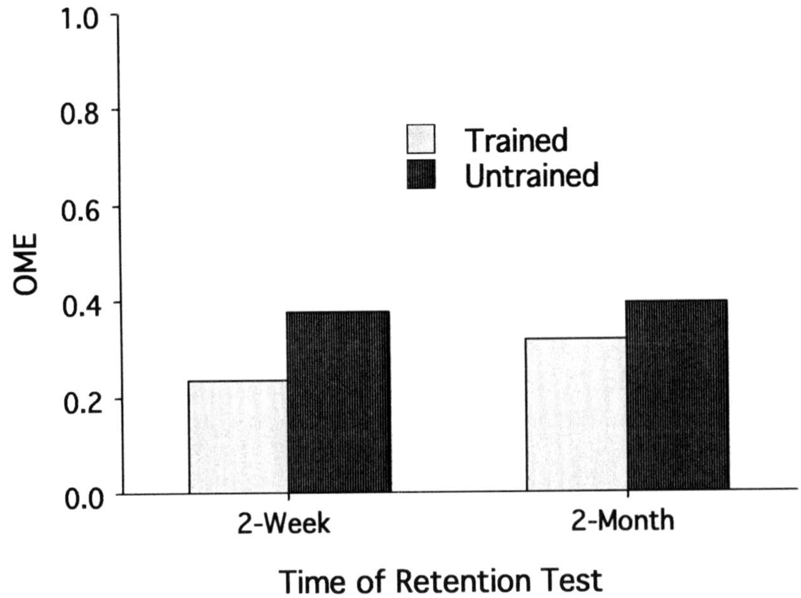

Figure 44. OME for the 8 subjects who returned at 2 months, as a function of time of retention test and training status in Experiment 1 (LaVoie et al., in press).

Our second experiment in this series (LaVoie et al., 1999, in press) investigated further the roles that specific and general knowledge may play in the seeding effect. However, instead of investigating the retention of seeding information, we focused our investigation on the use of seeding information while it is available. This experiment is similar to the first one with the subjects being pretested, given seeding information, and then posttested on different subsets of countries. There were six types of seeding information in this case, however, and the information was presented and left on the computer screen while the populations of the posttest countries were being estimated. In addition to six conditions determined by the type of information available, there was a control condition in which the subjects received no information at all, with 12 subjects in each condition.

Three of the experimental conditions were relevant to the explanation based on abstraction of general information. The subjects in these conditions were provided with either the mean population; the mean population and the range of populations; or the mean, the range, and a frequency distribution. The other three conditions were relevant to the explanation based on anchoring of specific information. The subjects in these conditions were shown either one country (the mean country in the seeding subset), with its name and population; the whole subset of 24 countries used in seeding, with their names and populations; or the seeding subset of 24 countries and their populations, but with no country names available to control for any real-world knowledge that the subjects could use.

As shown in Fig. 45, we found that in all six of the conditions in which subjects received additional information they showed improvement in estimations, as compared to the control condition in which they received no information. However, there were no differences in accuracy of estimations among the six experimental conditions. Hence, these results do not favor either the abstraction or the anchoring explanation, but they do indicate that presenting 24 countries with their names and populations is no better than presenting only a single population with no country name. This finding suggests that seeding may be accomplished with very minimal information.

In a third experiment designed to explore further the extent to which subjects' real-world knowledge influences seeding effects (Rutenberg, 1997), we examined the effects of seeding on 72 subjects' estimates of the radii of the 9 planets in the solar system. Subjects were given a pretest and then seeding information and a posttest. In the pretest and posttest, the names of each of the 9 planets in the solar system were listed in order of their distance from the sun, and subjects were asked simply to write the radius of each planet beside its name. The seeding information was provided on the same sheet as the posttest so it was immediately available to the subject and no memory for it was required. Three different between-subjects variables were crossed in our design: the number of seeds, seed specificity, and scale. Subjects were seeded with either 1 or 3 of the planets' radii; they were either given the names of the seeded planets or the names were withheld; and the seeding information and the estimates were either presented in a familiar scale (miles) or a less familiar scale (kilometers). As shown in Fig. 46, when we looked at the six planets not used in seeding, we found a large effect of seeding, so that posttest estimates were much more accurate than pretest estimates. Also, there was a marginally significant main effect of scale, with subjects' errors greater with the less familiar scale (kilometers) than with the more familiar scale (miles). However, the difference between the familiar and the unfamiliar scales was larger in the pretest than in the posttest after seeding. Furthermore, in agreement with our findings concerning seeding population estimates, there was no advantage on the posttest for providing three rather than just a single seed planet. Also in agreement with our findings concerning seeding population estimates, as shown in Fig. 47, for the planets not used in seeding there was no advantage on the posttest for providing the names of the seed planets, whether one or three planets were used in seeding. Thus, these experiments provide strong evidence that seeding information helps subjects simply by supplying them with information about the appropriate order of magnitude of their estimates and that the advantage of seeding does not depend in any way on subjects' real-world knowledge.

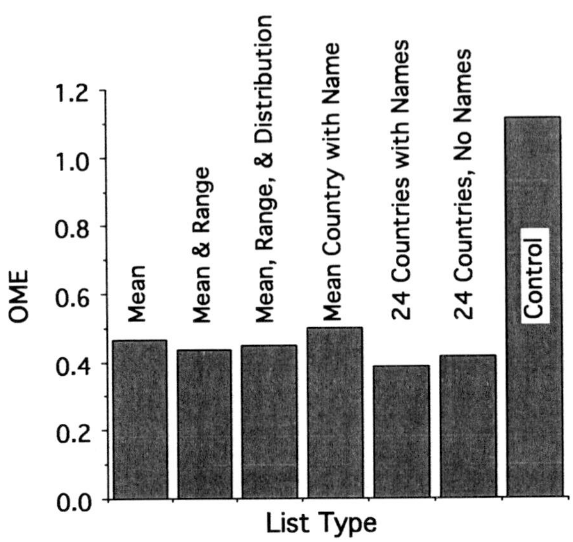

Figure 45. OME as a function of list type in Experiment 2 (LaVoie et al., in press).

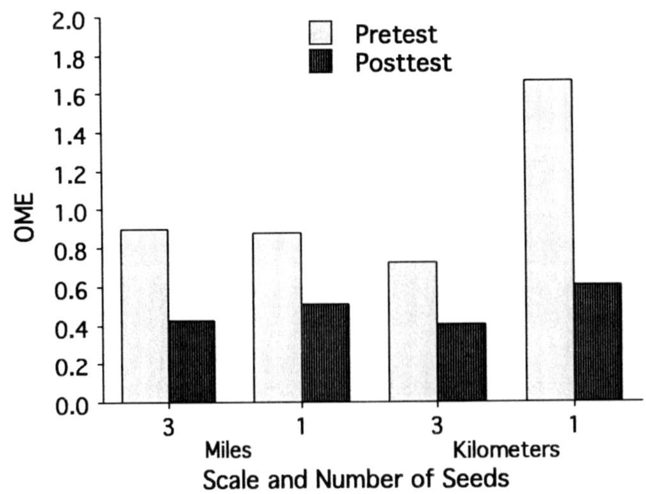

Figure 46. OME as a function of scale, number of seeds, and time of test in Experiment 3 (Rutenberg, 1997).

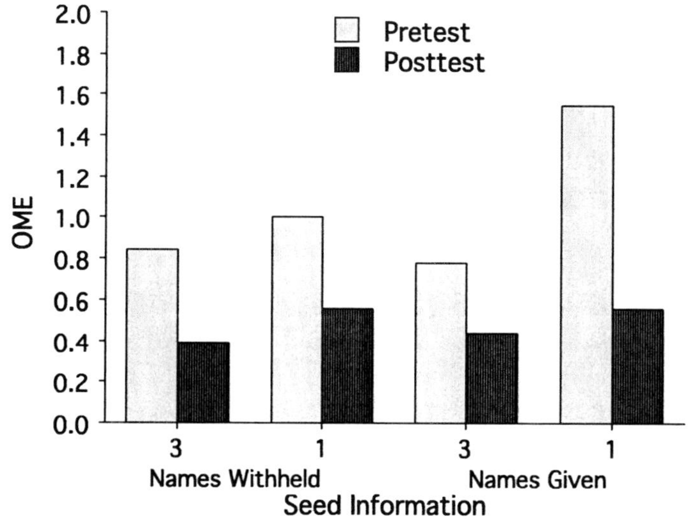

Figure 47. OME as a function of seed information and time of test in Experiment 3 (Rutenberg, 1997).

MANAGING RAPIDLY PRESENTED INFORMATION

In many communication situations, individuals are forced to perceive and comprehend large amounts of information in a relatively short period of time. Thus, the information is often presented at a faster than optimal rate. This problem is even more pronounced when the information is presented in a foreign language. Under these circumstances, even a normal conversational rate of speech can exceed the listener's ability to understand. One of the goals of our research is to develop training procedures to help individuals contend with rapid rates of information presentation.

Based on "above real speed" training used by the National Aeronautics and Space Administration (NASA) (Crane, Guckenberger, Schreiber, & Robbins, 1997), we conducted an experiment to test whether subjects could be trained to understand foreign language sentences presented at rapid rates (Bowles & Healy, 2000). Training at a speed faster than normal is inconsistent with most teaching practices (Derwing, 1990), because language instructors often try to help learners by speaking more slowly than normal. We hypothesized that learners would adapt to the speed of training, and that, at test, subjects trained with rapid (i.e., compressed) speech would find it easier to perceive foreign words spoken at normal conversational speed.

In our experiment, we tested subjects with no previous knowledge of Spanish and divided them into three training speed groups based on rate of stimulus presentation: slow (104 words per minute), medium (145 words per minute, which is approximately the normal conversational rate), or fast (203 words per minute). The stimuli were two- to four-word Spanish sentences pronounced by a native speaker and presented on a computer. The 48 stimulus sentences were created using words found in a first-year Spanish textbook. Example sentences are given in Fig. 48. The sentences were divided into four lists of 12. One base list was created (List 1) and from it three other lists were constructed. List 2 contained the same content words as List 1 rearranged into different sentences, with some sentences having different structures from those in List 1. List 3 was composed of sentences with the identical set of grammatical structures as those in List 1 but all new content words. List 4 was composed of sentences with the new set of content words from List 3 and the identical set of grammatical structures as used in List 2. Thus, each of the four lists was composed of sentences with the same content words as those in a second list but a different set of grammatical structures, the same set of grammatical structures as those in a third list but different content words, and both different content words and a different set of grammatical structures from those in a fourth list.

Each subject completed eight learning blocks for one of the lists of 12 sentences. Thus, each subject heard the same sentences repeated eight times. After hearing each sentence, the subject attempted to type it into the computer. The subject then received feedback on the display screen in the form of the correctly printed sentence.

Sample Sentences from Four Lists

List 1 (Base List):

Las mujeres cantan
Mis primos son tristes

List 2 (Same Content Words,
Different Set of Grammatical Structures):

Las mujeres son tristes
Mis primos cantan

List 3 (Same Grammatical Structures,
Different Content Words):

Las jovenes duermen
Mis nietos son fieles

List 4 (Different Content Words,
Different Set of Grammatical Structures):

Las jovenes son fieles
Mis nietos duermen

Figure 48. Sample sentences from the four different lists used (Bowles & Healy, 2000).

A test was given in the same session as training shortly after training was completed. During the test, each subject heard all of the 48 stimulus sentences from the four lists and attempted to type each of them into the computer. One third of the test sentences from each list was presented at the slow speed, one third at the medium speed, and one third at the fast speed. The assignment of speed to test sentences was counterbalanced so that across subjects in each training speed condition every sentence was heard equally often at each of the three testing speeds.

In summary, we examined the effects of two variables on performance during *training*: training speed (varied between subjects) and training block (varied within subjects). Further, we examined the effects of four variables on performance at *test*: training speed (varied between subjects), and testing speed, words, and structures (all varied within subjects). The variable of words compared sentences containing the same content words as those heard in training with sentences containing content words different from those heard in training. The variable of structures compared sentences containing the same grammatical structures as those heard in training with sentences containing different grammatical structures from those heard in training.

The results for training are summarized in Fig. 49 in terms of mean proportion correct as a function of training block and training speed group. It is clear that subjects in all three training speed groups showed large improvements throughout training, with better performance by the subjects in the slow and medium speed groups compared to those in the fast speed group. The difference between the three speed groups increased as training progressed; the interaction of training speed group and block was significant.

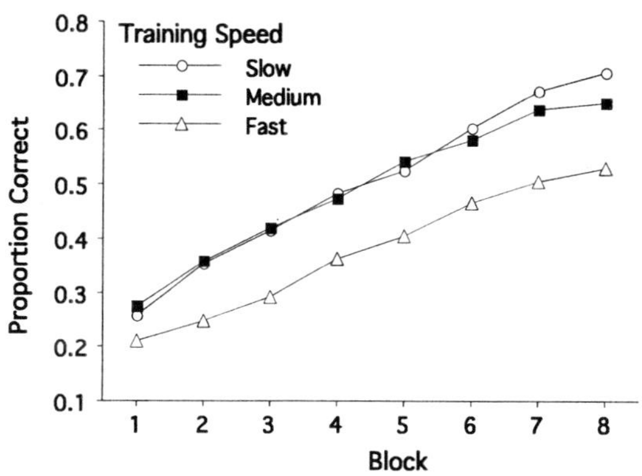

Figure 49. Proportion correct at training as a function of block and speed of presentation (Bowles & Healy, 2000).

Training speed group had a large impact on overall test accuracy. Subjects trained at the fast speed showed a significantly lower proportion of correct responses at test (.427) than did subjects trained at either the medium (.485) or slow (.496) speeds. Accuracy at test depended on test speed as well. Once again, subjects showed worse performance on the fast speed (.426) than on the other two speeds (medium = .490; slow = .491). The interaction of testing speed and training speed group was not significant. Nevertheless it is useful to examine this interaction, which is shown in Fig. 50, because it is clear that training with the fast speed did not aid test performance even on the fast test speed sentences.

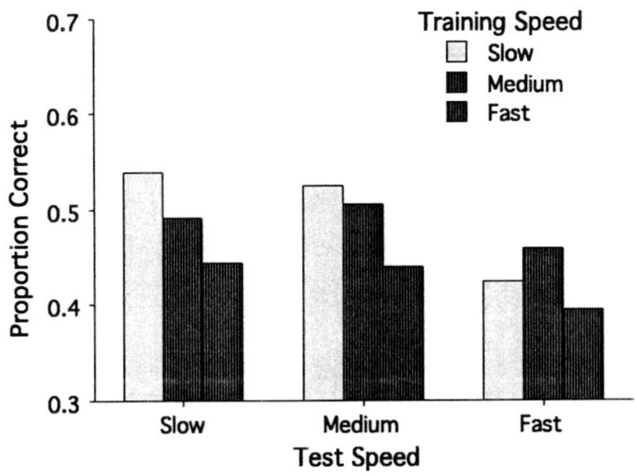

Figure 50. Proportion correct at test as a function of test speed and training speed (Bowles & Healy, 2000).

Accuracy at test also varied as a function of test list, as summarized in Fig. 51. Subjects performed best at test on the list used at training, but they also performed well on the list that had the same content words in different grammatical structures. In contrast, maintaining the same grammatical structures did not benefit performance when the content words were different from those used during training.

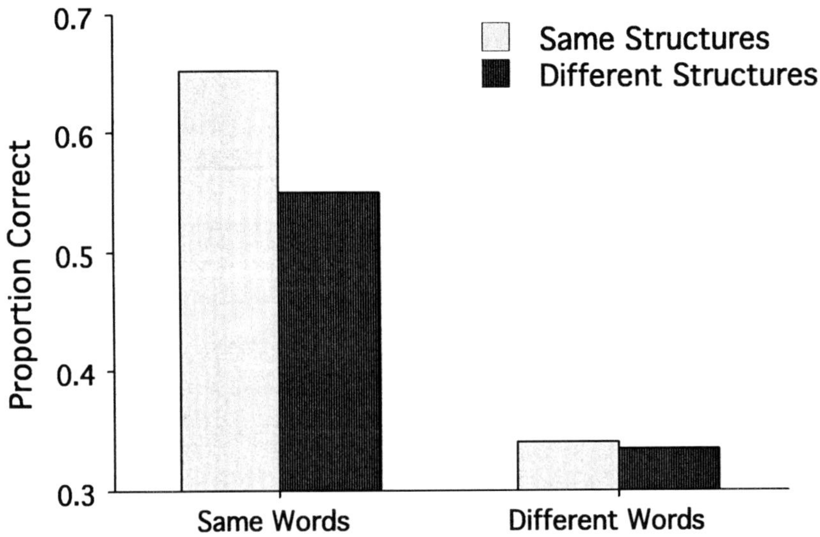

Figure 51. Proportion correct at test as a function of test list word similarity and test list structure similarity to training list (Bowles & Healy, 2000).

We expected that training at faster speeds would enhance performance on slower speeds. In fact, we found that subjects trained at the fastest speed showed the worst performance overall at test. These results suggest that training at fast speeds is not advisable, although training at speeds slower than normal is no better than training at the normal conversational speed. In addition, we found that subjects performed best at test on the sentences from the list given during training, and they performed better on the lists that included the same content words as those seen during training than on the lists that included new content words. However, even on the lists with new content words, subjects performed better at test than they had at the beginning of training (when the average proportion correct was less than .25). These results suggest that there were both specific and general effects of training.

In our current work, we are following up these findings by modifying our procedure in two important ways. First, all of the stimuli in our first experiment were originally produced by the native speaker at normal conversational speed with the intonation and pronunciation appropriate for that speed. In order to produce the sentences at the faster and slower speeds, we transformed the original sentences using a sound editing program on the computer either to compress or to expand the sentences in time without changing the pitch. This procedure may have provided an advantage to the medium speed because it was untransformed and, thus, perhaps more natural. For our second experiment, the native speaker was asked to produce all sentences at all three speeds, so that all sentences have a natural intonation and pronunciation, regardless of speed.

Second, because many of the errors reflected the subjects' unfamiliarity with the orthographic rules of Spanish rather than their ability to perceive the sounds, in our second experiment half of the subjects in each learning speed group are being pretrained on all the words that occur in their training list. Specifically, during pretraining, subjects see a list of Spanish words on a sheet of paper and are instructed to copy each word three times. Of interest will be the effect of pretraining on test performance and whether there is an interaction between pretraining and training speed. Perhaps we will be able to find circumstances in which there are advantages of high-speed training like that used by NASA.

CONCLUSIONS

Let us conclude by reviewing the issues we addressed, our findings concerning these issues, their implications for training, and some of our current and planned follow-up investigations.

The first issue concerned variations in contextual interference as a means to promote durability and generalizability of knowledge and skills. Our findings in the first two experiments were in accordance with predictions based on Battig's (1972, 1979) contextual interference principle. They imply that blocking material by category, which is frequently done in many types of training, may aid initial acquisition but may not yield optimal retention. If durable retention is the goal, contextual interference during training is advisable. In the third experiment, we found that increasing task difficulty during study aided transfer to a new task as well as long-term retention of the original task. The implication for training is that enhancing the difficulty of training exercises will lead to improved retention and transfer. In our future research, we plan to explore this phenomenon further by examining different methods for enhancing task difficulty during training.

The second issue concerned communication about navigating in space. We found that subjects' ability to repeat back and follow navigational instructions is affected by the length of the message and two aspects of the subjects' mental representation of the space through which they are navigating. First, there was a decrement in performance when the subjects had to move outside the picture plane. Second, there was a decrement in performance when the space was described as involving three dimensions. The implication for training is that additional practice should be given whenever a navigational task requires a three-dimensional mental representation, rather than simply a two-dimensional representation, or whenever the task involves movement along a plane not immediately shown. Our results imply further that three-dimensional displays, such as those found in virtual reality environments, are not sufficient to eliminate the need for this additional practice.

The third issue we addressed concerned response execution in a spatial array. We found that the simple task of response execution improved considerably with training, that both the identity and the location of the target affected speed of response, and that the two component response time measures—initiation and movement time—showed quite different patterns of results. These different patterns force us to conclude that there are two distinct processing stages involved in simple response execution. The implications are that the two components of response execution can be affected differently by training variables and may differ in their durability over time and generalizability over transfer conditions. Hence, any program that aims to maximize the efficiency of response execution must be sensitive to possible differential effects on these two components.

The fourth issue concerned the underlying causes of durability and specificity in digital data entry. We found a superiority of training with words even when the word format at training was faster than the numeral format because subjects were required to type the initial letter of the numbers rather than the digits. We conclude that the superiority for training with words is due to the fact that the word format at training leads to greater processing than does the numeral format. The implication for training is that using a format that forces greater processing should be encouraged whenever possible to promote durability and transferability.

The fifth issue we addressed concerned the relationship between the learning of examples and the learning of rules in conceptual tasks. We found strong support for the rote-and-analogy paradigm, as opposed to the rule-and-category paradigm, because we found that individuals' knowledge of the definite article pronunciation (/D´/ versus /Di/) distinction, unlike their knowledge of the indefinite article (*a*/*an*) distinction, does not conform exactly to the simple, general abstract rule included in the dictionary, but rather is more complex and involves information at various levels of abstraction. The implication for training is that individuals do learn complex rules involving different levels of abstraction when given sufficient specific examples but that they also benefit from explicit visual cues, reinforcement, and instruction concerning abstract, general rules. In our future research, we plan to examine the learning of novel linguistic rules so that we can develop general principles concerning the factors influencing the generalizability of known categories to new ones.

The sixth issue concerned types of strategies and strategy transitions in category learning tasks. We found that monotonic changes in overall measures of performance during practice masked the complexity of changes in the learner's basis of response or strategies employed. The implications for training are that instructors should monitor not only overall measures of performance but also the strategy employed by the trainee. When the most effective strategies are known, instructors would be advised to adopt procedures that can effectively bring these strategies forward earlier than usual in the training process. In our planned follow-up research, we seek to determine the role of prior knowledge in strategy selection.

The seventh issue we addressed concerned the durability and use of general and specific knowledge in quantitative estimation. We found that providing seeding facts, or training with a very small amount of general or specific information, led to large improvements in subjects' quantitative estimation skills. The implications for training is that seeding should be used wherever possible when quantitative estimation is required. In our future experiments, we plan to assess retention of quantitative estimation skills by exploring further the durability of improvements due to seeding. We also intend to examine the effects of seeding in other types of skills.

The eighth issue we addressed concerned managing rapidly presented information. We found that subjects with no prior knowledge who trained at the fastest speed showed worse performance than those who trained at normal or slow speeds. There was no difference between the normal and slow speed groups. The implication is that high-speed training is not recommended for language learning at least for introductory students. More generally, these results imply that training with normal speeds leads to performance at least as good as training with slow speeds and better than training with fast speeds, even when testing is done with a fast speed.

REFERENCES

Barshi, I. (1997). *Effects of linguistic properties and message length on misunderstandings in aviation communication.* Unpublished doctoral dissertation, Department of Linguistics, University of Colorado, Boulder.

Barshi, I. (1998). *The effects of mental representation on performance in a navigation task.* Unpublished doctoral dissertation, Department of Psychology, University of Colorado, Boulder.

Barshi, I., & Healy, A. F. (1997, November). *The cost of mentally representing three versus two spatial dimensions.* Paper presented at the 38th Annual Meeting of the Psychonomic Society, Philadelphia, PA.

Barshi, I., & Healy, A. F. (1998). Misunderstandings in voice communication: Effects of fluency in a second language. In A. F. Healy & L. E. Bourne, Jr. (Eds.), *Foreign language learning: Psycholinguistic studies on training and retention* (pp. 161-192). Mahwah, NJ: Erlbaum.

Barshi, I., & Healy, A. F. (2001). *The effects of mental representation on performance in a navigation task.* Manuscript submitted for publication.

Battig, W. F. (1972). Intratask interference as a source of facilitation in transfer and retention. In R. F. Thompson & J. F. Voss (Eds.), *Topics in learning and performance* (pp. 131-159). New York: Academic Press.

Battig, W. F. (1979). The flexibility of human memory. In L. S. Cermak & F. I. M. Craik (Eds.), *Levels of processing in human memory* (pp. 23-44). Hillsdale, NJ: Erlbaum.

Bourne, L. E., Jr., Healy, A. F., Parker, J. T., & Rickard, T. C. (1999). The strategic basis of performance in binary classification tasks: Strategy choices and strategy transitions. *Journal of Memory and Language, 41,* 223-252.

Bourne, L. E., Jr., Healy, A. F., Pauli, P., Parker, J. T., & Birbaumer, N. (2001). *The influence of stimulus array on training and transfer of a speeded response*. Manuscript submitted for publication.

Bowles, A. R., & Healy, A. F. (2000). Training and transfer of foreign word identification at three speeds. *Proceedings of the Twenty-Second Annual Conference of the Cognitive Science Society* (p. 1016). Mahwah, NJ: Erlbaum.

Brown, N. R., & Siegler, R. S. (1996). Long-term benefits of seeding the knowledge base. *Psychonomic Bulletin & Review, 3,* 385-388.

Buck-Gengler, C. J., & Healy, A. F. (2001). Processes underlying long-term repetition priming in data entry. *Journal of Experimental Psychology: Learning, Memory, and Cognition, 27,* 879-888.

Crane, P. M., Guckenberger, D., Schreiber, B. T., & Robbins, R. L. (1997). *Above real-time training applied to air combat skills*. Mesa, AZ: Hughes Training. (NTIS No. AD-A329 018/6INT).

Derwing, T. M. (1990). Speech rate is no simple matter: Rate adjustment and NS-NNS communicative success. *Studies in Second Language Acquisition, 12,* 303-313.

Fendrich, D. W., Gesi, A. T., Healy, A. F., & Bourne, L. E., Jr. (1995). The contribution of procedural reinstatement to implicit and explicit memory effects in a motor task. In A. F. Healy & L. E. Bourne, Jr. (Eds.), *Learning and memory of knowledge and skills: Durability and specificity* (pp. 66-94). Thousand Oaks, CA: Sage.

Fendrich, D. W., Healy, A. F., & Bourne, L. E., Jr. (1991). Long-term repetition effects for motoric and perceptual procedures. *Journal of Experimental Psychology: Learning, Memory, and Cognition, 17,* 137-151.

Healy, A. F. (1998). Optimizing the speed, durability, and transferability of training. *Proceedings of the 21st Army Science Conference: Science and technology for army after next* (pp. 445-454). Arlington, VA: Office of the Assistant Secretary of the Army.

Healy, A. F., Barshi, I., Crutcher, R. J., Tao, L., Rickard, T. C., Marmie, W. R., Schneider, V. I., Feldman, A., Buck-Gengler, C. J., Romero, S. G., Sherrod, N. B., Parker, J. T., & Bourne, L. E., Jr. (1998). Toward the improvement of training in foreign languages. In A. F. Healy & L. E. Bourne, Jr. (Eds.), *Foreign language learning: Psycholinguistic studies on training and retention* (pp. 3-53). Mahwah, NJ: Erlbaum.

Healy, A. F., & Bourne, L. E., Jr. (Eds.) (1995). *Learning and memory of knowledge and skills: Durability and specificity*. Thousand Oaks, CA: Sage.

Healy, A. F., & Bourne, L. E., Jr. (Eds.) (1998). *Foreign language learning: Psycholinguistic studies on training and retention*. Mahwah, NJ: Erlbaum.

Healy, A. F., Clawson, D. M., McNamara, D. S., Marmie, W. R., Schneider, V. I., Rickard, T. C., Crutcher, R. J., King, C. L., Ericsson, K. A., & Bourne, L. E., Jr. (1993). The long-term retention of knowledge and skills. In D. L. Medin (Ed.), *The psychology of learning and motivation: Advances in research and theory* (Vol. 30, pp. 135-164). New York: Academic Press.

Healy, A. F., Fendrich, D. W., Crutcher, R. J., Wittman, W. T., Gesi, A. T., Ericsson, K. A., & Bourne, L. E., Jr. (1992). The long-term retention of skills. In A. F. Healy, S. M. Kosslyn, & R. M. Shiffrin (Eds.), *From learning processes to cognitive processes: Essays in honor of William K. Estes, Volume 2* (pp. 87-118). Hillsdale, NJ: Erlbaum.

Healy, A. F., King, C. L., Clawson, D. M., Sinclair, G. P., Rickard, T. C., Crutcher, R. J., Ericsson, K. A., & Bourne, L. E., Jr. (1995). Optimizing the long-term retention of skills.

In A. F. Healy & L. E. Bourne, Jr. (Eds.), *Learning and memory of knowledge and skills: Durability and specificity* (pp. 1-29). Thousand Oaks, CA: Sage.

Healy, A. F., Parker, J. T., & Barshi, I. (2001). *The effects of mental representation on memory for verbal instructions*. Manuscript in preparation.

Healy, A. F., & Sherrod, N. (1994, November). *The/Thee pronunciation distinction: A local model of linguistic categories*. Paper presented at the 35th Annual Meeting of the Psychonomic Society, St. Louis, MO.

Homa, D., Sterling, S., & Trepel, L. (1981). Limitations of exemplar-based generalization and the abstraction of categorical information. *Journal of Experimental Psychology: Human Learning and Memory, 7,* 418-439.

Kosslyn, S. M. (1980). *Image and mind*. Cambridge, MA: Harvard University Press.

LaVoie, N., Bourne, L. E., Jr., & Healy, A. F. (1999, November). *Memory seeding: Representations underlying quantitative estimation*. Poster presented at the 40th Annual Meeting of the Psychonomic Society, Los Angeles, CA.

LaVoie, N. N., Bourne, L. E., Jr., & Healy, A. F. (in press). Memory seeding: Representations underlying quantitative estimations. *Journal of Experimental Psychology: Learning, Memory and Cognition*.

Pauli, P., Lutzenberger, W., Rau, H., Birbaumer, N., Rickard, T. C., Yaroush, R. A., & Bourne, L. E., Jr. (1994). Brain potentials during mental arithmetic: Effects of extensive practice and problem difficulty. *Cognitive Brain Research, 2,* 21-29.

Raymond, W. D., Fisher, J. A., & Healy, A. F. (in press). Linguistic knowledge and language performance in English article variant preference. *Language and Cognitive Processes*.

Raymond, W. D., Fisher, J. A., Healy, A. F., & Bourne, L. E., Jr. (1999, January). *Explaining language performance through rule interaction: English article variation in production and perception*. Paper presented at the Annual Meeting of the Linguistic Society of America, Los Angeles, CA.

Rickard, T. C. (1997). Bending the power law: A CMPL theory of strategy shifts and the automatization of cognitive skills. *Journal of Experimental Psychology: General, 126,* 288-311.

Rutenberg, B. D. (1997). *Immediate effects on seeding the knowledge base: The effects of seed number, seed specificity, and reference frame familiarity*. Unpublished senior thesis, University of Colorado, Boulder.

Schneider, V. I., Healy, A. F., & Bourne, L. E., Jr. (1998). Contextual interference effects in foreign language vocabulary acquisition and retention. In A. F. Healy & L. E. Bourne, Jr. (Eds.), *Foreign language learning: Psycholinguistic studies on training and retention* (pp. 77-90). Mahwah, NJ: Erlbaum.

Schneider, V. I., Healy, A. F., & Bourne, L. E., Jr. (in press). What is learned under difficult conditions is hard to forget: Contextual interference effects in foreign vocabulary acquisition, retention, and transfer. *Journal of Memory and Language*.

Schneider, V. I., Healy, A. F., Ericsson, K. A., & Bourne, L. E., Jr. (1995). The effects of contextual interference on the acquisition and retention of logical rules. In A. F. Healy & L. E. Bourne, Jr. (Eds.), *Learning and memory of knowledge and skills: Durability and specificity* (pp. 95-131). Thousand Oaks, CA: Sage.

Serwatka, M., & Healy, A. F. (1998). On the status of the count-mass distinction in a mental grammar. In A. F. Healy & L. E. Bourne, Jr. (Eds.), *Foreign language learning: Psycholinguistic studies on training and retention* (pp. 113-158). Mahwah, NJ: Erlbaum.

AUTHOR NOTES

Alice F. Healy, Department of Psychology; Carolyn J. Buck-Gengler, Department of Psychology; Immanuel Barshi, Department of Psychology (now at NASA Ames Research Center); James T. Parker, Department of Psychology (now at Knowledge Analysis Technologies, Boulder, Colorado); Vivian I. Schneider, Department of Psychology; William D. Raymond, Department of Linguistics (now at the Department of Linguistics, The Ohio State University); N. Noelle LaVoie, Department of Psychology; Anita R. Bowles, Department of Psychology; Paul Pauli, Institute of Medical Psychology and Behavioral Neurobiology (now at University of Würzburg, Germany); Julia A. Fisher, Department of Linguistics (now at Department of Psychology, University of Colorado at Denver); and Lyle E. Bourne, Jr., Department of Psychology.

This research was supported in part by Army Research Institute Contracts DASW01-96-K-0010 and DASW01-99-K-0002 (Alice F. Healy, Principal Investigator; Lyle E. Bourne, Jr., Co-Principal Investigator) and Army Research Office Grants DAAH04-95-1-0208 and DAAG55-98-1-0214 to the University of Colorado (Alice F. Healy, Principal Investigator).

Correspondence concerning this chapter should be addressed to Alice F. Healy, Department of Psychology, 345 UCB, University of Colorado, Boulder, CO 80309-0345 or electronically to ahealy@psych.colorado.edu.

Chapter 8

FOVEAL STIMULUS DEGRADATION DURING SCENE PERCEPTION

Paul M. J. van Diepen*

Laboratory of Experimental Psychology
University of Leuven, Belgium

ABSTRACT

Two experiments are reported using eye-contingent display changes during scene perception. Stimulus information was degraded within an ovoid window that was aligned with the fixation position, while outside of the window, an undegraded line drawing of a real-world scene was visible. Degradation types included: full masking by noise, partial masking by noise, and reduction of the foveal image contrast. Image degradation occurred after a preset delay from the onset of each fixation, ranging from 0 to 105 ms. In a control condition, the foveal ovoid was outlined by a red ellipse. It was shown that sufficient information could be extracted during the initial part of fixations, to accurately identify foveated objects. When the degradation occurred early after the fixation onset, fixations were sustained (with partial masking and contrast reduction), or repeated (with complete masking). Saccadic amplitudes, however, were always affected by foveal image degradation, regardless of the degradation delay, indicating that saccade target selection occurs later during fixations. The results support a local-to-global sequence of information processing during scene perception.

While exploring everyday life scenes, eye movements are characterized by an alternation of fixations and saccades. During fixations, the eye position is relatively stable, and information is extracted from the distal stimulus. Since retinal acuity decreases rapidly away from the fovea (e.g., De Valois & De Valois 1988), fine details can only be acquired from an area around the fixation position (i.e., the part of the stimulus that is projected on the high-acuity fovea). Saccades are quasi-ballistic eye movements that bring parts of the stimulus to the fovea. During saccades, visual resolution diminishes and sensitivity to stimulus changes

---

* Correspondence concerning this article should be addressed to Prof. Géry d'Ydewalle, Laboratory of Experimental Psychology, University of Leuven, Tiensestraat 102, B-3000 Leuven, Belgium.

decreases considerably (Ishida & Ikeda, 1989; Matin, 1974); therefore, no visual guidance is possible. Still, saccades are not directed to arbitrary stimulus locations: In scene perception, eye movements typically follow an object-to-object scanning pattern (Buswell, 1935; Henderson & Hollingworth, 1998; Yarbus, 1967), while in reading, fixations rarely occur on spaces between words (e.g., Rayner, 1979). Evidently, the destination of a saccade is determined during the preceding fixations. Saccadic amplitudes range from very small (directed to another part of the currently fixated object or word), to very large (crossing the entire scene or text). Apparently, both foveal and peripheral stimulus information is available to select potentially interesting stimulus locations.

In view of the above, two visual processes can be distinguished that occur during eye fixations. First, the foveated stimulus is analyzed in detail, and identified if possible. Second, the entire stimulus is analyzed at a rudimentary level, to select a sensible saccade target (i.e., a stimulus area that will be fixated next). Since information acquisition ends at the beginning of a saccade, extraction of information for the foveal identification process and the saccade target selection process must occur before the end of the fixation. Obviously, the actual saccade target selection and the associated oculomotor preparation have to be completed during fixations as well. Van Duren (1994) showed that some processes, such as integration of newly acquired information in the scene representation, and response selection, may continue during saccades. It is not clear, however, whether foveal identification can continue during saccades as well, or if it is bounded to the fixation interval.

The purpose of the present paper is twofold. First, it will be shown that image information for foveal identification is extracted during a relatively short period at the beginning of fixations. Second, it will be shown that fixation durations are sensitive to fovea image degradation, *only* during that period.

FOVEAL VERSUS PERIPHERAL INFORMATION PROCESSING

In this paper, the following model of information processing is adopted: For foveal stimulus identification, only the foveal image has to be available. During unrestricted scene exploration, relatively few intra-object saccades are observed. Hence, in most occasions, the saccade target selection process will exclusively use the peripheral image. Only when the foveal identification did not succeed, a new fixation on the same object is required, and foveal information is used for the saccade target selection. Consistent with several process models of eye movement control (e.g., Henderson's *sequential attention model*, 1992), it is hypothesized that a fixation starts with the identification of the foveal stimulus. Saccade target selection is initiated later during fixations, when the foveal identification process reaches a criterion of completion, or when a completion deadline is reached. According to this model, foveal information is required from the beginning of fixations, while peripheral information only has to be available later during fixations. Furthermore, given that relatively few intra-object saccades occur, absence of foveal information during the later part of fixations should hardly affect scene exploration. Hence, information is presumed to be extracted in a local-to-global sequence[1].

[1] In many studies, the terms "local" and "global" imply "fine" and "coarse", or "high and low spatial frequencies". While for the human visual system this might be correct (but see van Diepen, Wampers, & d'Ydewalle, 1998), in this paper they simply mean "foveal" and "foveal + peripheral", respectively.

The model above assumes that, generally, foveation is necessary to identify objects in scenes. Hence, multiple fixations are required to inspect a scene. Tachistoscopic studies, however, have shown that scene identity, as well as the identity of objects in scenes, can be determined from briefly presented stimuli (e.g., Antes, Penland, & Metzger, 1981; Biederman, 1981). In these "speeded object-verification tasks", typically an object name is presented, followed by the presentation of a scene that is masked after 100 to 200 ms. The mask contains a location cue, and it has been shown that participants often are able to decide if the named object has been present in the scene on that location. The findings suggest that both foveal and peripheral information can be extracted, with a high level of detail, within half the regular fixation duration. It has been argued, however, that attentional strategies and post-perceptual guessing explain at least part of the results obtained with speeded object-verification paradigms (Boyce & Pollatsek, 1992; De Graef, 1998). Probably a different level of information processing is addressed in tachistoscopic experiments, by attending to coarse aspects of the global stimulus. In most free-exploration studies, local stimulus detail is of greater importance, and object foveation is a necessity.

Shioiri (1993) studied recovery from saccadic suppression in scene perception. Participants explored photographs of natural scenes, in preparation for a recognition task. Eye movements were recorded, and during the initial 25, 50, 100, or 200 ms of each fixation, images were blanked or blurred. During the later part of fixations, the undegraded stimulus was visible again. In a control condition, no stimulus degradation took place. Fixation durations increased with approximately the duration of the degradation interval, relative to the control condition, both with blanking and blurring. Interestingly, presence of coarse stimulus information in the blurred condition did not improve performance on a subsequent recognition task, compared with the blanked condition. Apparently, high-resolution information is required even during the initial 25 ms of fixations, jeopardizing a coarse-to-fine hypothesis. However, since in this study the image manipulation affected the entire stimulus, no firm conclusions can be drawn with respect to foveal versus peripheral processing. The finding that the early presence of coarse information does not improve picture perception suggests that foveal encoding starts before peripheral processing. Note that here it is presumed that peripheral processes require coarse rather than fine information. Wampers, van Diepen, and d'Ydewalle (1997), however, showed that during scene exploration, fine peripheral information is more useful than coarse information (also see van Diepen, Wampers, & d'Ydewalle, 1998). Evidence against coarse-to-fine processing, therefore, does not automatically exclude the possibility of a global-to-local processing sequence.

Moving Mask and Moving Window Paradigms in Reading

Eye-contingent display change techniques have been used to manipulate part of a stimulus online, as a function of eye movement parameters. To unravel foveal and peripheral image acquisition, the so-called *moving mask* and *moving window* paradigms have been developed (McConkie & Rayner, 1975). Eye movements are monitored to manipulate parts of an image, depending on the concurrent fixation position and duration. In moving mask experiments, a small area around the fixation position (i.e., the foveal image) is replaced by a mask, while the remaining (peripheral) part of the stimulus remains visible. By delaying the mask appearance by a preset value from the beginning of each fixation, it is possible to estimate the required length of the foveal information extraction interval. Moving window studies manipulate peripheral, rather than foveal information. Within a window that is aligned with the fixation position, the undegraded stimulus is visible, while outside of the window the image is masked. By varying the size of the window, it can be determined how far from the fixation position useful information can be extracted from the stimulus.

Eye-contingent display changes have been employed frequently to investigate the time course of information extraction in reading. Below, I will briefly discuss two representative studies. For a more extensive overview, see Rayner (1998).

Rayner, Inhoff, Morrison, Slowiaczek, and Bertera (1981) studied foveal and peripheral information extraction in reading. Participants had to read individual sentences, displayed on a computer screen, to understand them. Foveal and/or peripheral text was masked during preset periods of fixations. It was shown that foveal vision is required for the identification of words. Peripheral information seems to be utilized as well, for other than semantic purposes. In one experiment, foveal text was masked after the onset of each fixation. Reading rate, fixation durations, and saccadic amplitudes asymptoted to levels that were observed in the control condition, when the mask delay exceeded 50 ms. The results indicate that for reading, sufficient foveal information can be acquired during the first 50 ms of fixations (see also Ishida & Ikeda, 1989). Peripheral, and full-line masking decreased reading performance, even at the longest, 150-ms delay. Apparently, peripheral information either has to be available *longer* than foveal information, or is required only *later* during fixations.

The finding that sufficient foveal information can be obtained during the initial 50 ms of fixations does not necessarily imply that foveal processing terminates thereafter. Blanchard, McConkie, Zola, and Wolverton (1984) indicated that foveal processing continues until the end of fixations. Participants had to read a sentence, while eye movements were being monitored. After a preset delay of 50, 80, or 120 ms from the beginning of each fixation, the sentence was masked for 30 ms, after which it appeared again. Each sentence contained a target word, that was replaced by a second target word after the mask presentation. The second target word differed in a single letter from the first target word. Both targets were consistent with the sentence. At the end of a fixation, the second target word was replaced by the original target word again. Participants often were able to identify both target words. When the mask delay increased, the probability of identifying the first and second target word increased and decreased, respectively. These findings clearly show that foveal information processing continues during the later part of fixations.

Scene Perception Research

To what extend can results from reading experiments be generalized to scene perception? Both scene perception and reading depend on the same visual system. With respect to low-level aspects such as information extraction, no differences might be expected. On the other hand, many differences can be pointed out that potentially disturb the comparability between scene exploration and reading. Regarding the stimulus, text is a densely packed image, following strict rules. An English text is written from left to right, and from top to bottom. Spaces mark the beginning of a new word, and words are not printed over each other. While fixating on a word, the next word is visible in parafoveal vision. Scenes can be thought of as consisting of a background, outlining the general setting, and a number of objects. Usually, considerably less objects can be found in a scene, than words in a text. The average distance between objects is larger than between adjacent words, but can also be so small that one object occludes another. Furthermore, no predictable saccadic pathway exists for scenes (e.g., see Mannan, Ruddock, & Wooding, 1996).

A clear illustration of the divergence between reading and scene perception is offered by research considering the span of effective vision. While reading English text, it has been shown that most information is extracted from four letters to the left of the fixation position, up to fourteen-fifteen letters to the right (McConkie & Rayner, 1976; Rayner et al., 1981). Text below or above the fixated line is hardly required (Pollatsek, Raney, LaGasse & Rayner, 1993). In scene perception, however, information is acquired from a window covering half the stimulus area, centered on the point of fixation (Saida & Ikeda, 1979).

Van Diepen, De Graef, and d'Ydewalle (1995) employed the moving mask paradigm to estimate the length of the foveal information extraction interval at the beginning of fixations during scene exploration. Participants explored scenes in the context of an *object-decision task* (see below), until they pressed a response button, indicating that they had completely explored the scene. Eye movements were registered to center an ovoid noise mask on the concurrent fixation position. Masks appeared on the fixation position after a delay of 15, 45, 75, or 120 ms from the beginning of fixations. In a control condition, no masking occurred. As in reading (Rayner et al., 1981), it was found that the scene inspection time asymptoted to the control level, when foveal masks were delayed by 45-75 ms or longer. For scene exploration, sufficient foveal information could be extracted during the initial part of each fixation.

A difference with Rayner et al. (1981) was that fixation durations did not asymptote to a control level, when the mask delay was increased. On average, fixation durations increased about equally in all mask conditions relative to the control condition. Fixation duration distributions revealed a time-locked effect of the *appearance* of the mask, comparable with the effect of a sudden text change, as described in McConkie, Underwood, Zola, and Wolverton (1985).

Rayner et al. (1981) argued that in their moving mask experiments, longer fixation durations were caused by cognitive factors. Specifically, early masking resulted in a degraded representation of the fixated stimulus, and increased fixation durations presumably reflected participant's increased activity and difficulty in identifying the foveal stimulus. Conceivably, in an object-decision task, it is harder to cognitively compensate for the lack of foveal stimulus detail: Word identification is aided by textual constraints, such as syntax and semantic context. Scene context aids object identification too, but in the object-decision task, only a sufficient amount of stimulus detail will result in an accurate verdict regarding the existence of objects. Maintaining fixations on completely masked objects is senseless, since longer fixations will not yield more stimulus detail. Therefore, foveal masking during scene perception could even be expected to *decrease* the duration of fixations on scenes, as the mask delay shortens.

In another experiment, van Diepen and Wampers (1998) manipulated the spatial frequency content of the peripheral image, while participants explored full-color, computer-generated images of real-world scenes. During the initial 150 ms of fixations, the peripheral image was low-pass, high-pass, or bandpass filtered, or blanked with the average scene color. During the later part, the image was undegraded. The foveal image was never degraded. In a control condition, no image degradation occurred. No reliable differences were found between the filter and blanking conditions, which could be interpreted as evidence for the hypothesis that peripheral information is not utilized during the initial part of fixations. However, in all degradation conditions, fixation durations and scene inspection times increased significantly, relative to the control condition. This effect could not be (exclusively) attributed to the massive stimulus change during fixations in the degradation conditions: In a second control condition, the luminance of the peripheral image was decreased by 50% during the initial part of fixations, inducing a stimulus change without changing the spatial frequency content. This luminance change did not significantly affect scene inspection time or fixation durations. Van Diepen and Wampers (1998) concluded that, while peripheral information can be extracted early during fixations, it is of minor importance for normal scene exploration.

THE PRESENT STUDY

In reading and free scene exploration, the hypothesis that information is acquired in a local-to-global sequence seems to be confirmed. One goal of the present study is to consolidate the conclusion that sufficient foveal information can be extracted during a short interval at the beginning of fixations.

A second, but not secondary goal is to show that fixation durations increase when the foveal information is degraded, but not completely removed. Fixation durations are commonly used as a measure of stimulus processing difficulty. For example, when participants explore real-world scenes, longer fixation durations are observed on objects that are inconsistent with the scene context. This finding is typically explained as reflecting shorter identification times for consistent objects, due to contextual facilitation of stimulus processing (e.g., De Graef, Christiaens, & d'Ydewalle, 1990). Other studies have shown that fixation durations increase when the stimulus contrast decreases, which has been explained in terms of a decreased information acquisition rate (e.g., Busey & Loftus, 1994; Loftus, Kaufman, Nishimoto, & Ruthruff, 1992).

Still, van Diepen et al. (1995) found no decrease in fixation durations when the foveal mask delay increased (see above). Scene inspection time, and correspondingly the number of fixations on a scene, did decrease as the mask delay increased. Foveal masking completely terminates the foveal identification process, which makes it necessary to increase the number of fixations on a given scene. In the present study it will be shown that, only when foveal information is completely masked during the early part of fixations, fixations are repeated. When foveal information is degraded, but not completely masked, fixations are sustained, instead of repeated. Degradation or complete masking during the later part of fixations will not affect fixation durations, or the number of fixations.

EXPERIMENT 1

Participants explored line drawings of real-world scenes in the context of the *object-decision task* (De Graef et al., 1990): Scenes could contain a number of meaningless, object-like volumes. Participants had to distinguish between these so-called *non-objects* and real objects as accurately and fast as possible. This task ensured that objects had to be processed sufficiently to identify them as existing objects, without the need for further processing, (e.g., to name the object). Stimuli were presented until participants pressed a response button, indicating that they had completely explored the scene. After each stimulus exploration, participants had to indicate how many non-objects had been present in the scene.

During the scene presentation, eye movements were registered to center an ovoid "manipulation area" on the concurrent fixation position. At the beginning of fixations, the undegraded scene was presented. After a preset delay from the fixation onset, the foveal image inside of the manipulation area was replaced by a degraded version of that image. Three degradation conditions were used in Experiment 1: In the *full-noise* condition, all pixels in the manipulation area were replaced by pixels with a randomly chosen grey-level. The full-noise condition completely masked the foveal image, as in van Diepen et al. (1995). In the *partial-noise* condition, only 40% of the foveal pixels were replaced by random grey levels, rendering the foveal stimulus partly visible. In the *low-contrast* condition, foveal image contrast was diminished by a factor of seven. The later two conditions presumably would slow foveal information processing, without completely terminating it, as in the full-noise condition. Consequently, fixations should be sustained, instead of repeated.

In a control condition, the image inside of the manipulation area was not degraded, but outlined by a red ellipse (along the border of the manipulation area). A pilot experiment had shown that the sudden appearance of the red ellipse induced a *stimulus-discontinuity effect* on fixation durations, as did the degradation conditions, without further disturbing scene perception: The sudden change of the foveal image sustained a number of fixations, resulting in a dip in the frequency distribution of fixation durations, and consequently increasing the average fixation duration. Therefore, the red-ellipse condition was considered a better control than the no-change control that was used in van Diepen et al. (1995).

According to the local-to-global processing model, foveal image degradation during the later part of fixations should not affect the foveal identification process. It might affect saccade target selection, though, since this process is presumed to occur later during fixations. Conversely, foveal degradation during the early part of fixations should affect foveal identification. In the full-noise condition, more fixations will be required to inspect a scene. In the low-contrast and partial-noise conditions, longer fixations can be used to deal with early image degradation.

Hence, in the 105-ms condition, image degradations were predicted not to affect fixation durations and scene inspection times. In the 15-ms full-noise condition, more fixations per scene, and correspondingly longer scene inspection times were predicted. In the 15-ms partial-noise and low-contrast conditions, longer fixation durations were expected. This potentially would increase the scene inspection time, but to a lesser extent than in the full-noise condition (assuming that repeating fixations is more time-consuming than sustaining a fixation). With respect to saccadic amplitudes, fewer small saccades were predicted in all degradation conditions, irrespective of the degradation delay, resulting in a larger average amplitude.

Method

Stimuli

Sixteen target scenes were selected from a library of black-on-white line drawings of real-world scenes (van Diepen & De Graef, 1994). The basic version of scenes showed the scene background, and occasionally a few typical objects. Two versions of each target scene were created. Six to 10 objects were added to the scenes. Many objects were used in both versions of the same scene, but arranged in a different way. In the second scene version, one of the objects was replaced by a non-object. All objects were easily recognizable, with presence, size, and position consistent with everyday experience. Non-objects were meaningless figures with a part-structure and size-range comparable with that of real objects (see Figure 1). Additionally, 12 filler scenes were assembled in two versions, containing several objects and one to four non-objects. In total, 32 target versions and 24 filler versions were used in the experiment.

Figure 1. Target stimulus containing one non-object (bottom-right). Here, the television is fixated, and locally the image has been partially masked (the television set remains recognizable, although the degradation in this printed version is probably worse than on the stimulus display).

On average, 0.5 non-objects were present in target scenes, and 2.3 non-objects in filler scenes. The limitation of the number of non-objects in target scenes minimized the effect of their presence on the scene exploration process, and maximized the number of fixations that could be used in the analyses. Filler scenes contained up to four non-objects, to maintain participants' commitment to extensively search each scene for non-objects.

Eight blocks of 28 trials were constructed, each containing all the 16 target scenes and the 12 filler scenes. Each of the two target versions and filler versions was used in half the blocks. Within a block, half the target scenes appeared with one non-object. Stimulus order in each block was randomized, with the restriction that the first and last scene in a block had to be filler scenes. Half the participants received the blocks and trials in reversed order.

The image within an ovoid foveal window was subject to four types of manipulation: In the low-contrast condition, image contrast within the window was diminished by a factor of seven. The full-noise condition completely masked the image inside of the window by replacing every pixel by a random value. In the partial-noise condition, only 40% of the pixels within the window were replaced by random grey-levels. Finally, in the control condition, the window image was outlined by a red ellipse. Two image-manipulation delays were used: A short 15-ms delay, and a long 105-ms delay. Window size was 2.5° x 1.9°. The average luminance of an undegraded scene was 1.8 cd/m^2. In the low-contrast condition, luminance within the window was slightly higher, because contrast was decreased by changing black contours to light-grey, while the background remained white. Luminance of the red ellipse was 1.1 cd/m^2. Noise-mask luminance approximated the scene luminance. Scenes subtended approximately 16° x 12°. Figure 1 displays a target scene that was used in the experiment.

Participants, Procedure and Design

Sixteen unpaid students from the University of Leuven participated in the experiment. All had normal or corrected-to-normal vision. None of the participants had experience with related experiments.

Upon arriving for the experiment, participants were told that they were to participate in a series of experiments on the speed and accuracy with which information could be extracted from complex visual displays. In the present experiment, they would have to search and count non-objects in line drawings of everyday scenes, that were presented on a display in front of them. The notion of non-objects was explained and illustrated with several examples. Participants were told that the number of non-objects would vary from stimulus to stimulus and that not all stimuli would contain non-objects. They were instructed to explore scenes as fast as possible, but it was emphasized that a high level of accuracy was required at all times. To help them maintain a good level of accuracy, feedback was provided following each trial. Participants were told that their eye movements would be recorded during the exploration of the scene, and that locally, the image could change. A printed example of a scene with the four image manipulations was shown. They were told to ignore the display changes, and to perform the task as quickly and accurately as possible.

Scene presentation was terminated by participants when they pressed the right response button, or automatically when the scene presentation time exceeded 20 s. The latter, however, hardly ever occurred. When the scene presentation was ended, the answering screen appeared, displaying two rows of five squares containing the numbers zero to nine. Participants could give the answer by looking at a number, which would then brighten, and by pressing the left response button. When the answer was correct, the word "correct" would appear below the rows for 0.5 s. When the answer was wrong, the correct answer was displayed for 1.7 s, together with the text "almost correct" (in the case that one non-object too much or too few was counted), or with the word "wrong" (otherwise).

Following these instructions, participants were seated at 125 cm from the stimulus display, with the heads stabilized by a headrest and a bite bar. Once the eye-tracker was successfully calibrated for nine points along the diagonals of the stimulus field, a block of five practice trials was initiated. Each trial consisted of the following events: First, a fixation spot appeared in the center of the display and participants were instructed to fixate this point. Once the eye-tracker determined that participants were steadily gazing at the fixation point, a stimulus was presented. Participants freely explored the stimulus, and terminated the display when they thought they had found all the non-objects. Displays were terminated by pressing the right of two response buttons, or, after 20 s, they finish automatically. Following this, the answering screen was displayed, until participants responded, and feedback was presented.

Conditions were counterbalanced across participants. Four different assignment sets were created, randomly assigning conditions to target trials with the following restrictions: Each set would be used by four participants, of which two received the reversed trial order. Within each block, each condition was assigned to two target scenes. Across blocks, every participant saw each target scene once in each of the eight conditions. Across all participants, every scene x condition combination was presented 16 times, twice in every block. Assignment of conditions to filler trials was identical for all participants, and occurred at random, with the restriction that each condition occurred an equal number of times. Between blocks, participants were given a brief rest. Typically, the whole experiment took about two hours.

Apparatus

Eye movements were recorded with a Generation-V dual-Purkinje-image eye-tracker (Crane & Steele, 1985). This system has an accuracy of less than 10 min of arc and a 1000 Hz sampling rate. It was interfaced to a 486 PC, storing every sample of the right eye's position. For each sample, the computer made an on-line decision about the eye state: fixation, saccade, blink, or signal loss. Eye state and position were fed into a second 486 PC, in control of stimulus presentation (for a description of this dual-PC eye-tracking system, see Van Rensbergen & De Troy, 1993; the eye-state discrimination algorithm is described in van Diepen, Ruelens, & d'Ydewalle, 1999).

The second PC contained two Truevision ATVista Videographics Adapter boards (henceforth called "Vista 1" and "Vista 2"), and was connected to a third 486 PC that contained a third ATVista board ("Vista 3"). The three vistas were synchronized, and connected to a video switcher (van Diepen, 1997). The output signal of Vista 1 was used as a key-signal for the video switcher, to select between the Vista 2 and Vista 3 image on a pixel-by-pixel basis. Vista 2 displayed the normal (undegraded) version of the stimulus. Vista 3 contained either the low-contrast version of the same stimulus (in the low-contrast condition), random noise (full-noise and partial-noise conditions), or a red image (control condition). Vista 1 contained a white ovoid on a black background in the full-noise and low-contrast conditions, and a white ellipse on a black background in the red-ellipse condition. In the partial-noise condition, Vista 1 contained a black background, while within an ovoid area, 40% of the pixels were set to white. White pixels of Vista 1 selected the pixels from Vista 3 that were superimposed on the Vista 2 image. The Vista 1 image could be repositioned very quickly to every screen position (van Diepen, De Graef, & Van Rensbergen, 1994). By moving the Vista 1 ovoid to an off-screen location, the undegraded stimulus was visible on the entire display.

Timing of the image manipulations was initiated whenever a new fixation was detected. At that time the computer determined during which video frame the manipulation area should be moved to the fixation position, in order to approximate the desired manipulation delay as closely as possible. During eye blinks and signal errors, no stimulus manipulation was applied. Stimuli appeared in 60-Hz interlaced NTSC mode on a Barco 6351 CRT with a 756 x 486 resolution, and subtended approximately 16° x 12°.

Dependent Variables

For each trial, seven dependent variables were calculated. Scene inspection time (i.e., the duration of the stimulus presentation, as determined by the participant) was regarded as a measure of overall stimulus processing difficulty. Another variable that could be affected by the task difficulty was the number of errors, defined as the absolute difference between the counted and the actual number of non-objects in a scene. By means of instructions,and feedback, it was attempted to maintain a low error level in all conditions.

Fixation durations were considered estimates of momentary visual processing speed. All fixations, with the exception of those on or near non-objects, were pooled in the (raw) average fixation duration for a given trial. At the level of objects, three related measures are included for comparability with other research: first fixation duration (FFD), gaze duration (GD), and total fixation duration (TFD). To calculate these measures, a number of square object regions was defined for each scene. The object regions tightly enclosed objects that were clearly distinguishable from the scene background, and did not overlap with other objects. The first fixation in an object region during a trial was taken as the FFD for that object. For each trial, an average FFD was calculated. The summed duration of all fixations on an object before leaving it for the first time was taken as the GD. The summed duration of *all* fixations on an object was defined as the TFD. As for FFD, all individual GDs and TFDs were averaged for each trial.

Finally, for each scene the average saccadic amplitude was calculated. This measure reveals changes in saccade target selection patterns. For example, in van Diepen et al. (1995), the average saccadic amplitude increased in the masking conditions, because the relative frequency of small, intra-object saccades decreased. Participants apparently had difficulty to saccade into noise. A similar pattern is expected in the present experiment.

Eye-movement Data Filtering

After the experiment, eye movement data were re-categorized offline as follows, in the given order. Fixations that contained eye position samples on locations more than 1° from the stimulus display (so-called false locks) for longer than 50 ms, or longer than 20% of the duration of the fixation, were marked as errors. The same was done for saccades with false lock durations exceeding 50 ms or 50% of the saccade duration. Occasional periods of missing data up to 15 ms were added to previous and/or subsequent events (e.g., when a few samples were missing after a fixation, and the following event was a fixation on approximately the same location, the two fixations were combined, extrapolating the missing samples, to one fixation).

Saccades preceding or following eye blinks were added to the blink, because eyelid movement often is erroneously categorized as a saccade, before and after the blink status signal is active. Fixations with durations shorter than 10 ms were categorized as saccades, and if possible added to preceding and/or following saccades. Then, saccades with durations shorter than 5 ms, peak amplitudes smaller than 0.10°, or mean velocities smaller than 10°/s were changed to fixations, and if possible added to preceding and/or following fixations.

Following the above data categorization, fixations and saccades were evaluated to decide which should be excluded from the analyses. Since only saccade-fixation and fixation-saccade transitions reliably marked fixation onsets and offset, all fixations that were not immediately preceded or followed by a saccade were rejected. Similarly, saccades were rejected when not directly preceded or followed by a fixation. Furthermore, fixations were rejected when a foveal image manipulation occurred more than 9 ms too soon, more than 9 ms too late, more than once, or when the undegraded image reappeared before the end of the fixation. All rejected saccades and fixations, as well as fixations on or near non-objects, were excluded from the analyses. For each trial, a data-loss ratio was calculated by dividing the total duration of rejected fixations, rejected saccades, and signal errors by the trial duration minus the blink duration. Data of filler trials, target trials shorter than 0.5 s, as well as target trials with a data-loss ratio exceeding 0.4, were entirely excluded from the analyses.

Results

The eye-movement data-filtering procedure (see above) excluded 23 of the 2048 target trials from the analyses. The average percentage of data loss of the remaining scenes was 3.8%. Furthermore, 242 of the 40643 saccades, and 1630 of the 39534 fixations were rejected and excluded from the analyses, together with 3342 fixations on or near non-objects. The mean image-manipulation delay was 15.2 ms and 105.1 ms in the 15 and 105 ms delay conditions, respectively, with a standard deviation of 4.9 ms. On average, 17.1 fixations were analyzed for each trial, of which 5.1 were identified as first fixations on objects.

Scene inspection time ranged from 1.4 s to 20.1 s, with a mean of 6.8 s. A log-transformation was used to normalize the positively skewed distribution of scene inspection times (Bush, Hess, & Wolford, 1993).

All data were entered in a Participant (16) x Manipulation Type (4) x Delay (2) repeated-measures ANOVA. The Huynh-Feldt estimator of sphericity (ε_{HF}) was used to correct $\underline{F}$-tests for violations of the circularity assumption, regarding within-subject manipulations with more than one degree of freedom (i.e., Manipulation Type and the Manipulation Type × Delay interaction). For the same reason, single degree contrasts were tested against contrast-specific error terms, instead of the omnibus error term. A family-wise alpha level of .05 was used in all tests.

The effect of the image-manipulation delay (i.e., the difference between the 15-ms and 105-ms delay conditions for a given image-manipulation) will be referred to as the *delay effect*. A priori, the delay effect in the ellipse condition was compared with the delay effect in the degradation conditions[2]. Furthermore, a two-tailed Dunnett test was used to compare saccadic amplitudes in the ellipse condition with the other image-manipulation types, averaged over the two delay levels. In Table 1, the ANOVA results are summarized. Condition means and delay effects can be found in Table 2.

[2] The Šidák procedure was used to control the familywise error rate of a set of a priori $\underline{F}$-tests at the .05 level. Reported $\underline{p}$-values represent the smallest *familywise* significance level at which the *individual* contrast would have been declared significant.

Table 1. ANOVA Results of Experiment 1

	Delay			Manipulation Type				Delay × Manipulation Type			
	MSe	$F(1,15)$	p	*MSe*	$F(3,45)$	ε_{HF}	p	*MSe*	$F(3,45)$	ε_{HF}	p
Inspection time (log(s))	0.0021	44.18	< .0001	0.0026	18.78	1.17	< .0001	0.0010	31.16	1.16	< .0001
Fixation duration (ms)	84.2	33.64	< .0001	178.0	3.75	1.19	< .02	81.6	4.43	1.10	< .009
First fixation duration (ms)	141.6	4.27	n.s.	238.7	8.56	1.02	< .0001	193.7	3.99	1.15	< .02
Gaze duration (ms)	416.7	4.13	n.s.	566.5	16.06	1.09	< .0001	345.9	4.76	1.14	< .006
Total fixation duration (ms)	1676	63.47	< .0001	3929	6.49	0.91	< .002	1834	14.23	0.93	< .0001
Saccadic amplitude (°)	0.023	3.60	n.s.	0.051	34.51	0.83	< .0001	0.017	4.01	0.88	< .02
Number of errors	0.010	9.39	< .008	0.011	7.89	0.99	< .0004	0.0061	3.01	1.06	< .04

Table 2. Condition Means and Delay Effects in Experiment 1

	M											
	Ellipse			Low Contrast			Partial Noise			Full Noise		
	15 delay	ms105 delay	msdelay effect	15 delay	ms105 delay	msdelay effect	15 delay	ms105 delay	msdelay effect	15 delay	ms105 delay	msdelay effect
Inspection time (log(s))	0.754	0.749	0.005_a	0.764	0.739	0.025_a	0.796	0.756	0.040_a	0.907	0.760	0.147_b
Fixation duration (ms)	243.7	244.1	-0.4_a	255.9	242.3	13.6_b	260.5	246.4	14.1_b	249.3	238.9	$10.4_{a,b}$
First fixation duration (ms)	265.7	271.1	-5.4_a	280.6	265.0	15.6_a	278.8	269.5	9.3_a	255.5	257.7	-2.2_a
Gaze duration (ms)	309.7	316.9	-7.2_a	314.1	294.8	$19.3_{a,b}$	319.6	299.6	20.0_b	274.8	277.6	$-2.8_{a,b}$
Total fixation duration (ms)	391.0	392.7	-1.7_a	413.1	377.5	$35.6_{a,b}$	453.8	390.6	63.2_b	519.4	385.9	133.5_b
Saccadic amplitude (°)	3.598	3.585	0.013_a	3.919	3.924	-0.005_a	3.890	3.884	0.006_a	4.257	4.069	0.188_a
Number of errors	0.107	0.067	0.040_a	0.112	0.091	0.011_a	0.150	0.122	0.028_a	0.265	0.141	0.124_a

Note. The *delay effect* for a given manipulation type is the difference between the 105- and the 15-ms condition for that manipulation type. Delay effects in the same row that do not share subscripts are significantly different from each other.

Regarding scene inspection time, the number of errors, fixation duration, FFD, and the TFD, no effect of manipulation type was present in the 105-ms condition: Performance was as good when foveal information was degraded during the later part of fixations, as in the ellipse condition. This was confirmed by non-significant simple main effects of manipulation type, for the 105-ms delay. Conversely, significant simple main effects[3] were found in the 15-ms condition, for scene inspection time, F (3,45) = 76.32, MSe = 0.0010, $p < .0001$, the number of errors, F (3,45) = 14.20, MSe = 0.0061, $p < .0001$, fixation duration, F (3,45) = 10.63, MSe = 81.6, $p < .0001$, FFD, F (3,45) = 11.58, MSe = 194, 1.15, $p < .0001$, and for TFD, F (3,45) = 27.7, MSe = 1834, $p < .0001$. With the 15-ms delay, performance was worse in the degradation conditions, relative to the ellipse condition. These results confirm that the foveal identification process extracts information early during fixations, and is hardly hampered by a later image degradation.

Note that the red ellipse could be expected to attract saccades, which could result in the present data pattern. However, inspection of the relative frequency distributions of saccadic amplitudes showed that this was not the case. Furthermore, a pilot study had shown that saccade amplitude distributions were very similar in the red ellipse condition and a no-change condition, where no image manipulation occurred.

The main goal of the present experiment was to show that fixations are sustained, when foveal information is degraded, but not completely removed, during the early part of fixations. Table 2 nicely confirms this position. For fixation durations, the delay effect was significantly larger in the low-contrast and partial-noise conditions, than in the ellipse condition. The same pattern was present for FFD, albeit not significantly. Fixation durations also increased in the full-noise condition, although the delay effect did not reliably differ from the ellipse condition. Rather than sustaining fixations, full masking after 15 ms increased the number of fixations needed to inspect a scene, as can be deduced from the increased inspection times in this condition.

A striking finding with respect to the fixation data was that the average fixation duration exhibited a 10-ms delay effect in the full-noise condition, while FFDs showed a negligible reversed delay effect. To determine whether this was due to the fact that for FFDs, fixations outside object regions were excluded, additionally the mean *object fixation duration* was examined, including both first and other fixations on objects. The data pattern for this variable took an intermediate position between the average fixation duration and first fixation data. A 5-ms delay effect was observed in the full-noise condition that did not reliably differ from the ellipse condition. The delay effects in the low-contrast and partial-noise conditions slightly increased, compared with the average fixation data, and significantly differed from the ellipse condition. These observations suggested that exclusion of other than first fixations, rather than other than object fixations, might have diminished the delay effect on first fixation durations in the full-noise condition. Therefore, *later fixation duration*, defined as the average duration of fixations on objects, excluding the first fixation of the first object entry, but including the first fixations of subsequent object entries, were examined. On average, 4.1 later fixations were analyzed for each scene. Delay effects of −15, 16, 36, and 24 ms were observed in the red-ellipse, low-contrast, partial-noise, and full-noise condition, respectively. The reversed delay effect in the ellipse condition differed significantly from the delay effect

[3] Simple main effects were considered to be significant when p was smaller than .15/6 = .025, according to Dunn's procedure (see Kirk, 1995, p. 381).

in the partial-noise and full-noise conditions, but not from the low-contrast condition. Apparently, foveal masking by noise, partially or fully, affected later fixations more than first fixations, whereas contrast manipulation affected all fixation duration measures about equally.

Discussion

The results largely confirm the prediction from the local-to-global processing model. Foveal image degradation reduces the likelihood of intra-object saccades, which increases the average saccadic amplitude, and decreases the number of fixations per gaze. This occurs irrespective of the degradation delay, confirming that the saccade target selection extracts image information during the later part of fixations. No other effects of foveal degradation are present, when the degradation is applied later during fixations. Foveal identification clearly acquires sufficient information during the early part of fixations. Early degradation of the foveal image disturbs the foveal identification process. When foveal information is degraded, but not completely removed, the visual system can deal with the degradation by sustaining fixations. If foveal information is completely masked at an early stage, fixations are not sustained, but repeated. In that case, participants typically saccade away from, and back to objects. This substantially increases the total time that is required to inspect a scene. When fixations are sustained, inspection times only increase little.

An interesting pattern of results was observed for first fixation durations, compared with average fixation durations. It appeared that first fixation durations were less sensitive to the delay manipulation, as opposed to other object fixations. Possibly, processing difficulty during the first fixation was carried over to subsequent object fixations. An implication could be that first fixation durations are primarily determined by visuo-motor constraints, and not by the ongoing stimulus processing, as has been argued by several authors (e.g., Vitu, O'Regan, Inhoff, & Topolski, 1995). On the other hand, first fixation durations might more purely reflect the foveal encoding speed, while later fixations are additionally affected by cognitive processing difficulties. The latter possibility is supported by the observation that in the full-noise condition, no delay effect was present on first fixations (i.e., first fixations were not sustained when foveal encoding terminated), whereas other fixation durations in that condition were significantly affected by the delay manipulation. Future research will focus on this issue.

Experiment 2

Experiment 2 was a replication of Experiment 1, employing shorter delays. A new eye-contingent display change algorithm was implemented that enabled a 0-ms delay condition. The 0-ms condition replaced the 15-ms delay condition of the previous experiment. Foveal image degradation, therefore, occurred from the start of fixations, fully eliminating the intake of undegraded foveal information. This was expected to considerably increase the impact of foveal contrast degradation and foveal masking. The long delay was conservatively set at 105 ms in Experiment 1, to ensure that sufficient object detail could be obtained during the undegraded interval. Earlier estimates with respect to foveal information encoding were much shorter. In the present experiment, a delay of 50 ms was employed instead of 105 ms, to provide a more stringent test of earlier claims.

As in Experiment 1, the low-contrast and partial-noise conditions were compared with the red-ellipse condition. The full-noise condition was not used. It was predicted that in the 0-ms condition, fixation durations would increase substantially when the foveal image was partially masked, or decreased in contrast, compared with the 50-ms condition. As in Experiment 1, the delay effect in the degradation conditions was compared with the corresponding difference in the red-ellipse condition, to control for the stimulus-discontinuity effect. Scene inspection times were also expected to exhibit larger delay effects in the low-contrast and partial-noise condition, compared with the ellipse condition, corresponding to the effects on fixation durations. Saccadic amplitudes were again expected to increase when foveal information was degraded, irrespective of the delay.

Method

Stimuli

Twelve of the 16 target scenes that were created for Experiment 1, as well as all the filler scenes, were used in Experiment 2. Furthermore, a new version of each of the 12 target scenes was created, by rearranging the objects of the target version that contained no non-objects. Where possible, new objects were added, sometimes replacing others. For each filler scene, a new version was created as well, containing several objects, and one to four non-objects. The three versions of a filler scene always contained different numbers of non-objects. The mean number of non-objects was 0.33 for target scenes, and 2.44 for filler scenes. In total, 36 target versions and 36 filler versions were used in the experiment.

Six blocks of 24 trials were presented to participants. Half the trials in each block were filler trials. Within each block, one third of the target scenes contained one non-object. Across blocks, each of the target and filler versions was presented twice to every participant. Stimulus order within each block was randomized, with the restriction that the first and last scene in a block had to be filler scenes. Half the participants received the blocks and trials in reversed order.

Three types of stimulus manipulation were applied to the image within an ovoid foveal window. In the low-contrast condition, foveal image contrast was decreased by a factor of seven. In the partial-noise condition, half the pixels in the foveal window were replaced by noise. The window was outlined by a red ellipse in the control condition. The delay was 0 ms or 50 ms. Window size was 2.5° × 1.9°. Average undegraded scene luminance was 1.8 cd/m^2, which approximated the luminance of the noise mask. Within the low-contrast window, scene luminance was slightly raised (see Experiment 1). Luminance of the red ellipse was 1.1 cd/m^2.

Participants, Procedure and Design

Twelve unpaid students from the University of Leuven participated in the experiment. All had normal or corrected-to-normal vision, and had no previous experience with related experiments.

The procedure was identical to Experiment 1. Half the participants saw six trial-blocks in the same order, while the blocks and trials were reversed for the other six participants. The Manipulation Type × Delay conditions were counterbalanced across participants. Three sets of condition-to-trial assignments were created at random with the following restrictions: Each set would be used by four participants, of which two received the reversed trial order. Within each block, each condition was assigned to two targets scenes for each participant. Across blocks, every participant saw each target scene six times, once in each condition. Overall, every scene x condition combination appeared 12 times, twice in every block. Assignment of conditions to filler trials was identical for all participants, and occurred at random, with the restriction that each condition occurred an equal number of times. Between blocks, participants were given a brief rest. Typically, the experiment took about 90 minutes.

Apparatus

The experimental setup was identical to the previous experiment. In the 0-ms condition, eye movements were monitored during saccades to update the window position. Consequentially, at the beginning of a fixation that was preceded by a saccade, the window position approximated the fixation position. When at the beginning of the fixation the window position deviated more than 0.1° from the horizontal or vertical fixation coordinate, the window was repositioned to the correct position. Note that, even with position deviations up to 0.5°, the image close to the fixation position would still be affected by the stimulus manipulations, from the beginning of the fixation. Only the window border would shift as a result of the repositioning. The 50-ms condition was created in a similar way as the 15- and 105-ms conditions of the previous experiment.

Eye-movement Data Filtering

The same data reduction procedure was used as in Experiment 1, with the following exception: In the 50-ms delay condition, the window was allowed to appear 7 ms too soon, or 11 ms too late. In this way, the number of accepted fixations was maximized, without changing the range of accepted delays. In the 0-ms condition, the window was required to be present from the beginning of fixations. At most, one position correction was allowed, occurring within 30 ms after the fixation onset.

Results

Of the 864 target trials, 58 had to be excluded. The data-loss percentage of the remaining trials was 9.8% on average. Here, 452 of the 14193 saccades, and 1059 of the 13711 fixations were rejected. Additionally, 789 fixations on or near non-object were excluded from the analyses. In the 0-ms condition, a window position correction occurred during 83% of the fixations. On average, the window position was adjusted 13.2 ms after the beginning of the fixation ($\underline{SD}$ = 6.5 ms). For the remaining fixations, the window was correctly positioned from the beginning of the fixation. Taken together, the correct window position was achieved after 11.0 ms on average ($\underline{SD}$ = 7.7 ms). Note that incorrectly positioned windows usually were already close to the actual fixation position. Image degradations thus affected foveal information from the onset of fixations, even though a position correction followed. In the 50-ms condition, the actual delay was 52.1 ms on average, with a standard deviation of 4.8 ms.

Scene inspection time ranged from 2.2 to 21.4 s, with a mean of 6.4 s. A log-transformation was used to normalize the distribution of scene inspection times.

All data were entered in a Participant (12) x Manipulation Type (3) x Delay (2) repeated-measures ANOVA. As in Experiment 1, the Huynh-Feldt estimator of sphericity (ε_{HF}) was used to correct F-tests, and single degree contrasts were tested against contrast-specific error terms. A priori, the delay effect (i.e., the difference between the 0- and 50-ms conditions for a given manipulation type) was compared for all pairwise combinations of manipulation types. A family-wise alpha level of .05 was used in all tests. Table 3 summarizes the ANOVA results, while Table 4 reports the condition means, delay effects, and the results of the pairwise comparison of the latter.

Significant Manipulation Type × Delay interactions were found for all dependent variables, with the exception of saccadic amplitude and the number of errors. Table 4 indicates that these interactions are due to the fact that foveal image degradation only affected performance in the 0-ms conditions. Correspondingly, the simple main effect of manipulation type was significant only for the 0-ms delay: $F\ (2,22) = 31.95$, $MSe = 0.0024$, $p < .0001$, for scene inspection time, $F\ (2,22) = 17.60$, $MSe = 192$, $p < .0001$, for fixation duration, $F\ (2,22) = 12.70$, $MSe = 275$, $p < .0002$, for FFD, $F\ (2,22) = 5.42$, $MSe = 526$, $p < .02$, for GD, and $F\ (2,22) = 46.53$, $MSe = 2595$, $p < .0002$ for TFD. With the 50-ms delay, the simple main effect of manipulation type was not significant for any of these dependent variables.

While the above findings confirm that foveal information can be extracted during the initial part of fixations, the main effect of Manipulation Type on the number of errors indicates that some disturbance of foveal image degradation was present in the 50-ms condition. A Tukey test showed that the error rate significantly increased in the partial-noise condition, relative to the ellipse condition, $q\ (11) = 6.09$, $MSe = 0.066$, $p < .004$, and the low-contrast condition, $q\ (11) = 4.06$, $MSe = 0.057$, $p < .04$. Table 4 shows that this increase is smaller with the 50-ms delay, than with the 0-ms delay. The Manipulation Type × Delay interaction, however, did not reach significance.

Saccadic amplitudes were significantly affected only by Manipulation Type, $F\ (2,22) = 22.65$, $MSe = 0.032$, $\varepsilon_{HF} = 1.09$, $p < .0001$. A Tukey test revealed that saccadic amplitudes in the ellipse condition were significantly smaller compared with the low-contrast condition, $q\ (11) = 8.98$, $MSe = 0.12$, $p < .0002$, and the partial-noise condition, $q\ (11) = 6.85$, $MSe = 0.16$, $p < .002$. As in Experiment 1, the saccade target selection process seemed to exclude targets from the foveal area more often when the image was degraded, than in the ellipse condition, irrespective of the level of delay.

Discussion

In Experiment 2, detrimental effects of foveal degradation started to show up when undegraded information was present for only 50 ms at the beginning of fixations: More errors were made in the 50-ms degradation conditions than in the control condition. Fixation durations and scene inspection time, however, were hardly affected by foveal image degradation after 50 ms. Evidently, the 50-ms delay time approximated the length of the foveal information extraction interval. Conceivably, the length of this interval will be different when other stimuli than line drawings are used.

Table 3. ANOVA Results of Experiment 2

	Delay			Manipulation Type				Delay × Manipulation Type			
	MSe	*F*(1,11)	*p*	*MSe*	*F*(2,22)	ε_{HF}	*p*	*MSe*	*F*(2,22)	ε_{HF}	*p*
Inspection time (log(s))	0.0017	37.38	< .0001	0.0020	30.33	1.02	< .0001	0.0024	9.18	1.01	< .002
Fixation duration (ms)	69.6	53.77	< .0001	202.7	12.31	1.08	< .0003	192.4	5.94	0.85	< .02
First fixation duration (ms)	258.8	7.57	< .02	346.1	6.29	1.22	< .007	275.0	6.65	1.00	< .006
Gaze duration (ms)	447.0	8.50	< .02	474.6	0.89	1.16	n.s.	526.3	6.12	0.90	< .01
Total fixation duration (ms)	3554	32.81	< .0001	2090	38.21	1.15	< .0001	2595	16.99	1.15	< .0001
Saccadic amplitude (°)	0.031	0.24	n.s.	0.032	22.65	1.09	< .0001	0.032	0.71	0.86	n.s.
Number of errors	0.0074	9.73	< .01	0.015	10.75	1.19	< .0006	0.015	2.08	0.80	n.s.

Table 4. Condition Means and Delay Effects in Experiment 2

	M								
	Ellipse			Low Contrast			Partial Noise		
	0 ms delay	50 ms delay	delay effect	0 ms delay	50 ms delay	delay effect	0 ms delay	50 ms delay	delay effect
Inspection time (log(s))	0.711	0.710	0.001_a	0.761	0.706	$0.055_{a,b}$	0.867	0.745	0.122_b
Fixation duration (ms)	241.9	243.4	-1.5_a	267.6	244.5	23.1_b	273.5	251.9	21.6_b
First fixation duration (ms)	257.7	266.9	-9.2_a	284.3	260.1	24.2_b	289.5	273.2	16.3_b
Gaze duration (ms)	288.3	300.5	-12.2_a	313.1	284.8	28.3_b	316.5	289.0	27.5_b
Total fixation duration (ms)	341.3	342.6	-1.3_a	414.0	341.1	72.9_a	539.6	369.9	169.7_b
Saccadic amplitude (°)	3.486	3.534	-0.048_a	3.850	3.799	0.051_a	3.762	3.824	-0.062_a
Number of errors	0.074	0.066	0.008_a	0.150	0.112	0.038_a	0.302	0.158	0.144_a

Note. The *delay effect* for a given manipulation type is the difference between the 0-ms and the 50-ms delay condition for that manipulation type. Delay effects in the same row that do not share subscripts are significantly different from each other.

In the 0-ms condition, large degradation effects were present on all dependent variables. Fixation durations increased substantially in both the partial-noise and low-contrast conditions. Compared with the 15-ms condition in Experiment 1, loss of the initial 15-ms undegraded view of foveal information had relatively large effects on fixation durations and scene inspection times. Relative to the ellipse condition, fixation durations in the 0-ms degradation conditions increased about 16 ms *more* compared with the corresponding increase in the 15-ms condition of Experiment 1. This should not be interpreted merely as an additive effect of the duration of the undegraded stimulus, because the present image degradations slow down, rather than terminate, foveal information encoding. Apparently, a 15-ms undegraded view of the foveal image in Experiment 1 greatly enhanced perception.

Fixation durations were about equal in the 0-ms and 50-ms ellipse conditions. This indicates that a stimulus-discontinuity effect was still present in the 0-ms condition, presumably caused by the window position adjustment at the beginning of most fixations in that condition. Although this could have been avoided by suppressing position adjustments during fixations in the 0-ms conditions, this would result in a less optimal manipulation-area position.

Saccade programming again was affected about equally in both delay conditions, confirming the hypothesis that saccade target selection occurs during the later part of fixations.

General Discussion

The present study examined foveal information extraction during scene exploration. Two moving window experiments confirmed the hypothesis that sufficient foveal information for object identification can be acquired in about 50 ms following the fixation onset, and that fixation durations can be affected by foveal image degradation.

When foveal information is degraded early during fixations, the visual system has two means to compensate for the stimulus deterioration. First, additional fixations can be made to obtain a number of snapshots of the foveal stimulus. This method especially is used when foveal information is completely masked. Although small, intra-object saccades in principle could be used to increase the number of snapshots, participants generally saccade away from an object, to fixate it again later. Apparently, it is difficult to program intra-object saccades once the object has been replaced by random noise. A second method to deal with degraded foveal information is to increase fixation durations. When foveal information is completely removed, sustaining fixations is not very helpful. However, when some information remains, as is the case in the partial-noise and low-contrast conditions, increasing the fixation duration is a more efficient method to deal with the degradation than the refixation method.

Foveal identification can be achieved when foveal information is present only for a short period at the beginning of fixations. Saccade programming, however, is affected by the quality of the foveal image later during the fixation. Clearly, registration of the foveal image does not stop after the initial 50 ms. This raises the question whether information extraction for object identification should mandatorily occur at the beginning of fixations. Some research indicated that saccadic suppression continues up to 50 ms after saccades terminate (for a review, see Shioiri, 1993). Accordingly, a more efficient foveal information extraction period could exist later during fixations. However, Shioiri (1993) himself showed that stimulus degradation during the early part of fixations definitely affected picture perception. In future experiments, the moving window technique will be used to mask foveal information during a preset period at the *beginning* of fixations. If foveal information extraction starts immediately following the start of fixations, initial masking should slow performance.

The present results favor a local-to-global hierarchy of information processing during scene exploration. Other studies, however, have argued for a global-to-local sequence. For example, Navon (1977) showed that global information dominated local information (e.g., a large letter H that consists of small Es is processed faster than its local components). Hughes, Nozawa, and Kitterle (1996) accepted *global precedence* as a principle of human pattern perception, and presumed a corresponding low spatial frequency dominance. The latter was confirmed in two experiments, that presented sine wave gratings of high and low spatial frequencies. Schyns and Oliva (1994) presented scenes that were either low- or high-pass filtered, and "hybrids", that combined the low frequencies of one scene with high frequency information of another. Presentation time was 30 or 150 ms. After each presentation an undegraded scene was presented, and participants had to indicate if this scene matched to previous presentation. In the 30-ms condition, performance was much better when target scenes matched the low-frequency component of hybrid scenes. In the 150-ms condition, better performance was obtained when hybrid and matching stimuli shared high-frequency information. These results support also a coarse-to-fine hierarchy of information processing. In a more recent study, however, Schyns and Oliva (1997) show that, depending on task constraints, the visual system can select spatial scales in a flexible, rather than a fixed way.

The above findings apparently are at odds with the conclusions from the present study. How can the visual system both show a global-to-local and a local-to-global processing hierarchy? To answer this question, the differences between the studies have to be considered. In the present experiments, participants were engaged in a free exploration task, where eye movements were possible, with realistic stimuli, in search of non-objects. Results from this experiment should be interpreted as the "normal" viewing behavior in natural environments. The experiments discussed in the previous paragraph explore the limits of the visual system. Participants perform unnatural tasks, with briefly presented and often meaningless stimuli. Furthermore, task demands (e.g., scene matching) often stimulate the use of global, coarse stimulus characteristics. The present study favors the local processing of objects in scenes, but also required global processing (e.g., for saccade programming), as is the case in many real-life situations.

To conclude, with the above considerations in mind, the present study indicates that during normal scene exploration, a local-to-global processing scheme is adopted, even though in some occasions, when pushed to its limits, the visual system can use a global-to-local sequence.

ACKNOWLEDGEMENTS

This research was supported by the Programma on Interuniversity Poles of Attraction contract No. 31 and the G.O.A. Convention No. 93/1. The author wishes to thank Géry d'Ydewalle, Johan Wagemans, Jacqueline van Diepen, Martien Wampers and Keith Rayner for their comments on an earlier draft.

Additional information about ongoing scene perception research can be found on the web: http://www.psy.kuleuven.ac.be/~paul/.

REFERENCES

Antes, J. R., Penland, J. G., & Metzger, R. L. (1981). Processing global information in briefly presented pictures. *Psychological Research, 43,* 277-292.

Biederman, I. (1981). On the semantics of a glance at a scene. In M. Kubovy, & J. R. Pomerantz (Eds.), *Perceptual organization* (pp. 213-253). Hillsdale, NJ: Erlbaum.

Blanchard, H. E., McConkie, G. W., Zola, D., & Wolverton, G. S. (1984). Time course of visual information utilization during fixations in reading. *Journal of Experimental Psychology: Human Perception and Performance, 10,* 75-89.

Boyce, S. J., & Pollatsek, A. (1992). Identification of objects in scenes: The role of scene background in object naming. *Journal of Experimental Psychology: Learning, Memory, and Cognition, 18,* 531-543.

Busey, T. A., & Loftus, G. R. (1994). Sensory and cognitive components of visual information acquisition. *Psychological Review, 101,* 446-469.

Bush, L. K., Hess, U., & Wolford, G. (1993). Transformations for within-subjects designs: A Monte Carlo investigation. *Psychological Bulletin, 113,* 566-579.

Buswell, G. T. (1935). *How people look at pictures.* Chicago: University of Chicago Press.

Crane, H. D., & Steele, C. M. (1985). Generation-V dual-Purkinje-image eyetracker. *Applied Optics, 24,* 527-537.

De Graef, P. (1998). *Object identification in real-world scenes: Parallel processing versus selective presaccadic encoding* (Psych. Rep. No. 231). Laboratory of Experimental Psychology, University of Leuven, Leuven.

De Graef, P., Christiaens, D., & d'Ydewalle, G. (1990). Perceptual effects of scene context on object identification. *Psychological Research, 52,* 317-329.

De Valois, R. L., & De Valois, K. K. (1988). *Spatial vision.* New York: Oxford University Press.

Henderson, J. M. (1992). Visual attention and eye movement control during reading and picture viewing. In K. Rayner (Ed.), *Eye movements and visual cognition: Scene perception and reading* (pp. 260-283). New York: Springer-Verlag.

Henderson, J. M., & Hollingworth (1998). Eye movements during scene viewing: An overview. In G. Underwood (Ed), *Eye guidance in reading and scene perception (pp. 269-293).* Oxford: Elsevier.

Hughes, H. C., Nozawa, G., & Kitterle, F. (1996). Global precedence, spatial frequency channels, and the statistics of natural images. *Journal of Cognitive Neuroscience, 8,* 197-230.

Ishida, T., & Ikeda, M. (1989). Temporal properties of information extraction in reading studied by a text-mask replacement technique. *Journal of the Optical Society of America, 6,* 1624-1632.

Kirk, R. E. (1995). *Experimental design: Procedures for the behavioral sciences.* Monterey, CA: Brooks/Cole.

Loftus, G. R., Kaufman, L., Nishimoto, T., & Ruthruff, E. (1992). Effects of visual degradation on eye-fixation durations, perceptual processing, and long-term visual memory. In K. Rayner (Ed.), *Eye movements and visual cognition: Scene perception and reading* (pp. 203-226). New York: Springer-Verlag.

Mannan, S. K., Ruddock, K. H., & Wooding, D. S. (1996). The relationship between the locations of spatial features and those of fixations made during visual examination of briefly presented images. *Spatial Vision, 10,* 165-188.

Matin, E. (1974). Saccadic suppression: A review and an analysis. *Psychological Bulletin, 81,* 899-917.

McConkie, G. W., & Rayner, K. (1975). The span of the effective stimulus during a fixation in reading. *Perception & Psychophysics, 17,* 578-586.

McConkie, G. W., & Rayner, K. (1976). Asymmetry of the perceptual span in reading. *Bulletin of the Psychonomic Society, 8,* 365-368.

McConkie, G. W., Underwood, N. R., Zola, D., & Wolverton, G. S. (1985). Some temporal characteristics of processing during reading. *Journal of Experimental Psychology: Human Perception and Performance, 11,* 168-186.

Navon, D. (1977). Forest before trees: the precedence of global features in visual perception. *Cognitive Psychology, 9,* 353-383.

Pollatsek, A., Raney, G. E., Lagasse, L., & Rayner, K. (1993). The use of information below fixation in reading and in visual search. *Canadian Journal of Experimental Psychology, 47,* 179-200.

Rayner, K. (1979). Eye guidance in reading: Fixation locations within words. *Perception, 8,* 21-30.

Rayner, K. (1998). Eye Movements in Reading and Information Processing: 20 years of research. *Psychological Bulletin, 124,* 372-422.

Rayner, K., Inhoff, A. W., Morrison, R. E., Slowiaczek, M. L., & Bertera, J. H. (1981). Masking of foveal and parafoveal vision during eye fixations in reading. *Journal of Experimental Psychology: Human Perception and Performance, 7,* 167-179.

Saida, S., & Ikeda, M. (1979). Useful visual field size for pattern perception. *Perception & Psychophysics, 25,* 119-125.

Schyns, P. G., & Oliva, A. (1994). From blobs to boundary edges: Evidence for time- and spatial-scale-dependent scene recognition. *Psychological Science, 5,* 195-200.

Schyns, P. G., & Oliva, A. (1997). Flexible, diagnosticity-driven, rather than fixed, perceptually determined scale selection in scene and face recognition. *Perception, 26,* 1027-1038.

Shioiri, S. (1993). Postsaccadic processing of the retinal image during picture scanning. *Perception & Psychophysics, 53,* 305-314.

van Diepen, P. M. J. (1997). A pixel-resolution video switcher for eye-contingent display changes. *Spatial Vision, 10,* 335-344.

van Diepen, P. M. J., & De Graef, P. (1994). *Line-drawing library and software toolbox* (Psych. Rep. No. 165). Laboratory of Experimental Psychology, University of Leuven, Belgium.

van Diepen, P. M. J., De Graef, P., & d'Ydewalle, G. (1995). Chronometry of foveal information extraction during scene perception. In J. M. Findlay, R. Walker, & R. W. Kentridge (Eds.), *Eye movement research: Mechanisms, processes and applications* (pp. 349-362). Amsterdam: Elsevier.

van Diepen, P. M. J., De Graef, P., & Van Rensbergen, J. (1994). On-line control of moving masks and windows on a complex background using the ATVista Videographics Adapter. *Behavior Research Methods, Instruments, & Computers, 26,* 454-460.

van Diepen, P. M. J., Ruelens, L., & d'Ydewalle, G. (1999). Brief foveal masking during scene perception. *Acta Psychologica, 101,* 91-103.

van Diepen, P. M. J., & Wampers, M. (1998). Scene exploration with Fourier filtered peripheral information. *Perception, 27,* 1141-1151.

van Diepen, P. M. J., Wampers, M., & d'Ydewalle, G. (1998). Functional division of the visual field: Moving masks and moving windows. In G. Underwood (Ed.), *Eye guidance in reading and scene perception* (pp. 337-355). Oxford: Elsevier.

van Duren, L. (1994). *Signal Processing During Eye Fixations and Saccades.* Unpublished doctoral dissertation, Vrije Universiteit Amsterdam, Amsterdam.

Van Rensbergen, J., & De Troy, A. (1993). *A reference guide for the Leuven dual-PC controlled Purkinje eyetracking system* (Psych. Rep. No. 145). Laboratory of Experimental Psychology, University of Leuven, Belgium.

Vitu, F., O'Regan, J. K., Inhoff, A. W., & Topolski, R. (1995). Mindless reading: Eye-movement characteristics are similar in scanning strings and reading texts. *Perception & Psychophysics, 57,* 352-364.

Wampers, M., van Diepen, P. M. J., & d'Ydewalle, G. (1997). *The use of coarse and fine peripheral information during scene perception* (Psych. Rep. No. 212). Laboratory of Experimental Psychology, University of Leuven, Belgium.

Yarbus, A. L. (1967). *Eye movements and vision.* New York: Plenum Press.

Chapter 9

PREDICATES, EVENTS AND SITUATIONS: A CROSS-COGNITIVE ANALYSIS AND RECONSIDERATION OF SOME BASIC NOTIONS OF DISCOURSE COMPREHENSION

Jean-François Le Ny*

University of Paris-Sud,
Orsay, France

A comprehensive study of discourse comprehension should take into account the contributions of many cognitive disciplines, including logic, linguistics, artificial intelligence, cognitive psychology and neurobiology. This type of cross-cognitive approach should also be applied to many other high-level intelligent activities. The following general remarks about language and thought in discourse comprehension are designed to include the contributions of such disciplines.

One of the main theoretical notions used in this domain is that of proposition. The modern view of proposition, introduced by Frege (1892), is that it is first defined in terms of truth, as the smallest portion of discourse that can have a truth value, True or False. The analysis of proposition into two components, a Fregean predicate and its argument(s), is a major part of this modern view, and in today's logic, proposition calculus and predicate calculus, most often restricted to the first order, derive from these main orientations.

However, although Frege's founding views basically involved a sound anti-psychologism, a major change in this field is today's effort to naturalize its main notions, and to relate them to psychology. Propositions, predicates and their arguments, and other abstract entities were not only non-mental for Frege, but were also endowed with a specific mode of ideal existence in a Platonic world. Some followers of Frege, in particular mathematicians, still adhere to this type of neo-platonism, whereas others only believe that propositions, as logical entities, are strictly independent of the mind. One aspect of this idea, that true

---

* LIMSI-CNRS. Groupe Cognition Humaine. Universite de Paris-Sud. BP 133. - F 91403 Orsay Cedex.- leny@limsi.fr

statements and thoughts are independent of particular minds, must certainly be included in a cross-cognitive framework, but it must be examined in connection with more recent concepts.

A propositional view of human cognition became part of psychology only about 70 years after the publication of Frege's main work, which was largely ignored or underestimated for a long time. This viewpoint became part of the new topic of cognitive psychology, mainly in connection with new research in artificial intelligence. It was based on an analogy between how the mind and computers process reasoning and problem solving (notably Newell and Simon, 1972; Norman and Rumelhart, 1975; Schank, 1975), but it has also been applied to the psychological study of discourse comprehension in connection with linguistics (FranÁois, 1991; Kintsch, 1974; Kintsch and van Dijk, 1978; Le Ny, 1979). It has since proved to be most useful for interpreting experimental data. All psychologists do not have exactly the same concept of propositions so that the relationships between logical and psychological views, together with the linguistic and artificial intelligence concepts, deserve further theoretical exploration. A fortiori, enlarging them to the point of assuming they have an underlying neurobiological structure may be useful for constructing psychological models of discourse comprehension.

I will thereafter examine certain aspects of the notion of proposition, and its core, the notion of predicate, in connection with the rational representation of sentences reporting events, to identify their basic nature. This will be equated with the idea of semantic predicative structure in language and thought. "Semantic" here means that this structure is not purely syntactic, but represents some structure of the world. I will use the conclusions of this analysis to produce a united view of predicate and concept, and a view of discourse comprehension based on the notions of situation and situation representation.

DIFFERENCES IN THE NOTION OF 'PROPOSITION' IN LOGIC AND COGNITIVE PSYCHOLOGY

The use of the notion of 'proposition'[1] in cognitive psychology differs in several respects from its use in logic and, consequently, artificial intelligence and formal linguistic. But these apparently large differences do not totally obscure the underlying unity.

One difference, perhaps surprising in view of the above definition, is the role of the truth criterion in describing a proposition. According to this, propositions are fragments of statements that are defined as a function of logical requirements, analysis of inferences to make them necessarily valid, to guarantee that truth is conserved when thought proceeds from a statement given as true to a derived one. This objective, to clearly separate truth-conserving inferences from others, concerns not only human minds but also automatic systems on computers. Cognitive psychology, by contrast, tries to determine how human thought works naturally, irrespective of the true or false status of its content or of any conservation or loss of truth in the process. Naturalization of the 'proposition' notion, and some others as 'predicate', thus consists of no longer considering their properties as being laid down by axiomatic thought, but as being discovered by empirical and theoretical research. These notions then become descriptions of actual units in the natural, psychological, activities of discourse and

[1] I insert between single quotes words referring to mental entities: meanings, notions, etc., and between double quotes, as usual, words referring to words or statements.

thought, and the contents of their logical or formal linguistic counterparts become idealizations of these natural entities.

Another, related difference is the apparently poor use made by cognitive psychology of a major property of thought - reference - when studying statements and the way they are processed by human minds. For example, cognitive psychologists often disregard their reference in discourse comprehension, and are only interested in meaning as it is in the mind. The word "semantic" does not require a reference basis in psychology, as it does in logic and related disciplines. This can be justified by the fact that human micro-cognitive activities work in exactly the same way whether subjects process false or true contents. Thus, sentences used as linguistic material in psychological experiments are built to satisfy the requirements of the particular experiment, but not for their truth. An experiment subject need not be concerned about the underlying fact when reading that "a bus run into a sedan last night...". Nevertheless, these sentences are partially truth dependent, or "pseudo-true", in that reference is as a rule present in the background of the experiments, and truth is always pretended in declarative discourse. Experimenters always seek to use plausible sentences, and to abide by the general properties of the reference reality, in particular probabilistic ones. I will use this notion of reference in analysis of verbs below, in the belief that the relationship to reality must be fundamental to any attempt to naturalize the mind (Dretske, 1995), since evolution has presumably made humans and their brains able to represent the external world in an appropriate, if approximate, way.

The last major difference between standard logic and cognitive psychology in this domain that I will mention is the separation of predicates and quantifiers. Whereas quantifiers are basic components of first order predicate calculus, and this calculus cannot be used without quantifiers, cognitive psychologists study them only in research on modes of reasoning that explicitly use them. They are seldom to the fore in studies on pragmatic inferences during discourse comprehension (Campion and Rossi, 1999; Fincher-Kiefer, 1993; Graesser and Bower, 1990; Graesser, Singer and Trabasso, 1994; McKoon & Ratcliff, 1992; Martins & Le Bouédec, 1998; Walker and Yekovich, 1984, 1987), and they are often omitted from formal notations, as will be seen below.

PREDICATIVE STRUCTURE

It should be possible to find a cross-cognitive common content of the notions of 'proposition' and 'predicate' if we view them as describing general properties of the representational content of language and thought, i. e. as syntactical-semantic structures. This section examines the structure of many-place predicates by analysing a major class of them, and shows that extending the analysis proposed by Davidson (1980) for actions or events[2] is highly illuminating and has wide-reaching consequences. I will use the word "predicate", and the corresponding notion, as an equivalent of "predicator" (Carnap, 1947), referring to an abstract type of unit, the tokens of which are identifiable in sentences, and the content inferred to be their "logical meaning". This is clearly different from a logical-behavioristic view, such as that of Quine (1953, 1960, 1970). Naturalization of this notion by cognitive psychology leads to the idea that propositions are actual states of individual thought and

[2] Take actions as a subset of events and thus, thereafter, only mention the latter ones.

discourse, while predicates, as constituents of propositions, are another kind of mental state. I will return to this question later.

I will first examine the symbolic description of propositions, i. e. of discourse and thought contents, as it is expressed in symbolic notations, or in rational representation, as opposed to natural representations. In the conventional symbols of logic one-place predicates are abstractly noted as P, and simple propositions as P (a), or more explicitly as P (x), or even P ( ), for example, DOG (x), or BLOND ( ), in which x is a variable, or the space between the brackets the place for the argument. The latter notations already show the basic predicative structure and high level properties of one-place predicates I will use more widely later; they are pieces of thought content; they contain one or more empty place(s) and are, thus, incomplete entities, "unsaturated" in Frege's words; and each of these places has a specific nature that allows it to be filled only by other determined pieces of content.

This structure is what makes predication possible. Predication in discourse can thus be described in production or in comprehension. Let us begin with the latter:

Predication in comprehension is a cognitive, mental act by which comprehenders who have heard or read two appropriate successive words, and accessed their meanings, mentally assign the meaning of the predicate word to that of the argument word, or vice versa. In this way they build a mental proposition as a part of the meaning of the current sentence.

Similarly, predication in production is the cognitive, mental, act by which speakers first mentally assign a predicate to an argument, or vice versa, so building a mental proposition they must then express by enunciating the corresponding words to form the sentence or sentence part. The true or false value of the proposition thus becomes a posterior property, assigned to it as the result of an actual or idealized judgement. The sentence "Socrates is a man" must be given a meaning before any truth can be assigned and it treated logically.

I will return to the predicative structure later.

Many-Place Predicates of Events and Actions

The conventional classification of predicates based on the number of places they offer, their arity or "adicity", seems to yield a slightly different abstract structure for two-place predicates, for example P (x, y). This becomes different after the major change recently introduced by Davidson (1980) for the "logical form" of event (including action) sentences.

Davidson took as an example the sentence:

"Smith buttered the toast in the bathroom with a knife at midnight".

Its conventional logical analysis would be based upon a polyadic, five-argument predicate, BUTTERED (a, b, c, d, e), containing argument places that are usually filled by singular terms or bound variables, here 'Smith', 'toast', 'bathroom', 'knife', 'midnight'. Davidson states that event (action) verbs are predicates that actually contain one more place, to be filled by the particular action or event referred to by the sentence concerned. This must thus be represented by a special letter in the notation.

Let us use a simpler example with only two arguments

"John hit Paul"

and compare its possible notations. In the conventional notation adopted in cognitive psychology (Kintsch, 1974, 1998; Le Ny, 1979; van Dijk and Kintsch, 1983), the predicate, HIT, has two places, HIT (x, y), and the notation will be, disregarding quantifiers:

HIT (John, Paul)

By contrast, a notation derived from Davidson's analysis would take the predicate as a three-place one, HIT (e, x, y), yielding the notation:

HIT (e, John, Paul)

which explicitly expresses the idea that there was a particular event, denoted e, with a determined content (the hitting), in which John and Paul were concerned. Davidson (1980) showed the logical advantages of this solution. It permits to quantify on events, as standard logic does for individuals, and deductive inferences to be formally drawn from a complex sentence to a simpler one contained within it, for example: IF "Smith buttered the toast in the bathroom with a knife at midnight", THEN " Smith buttered the toast". It also allows reference to the mentioned action in discourse, for example by adding a sentence (5) to my sentence 2: "....and this was bad".

These advantages can be interpreted as implying that speakers, when they enunciate an event sentence, such as 1. or 2., or comprehenders, when they process them, make a mental reference to the particular event concerned, in the same way as they do to objects or protagonists. The possibility of making reference linguistically noticed by Davidson, and exemplified in my sentence 5. is behavioral evidence of this. The advantages cited by Davidson can thus be viewed as being both logical and cross-cognitive .

The superiority of this notation has many consequences. Before discussing them, it may be useful to examine exactly how cognitive psychologists can use such notations, and why some notations are better than others. An adequate notation of a statement is, as I indicated above, a rational representation that allows to describe appropriately the content of a statement to be conveyed from a speaker to a comprehender. This conveyed content cannot in general be better assessed than by observing/inferring its effect on the mind of the comprehender. In other words, the "semantic post-representations" (Le Ny, 1997; see below) it produces in their mind/brain. Cognitive psychologists have developed several techniques collecting behavioral data after comprehension of a statement and inferring from them the corresponding semantic representation and its properties. An adequate notation of a statement in an appropriate symbolic language is thus equivalent to one designed to describe the contents of the semantic representation built in the mind/brain of an ideal comprehender who has understood the statement concerned. For ordinary statements, this ideal semantic representation would not be too different from those built by average people comprehending the same statement. The subjects of a cognitive psychology experiment are assumed to be a representative sample of such people.

The terms used in such a notation have thus an inferred, mental but real, reference. For example, the term "toast" in the notation of 6.1. (below) will mean: "the mental representation of the actual object toast, as shared by all (excepting a very few) English speakers". Between-speaker variations in the representation is assumed to be small in this particular case, but it can be large in others, making the attribution of meaning more risky. I believe that requirements for notations of cognitive psychology are convergent with those of other cognitive disciplines, so that adequate cross-cognitive notation is possible.

Let us now return to example 1. The conventional psychological notation of it would take several lines:

6.1. BUTTER (John, toast)
6.2. IN (bathroom, (1))
6.3. WITH (knife, (1))
6.4. AT (midnight, (1))

Preposition meanings are used as predicates in this notation, and a proposition, noted (1), is taken as an argument of these predicates. This second feature is contrary to standard logic, and thus inappropriate from a cross-cognitive point of view. Its use in cognitive psychology is due to the notion of 'atomic proposition' (van Dijk and Kintsch, 1983), viewed as a minimal proposition, especially in long clauses or sentences. This is an excellent notion that makes it unnecessary to use many-place polyadic predicates.

By contrast, the conventional logical notation of sentence 1. would be:

BUTTERED (Smith, toast, in the bathroom, with a knife, at midnight)

This makes a single proposition, and the last three arguments are inserted into it in exactly the same way as the first two. This notation allows the unlimited inclusion of more arguments. Neither does it seem to give an appropriate account of multiple-adjunct sentences like 1. It thus seems preferable to use atomic propositions, especially as each one can be separately negated, and so has its own truth value. This confirms that 6.1. actually notes a central atomic proposition, equivalent from a syntactic point of view to the meaning of the clause's syntactical core. This proposition is also precisely the one logicians wish to be able to deduce from a complex proposition such as 7., if we choose this notation. The idea that particular predicates as prepositions can directly take propositions as arguments is unsatisfactory. This possibility should be reserved for relationships that are clearly inter-propositional, to be denoted R (p1, p2), as AND, OR, BEFORE, AFTER, BECAUSE, and other connectives.

The notation presented by Davidson (1980), seems to be very satisfactory for resolving these problems. By introducing a particular argument intended to separately note the event denoted by "buttered" in the sentence, it has the major advantage of avoiding both notations 6. and 7. It gives (notation mine, JFLN):

8.1. BUTTERED (John, toast, e)
8.2. IN (bathroom, e)
8.3. WITH (knife, e)
8.4. AT (midnight, e)

Here, 8. 1. notes the main atomic proposition in the sentence, the meaning of its syntactical core, and e the particular event in it, the action of buttering. 8.2. notes a second atomic proposition, in which the predicate is IN, and has two arguments: the first, "bathroom", represents the content of a particular adjunct, a particular location, and the second, e, again represents the event. Similarly, 8. 3. notes a third atomic proposition, in which WITH is the predicate, also with two arguments, the first, "knife", representing a particular instrument. Lastly, 8. 4. notes a fourth atomic proposition, in which AT (in a determined sense) is the predicate, still with two arguments, the first, "midnight" representing time. The use of prepositions as predicates can be justified in this context by the same reasons as for events as BUTTERED. A more sophisticated notation would thus have to introduce a letter noting instantiation of the relation they express, but this matter cannot be discussed here. I will also not examine the consequences for sentence interpretation that would result from introducing an existential quantifier.

VERBS

I have followed Davidson's notation in 7. and 8. 1. in using BUTTERED to note the first predicate, but I believe it is preferable to write BUTTER, the infinitive form of the verb, rather than a conjugated form such as BUTTERED. This choice of course implies that a mark for 'past' should be separately introduced if we wanted a complete notation. But it seems more appropriate, from a cross-cognitive point of view, to use high level type lexical units, the infinitive being assumed to carry the conceptual content of the word more centrally than inflexed forms.

BUTTER is not an ideal example of predicative structure. Thus, before generalizing Davidson's analysis, I will apply it to verbs that conventionally are one-place event predicates. RUN (x) is an example of a conventional one-place event predicate, P (x), but must be viewed as being two-place, RUN (e, x), belonging to the sub-class P (e, x), in exactly the same way as BUTTER, that previously appeared to be an instance of a two-place predicate, P (x, y) under the notation BUTTER (x, y) proved to be three-place, BUTTER (e, x, y), belonging to the sub-class P (e, x, y). A notation of the sentence:

"Mary runs"

thus should not be written:

RUN (Mary), but
RUN (e, Mary).

The underlying cognitive interpretation is that verb predicates are given to speakers as pre-existent type linguistic units, comprising a set of word forms and their common central meaning, shared by a language community, and generally individually stored by the community members in their long-term memory as words of their mother tongue (or any other learned language). Word form and meaning can exceptionally be dissociated (Cordier and Le Ny, 1998, 1999).

The typical classes of word predicates we usually meet in general lexical studies are nouns, and sometimes adjectives, and they are commonly considered to be psychologically used in discourse through an instantiation process. The above discussion basically shows how verbs can be viewed as similar lexical units. They can thus be characterized (Le Ny, 1998) as: 1. a class of word forms (or, taking their inflexions into consideration, of sub-classes of word forms), 2. the function of which is to convey information on certain world entities, 3. which work in comprehension by first activating token representations in a comprehender's working memory from a corresponding class of long term representations stored in memory, and later allowing predication as shown above.

But which entities in the actual world are specifically represented by the linguistic class of verbs? What are verbs about? This question is similar to the one that can be asked about nouns, or adjectives. The response is that nouns typically represent things (and more generally entities viewed as things), and adjectives their properties. I write "typically" because this attribution of reference cannot apply to all instances of the class.

Several empirical studies, using various, linguistic, philosophical or psychological methods, have been recently devoted to the semantic classification of verbs, in particular in English and French (e. g. Baudet, Cordier and François, 1997; Cordier, 1987, 1993; Desclés, 1990; Desclés, Flageul, Kekenbosch, Meunier and Richard, 1995; Fiez and Tranel, 1997; François, 1997a, 1997b; François and Tapiero, 1997; Gentner, 1981; Jackendoff, 1992; Langacker, 1987, 1991, 1992; Miller and Johnson-Laird, 1976; Récanati and Récanati, 1997; Tijus, Legros, Hamilton and Richard, 1998; Vendler, 1987). They show that differences among verbs can be characterized in terms of a few semantic classifiers, despite the great heterogeneity of this lexical class. There are two main groups of these. First, change, causativity, agentivity or control by humans, time, aspect and mode of progress in time, and secondarily, the presence or absence of an end, attainment of this end, homogeneity or heterogeneity of change, nature of states of no-change during the process, opposition between process in progress and the rest of it, in short the multiple potential constituents of an event or process. Second, the role or participation of the entities or protagonists submitted to or bringing about this event or process, as studied in linguistic research on actantial roles (in the words of Tesniére, 1959) or casual roles (Fillmore, 1968). These dimensions of difference may be interpreted as semantic features, receiving various degrees of salience in the meanings concerned. These analyses can also enrich this notion of a 'semantic feature' (Le Ny, 1976, 1979, 1998).

The results of these linguistic and psycholinguistic studies support the basic characterization of general verb reference presented above. They typically convey change meanings, i. e. the result of their processing in sentence comprehension is the activation of mental representations of physical changes or processes in the world flow. I will return to the nature of these world processes later.

Unification and Superunification of the Notions of Predicate and Concept

The notational change I have described above, based on that of Davidson, is not only a symbolic, purely formal one. It expresses a mental view of event predicates that deserves to be extended, with major consequences, to any cognitive theory of language and thought.

The first consequence is that it leads to a unification of the notion of predicate that makes it somewhat different from the conventional one. This is best shown by the way reference is made in event sentences or propositions to the event concerned. This particular reference, noted above by e, is of exactly the same nature as the usual reference made to an object, let us say a, in the formulas:

12.1. TOAST (a)
12. 2. BUTTER (e, Smith, a)

This reference to the event can explicitly appear in discourse, as in sentence 5 above, which used an indexical, "this (was bad)", to express it. Another linguistic device that is more general and has the same function is the nominalization of verbs. This makes predication of events easier. It may have various forms, some directly related to the verb form as in:

"the buttering of the toast by Smith in the bathroom at midnight was jerky, (because he was angry)"

or, with a metonymy,

"the buttering of the toast by Smith in the bathroom at midnight was angry".

Nominalization may also be independent of the verb form, as in:

"the blow of John at Paul was violent", corresponding to 2., 3., 4., or
"Mary's race was feline", corresponding to 9., 10., 11.

Some languages have linguistic devices that make it easy to nominalize all verbs, as in German with "das + infinitive", or English with "the + gerund", with, in addition, specific names of events, as "blow" and "race" above. Other languages, such as French, have no handy device for designating events and actions from verbs, and use various types of names. Besides, some languages, such as Russian, like to use nominalizations where others prefer to use verbs, but reference is made similarly. The existence of adjectives that apply to events, and in particular to actions, like "slow" and "fast ", "jerky", "violent", "feline", conveying meanings that are general properties of events or actions, not of objects, is also linguistic evidence of this unified structure of predicates, however many places they have.

The main general property of predicates that stands out from this unified structure is their capacity to be instantiated. Whatever their number of places, the content that must be inserted into one of these first refers to an individual entity. This can be an object, property, or event, depending on the semantic class of the word predicate.

A second major consequence of the analysis of event sentences, and a sub-consequence of the previous consequence, is the possibility of unifying the notions of 'predicate' and 'concept'. These are generally encountered in different contexts of cognitive science, 'predicate' more often in logic, or logic-based philosophy and linguistics, 'concept', with its brother 'meaning', more often in philosophy of the mind and cognitive psychology. For example, the nature of concepts can be thoroughly analysed without reference to the notion of 'predicate' (Fodor, 1998).

But if we welcome the notions of 'predicate' and 'concept' in a cross-cognitive framework, we must determine their respective mental forms, either as mental realizations of abstract, non-psychological entities, or the basic contents of mind/brain states, of which the abstract entities handled by other cognitive disciplines are idealizations. We then see that mental predicates and mental concepts cannot be two distinct categories. Similarly, if we wanted to implement a cognitive system able to process a wide range of propositions or sentences, such as a natural language processing machine, we would have to build a lexicon, with meanings associated with word forms (Le Ny, 1995), and it would be preposterous to create two distinct categories of units, each with its own processing procedures.

It is therefore rational to unify the two notions, and believe there is only one category of mental semantic representations in long-term memory[3], that are active in discourse comprehension or production as in internal speech and thought. The foregoing analysis shows that predicates have a basic property, in common with concepts, the possibility to be instantiated. A natural concept is a particular type of mental representation in a subject, and a predicate is exactly the same reality.

The notion of 'predicate' is used in very abstract contexts, and the corresponding entities themselves are often assumed to be very abstract, whereas mental concepts or meanings have been shown by cognitive psychology research to have various degrees of abstractness/concreteness and imageability. But this does not justify a particular category for the basic, natural reality of predicates. The notion of 'predicate' focuses on one aspect of the semantic representations, as stressed by Frege, their "unsaturated", incomplete, character, so that they need completion by perception or discourse. 'Dog', or 'to hit' are generic representations that can apply to any instance of the infinite set of dogs, or hittings, as long as the mental activity involving this application allows us to disregard many attribute-values as irrelevant in the current context. But when one perceives a dog, or someone (something) hitting someone (something), many more attribute-values, mainly figurative, are reinstated. Similarly, when "dog" or "hit" are mentioned in a sentence, many attribute-values of the particular animal or event referred to are unmentioned, and the relevant ones must be reinstated by the remainder of the sentence or discourse.

Thus, these representations "wait for", rather than "need", completion, which paves the way to a transposition of this property to a neurobiological interpretation of representations' substrates (Le Ny, 1998): I will not elaborate on this transposition here.

Discourse Comprehension and Semantic Post-Representations

Discourse comprehension can be viewed (Britton and Graesser, 1995; Gernsbacher, 1990, 1997; Just & Carpenter, 1992; Kintsch, 1988, 1998; Kintsch and van Dijk, 1978; Le Ny, 1989, 1997; Sharkey, 1990; van Dijk and Kintsch, 1983) as basically consisting of the construction of transient representations in working memory, semantic post-representations, that can be conserved later, perhaps in a long-term working memory (Ericsson and Kintsch, 1995), or in conventional long-term memory. As comprehension is generally believed to

[3] I make here no judgement whichsoever on the format of these representations, and the role of mental images, which is a matter of psychological study.

proceed sentence by sentence (provided these are of medium length and difficulty), and event sentences are the basis of narrative discourse, then how are semantic post-representations generated by these sentences? I will disregard the more complex issue of larger semantic representations, such as those generated by comprehension of a whole story, which are obviously built by integrating several successive one-sentence semantic post-representations.

Research in this domain has introduced the notion of 'situation representation'. A series of experimental studies developed it under several denominations as "mental models" (Albrecht and O'Brien, 1993; Ehrlich, Tardieu and Cavazza, 1993;Glenberg, Meyer and Lindem, 1987; Johnson-Laird, 1983), "situation models"[4] (Carreiras, Carriedo, Alonso and Fernandez, 1997; Fincher-Kiefer, 1993; Morrow, 1985; Morrow, Bower and Greenspan, 1989; Morrow, Greenspan and Bower, 1987; Rinck and Bower, 1995; Rinck, H"hnel, Bower, and Glowalla, 1997; Rinck, Williams, Bower, and Becker, 1996; Wilson, Rinck, McNamara, Bower, and Morrow, 1993; Zwaan, Langston and Graesser, 1995; Zwaan, Magliano and Graesser, 1995; see Zwaan and Radvansky, 1998, for a recent review). One result of these studies is that comprehension of an event sentence (and of several successive ones dealing with the same topic) produces a representation of the situation described by the sentence, which contains a large set of organized semantic constituents.

Zwaan, Langston and Graesser (1995), Zwaan and Radvansky (1998) have developed an event-indexing model in which they try to encompass the results of these experimental studies. This model is based on the idea that "events [i. e. their representations, JFLN] are the building blocks of integrated situation models", and has many similarities to the analysis of events and verbs presented above. In particular, this model involves five "dimensions" of situations: space, causation, intentionality, protagonists and objects, and time, that are very similar to the semantic components of verbs mentioned above.

I will be mostly interested in another major result of studies on situation representations - the focus of their constituents varies, and can be experimentally assessed. Denis and Le Ny (1986; experiment 1) used pairs of sentences containing the same word referring to an object, for example "church". The first sentence of this pair induced comprehenders to focus on a part of this object, such as the church porch: "every Sunday morning at Cormainville church a beggar stood, holding out his hand to people coming out after the mass". The second sentence, by contrast, induced comprehenders to focus on another part of the same object, the church bell-tower: "as Cormainville came into view, the first building one could see was the church, proudly dominating the roofs of the village". The relevant part of the object was not mentioned. Subjects were instructed to read their sentences, and immediately shown drawings of one or the other part of the object. They had to indicate whether this drawing illustrated the scene correctly or not by pressing "yes" or "no" buttons. The responses were always correct, but the response times were shorter when the subjects were given a drawing for the object part focused on in the sentence.

Glenberg, Meyer and Linden (1987) obtained similar results from subjects shown one of two short stories mentioning the presence or absence of an object, for example a sweater, that was kept on or taken off by a running man in the story. "Yes" or "no" recognition responses

[4] Use of the word "model" in this sense and this context seems to me to be infortunate, as is shown up, for example, by the fact that the model presented by Radvansky and Zwaan (1998) has to be a "model of situation models", with two different senses of the word '"model" in the same phrase. It seems preferable to reserve the word "model" to the sub-category of systematic, rational, representations built by scientists. scientists, and use "representations" in the general case.

to probe words, for example "sweater", were given after reading. Response times were shorter when the story said the object was present in the situation than when it was absent.

These results were interpreted as evidence that the representation of the church porch or bell-tower, in Denis and Le Ny's experiment, the sweater in that of Glenberg and al., which is assumed to be present in subjects' working memory, is more accessible when it is addressed via the picture or name of the object in the subsequent test. The results of other experiments on representations of object locations in the situation (Morrow, 1985; Morrow, Bower and Greenspan, 1989; Morrow, Greenspan and Bower, 1987; Rinck and Bower, 1995), its time (Carreiras, Carriedo, Alonso and Fernandez, 1997), and other constituents of it, as its protagonists (Morrow, 1985), its cause, etc., are in general agreement with the previous ones.

The general fact evidenced by these studies can be described as the role of attentional focusing in discourse comprehension and building of the semantic post-representation. Metaphorically, the constituents concerned, the church porch or bell-tower, or the sweater, are said to be "in the foreground" (Carreiras and al, 1997; Glenberg and al., 1987; Hopper, 1979; Whitney, Ritchie and Crane, 1992; Zwaan and Radvansky, 1998) of the situation representation. Another interpretation, given in a more audacious cross-cognitive framework[5] related to neurobiology (Le Ny, 1997), is that attentional focusing brings about greater activation of the concerned representational constituent in the whole semantic post-representation, more precisely of its substrate in the subjects' mind/brain. These results show the importance of focusing in discourse comprehension.

THE FOCUSED SITUATION

A situation representation generated by a sentence is, as a rule, a representation of a short situation, and thus identical to an event representation, but it takes the setting of the event into greater consideration. Adding the idea of focussing to our views of event and situation representations can help to enrich them and understand them more clearly. The situation representation is considered to have "semantic constituents", or structured parts referring to world elements in a representational structure, rather than "dimensions" (Zwaan and Radvansky, 1998), which concern space and time. All are differently focussed.

The setting is expressed in a sentence by the adjuncts, which describe the circumstances of the event. My Davidsonian analysis in the beginning of this paper shows this structure of representations generated by event sentences in a formal way, based on a core/circumstances, sun/satellite relationship, but it is more accurately/thoroughly described as a situation representation. Semantic analysis of verb contents shows this as a latent structure of a verb's meaning. Psychological experiments have largely confirmed this analysis and added that focusing, or the activation level in another framework, varies considerably among the semantic constituents in the comprehender's mind/brain.

[5] I do not elaborate this transposition here. See Le Ny, 1998.

A General Schema of Situation Representations

These analyses suggest that people not only build token representations of situations, including attentional focusing of their constituents, as a result of their comprehension of event sentences, but also have a general schema that guides that construction.

A situation representation can be viewed as the representation of a spatio-temporal slice of the world, cut out from the total fabric of the flowing reality, and containing a set of differently focused parts. It is analogous to a frame of a motion picture.

Any token situation has its limits, a priori undetermined, and a central area with a more focused content. In comprehension, degrees of focusing, or activation levels, as assessed by experimental methods, such as measuring response times in probing tasks as shown above, seem to be attributed to the various semantic constituents of the situation representation as a function of their cognitive importance (Le Ny, 1997). This view is close to the description of the final outputs of processing introduced by connectionist models or attractor networks (Hinton and Shallice, 1991; Rumelhart and McClelland, 1986; Sharkey,1990). It can be illustrated metaphorically by a mountain landscape with peaks and valleys. The constituents corresponding to the verb and the protagonists would usually be on the highest peaks, and those corresponding to sentence adjuncts, or products of pragmatic inferences, would be lower, with heights corresponding to their current activation level.

This view throws light upon the abstract schema analyzed above, which can be made generally representative of these degrees of focusing and activation. In an abstract notation summarizing 8., we have:

17.1. P (e, a, b)
17.2. Pp1 (e, c)
17.3. Pp2 (e, d)
17.3. Pp3 (e, f)
.....

in which e notes the verb meaning, Pp1, Pp2 and Pp3 the meanings of three prepositions, a and b refer to the protagonists, c, d and f to particular circumstances. It is clear that e, a and b would be the most focused constituents, c, d and f the lesser ones.

Focusing in Language and Perception

I have claimed that verbs, with their wide range of contents, are abstract units of the vocabulary, stored in a speakers' long-term memory, that very often focus the attention highly. But focusing is not only active during sentence comprehension, it is also active in individual development, and presumably in the learning of meaning/concept/predicate units.

We can obtain a fair idea of how focusing can determine situation representations by comparing language structures with perceptive ones. Let us begin with the limits of a situation. Perception, in particular vision, cuts out perceptual slices, token «perceptions», from individual environments The contents of these slices are "states of affairs" or, as here, "situations". These perceptions are necessarily centered on the perceiver him/herself, and are first delimited in the immediate environment by the reach of the individual's sense organs, his orientation and point of view. Both the spatial limits of the perceived situation and salience of its constituents are determined and modulated by the focus of the perceiver's attention on the current environment, which controls where and what the subject looks at, or hears, and is retroactively controlled by it. Eye movements and fixations are the major perceptive instruments of this in vision, ear orientation in hearing. The time limits of a momentary situation are less strictly given by nature and management of sense organs: they may vary considerably, depending on eye movements and the duration of fixation. But a basic common characteristic of perception and discourse must be stressed: as perception pastes together successive short frames resulting from successive fixations to construct an apparently continuous view of the world, discourse comprehension, best illustrated by the comprehension of a story, pastes together and integrates successive short situation representations, each one built separately after processing an event sentence.

An essential function of discourse, and presumably one of the first to evolve in humans, is for a person perceiving an interesting event to pass the information about it to another person who did not directly perceive it. Thus, a semantic post-representation generated after an event sentence is a substitute for a perceptual representation constructed when regarding the event itself. The major difference is that language suppresses a number of figurative aspects present in perception. But a major finding of the studies on situation representations mentioned above is that these retain more cognitive properties from perception than was previously believed.

For example, space and time proximity relationships are conserved within certain limits. More figurative aspects can also be activated in ordinary discourse comprehension, provided an extra imaging cognitive effort is made, with a time cost (Denis and Cocude, 1992). But in general, many figurative features are irretrievably lost by language, though this varies greatly from one individual to another.

Change and Permanence

It is not surprising that many of the representational pieces stored by human minds/brains, chiefly verbs, are used to report events. In learning a language, verbs are pieces of meaning that come mainly from extracting knowledge of changes from the flowing world, and are crystallized into words of this particular semantic category. Nouns and adjective, by contrast, are primarily words that result from permanence extraction and crystallization. Some verb meanings encode basic changes of the world at a high or medium high level of abstraction, such as 'to change', 'to move' or 'to run', but there are many more particular verbs, some very particular, of which 'to butter' is a good example. Details of the formation of these pieces of meaning is relevant to human life. These may encode minor changes, if they are interesting with regard to human life.

Focusing, or attention, can thus play a major role in learning the meanings of words, not only their use in discourse. Attention is mostly captured by changes in world parts or aspects, and interest is commonly judged to be associated with an optimal amount of change, as was first shown by Berlyne (1960). Relationships between these changes and objects or individuals that change, i. e. to which something happens, are precisely the core of event sentences, and of actions sentences, when changes or moves are caused by a human.

The contrast between change and permanence, in connection with another pair of contrasts, same and different, is a major basis of cognition. The complex role of perceptual change in capturing the attention of very young babies and giving place to habituation is the first evidence of this very early role. These phenomena are now used as methods for the study of cognition in infancy (Spelke, 1986). Further evidence is the importance of the 'permanent object' schema as a cornerstone of cognitive development in infants, as stressed by Piaget (1936, 1937). Subsequent studies (e. g. Baillargeon, 1987; Baillargeon and de Vos, 1991; Baillargeon, Spelke, and Wasserman, 1985; Lécuyer, Pécheux and Streri, 1994; Spelke, Breilinger, Macomber and Jacobson, 1992) specified and sometimes corrected Piaget's views, showing that the "permanent object" schema appears earlier, is presumably based on a perceptual rather than action cognitive activity, and may be even more basic than Piaget believed. But they did not diminish their importance. Perception and recognition of particular objects, their properties, and their changes must be considered to be inserted into this perceptive and conceptual scheme, or schema, and classes of lexical meanings can be viewed as largely based on the underlying contrast of permanence and change. That does not mean that the world is perceived and thought of in cognitive "categories" superimposed on it in a transcendental way, as in Kant's view (Kant, 1781), but that perception, thought and language are controlled by representational schemas suited to world structures, presumably as a result of evolution.

Change (or non-change when change is expected) is a basic property of the world represented in verb meanings. But it follows from this that verbs can only be meaningful if they are applied - predicated - to non-changing entities. These entities are objects, including living and human ones, that have a permanent object status, and are thought of as invariant with respect to many attributes. Actantial or casual roles are identified, and attributed to these entities during discourse processing as a function of a set of implicit general questions: what is said to change? what, in this, remains correlatively permanent? in respect to what does this change of permanent entities take place? under which circumstances (which conditions of external changes and permanence, associated with the main change) does this take place? Some of the responses to these questions are conveyed in the message, and given to the comprehender, but others remain implicit, and must be reconstructed by the comprehender's inferences. The information contained in an event sentence thus leads to representation of a whole situation.

However, while verbs represent change, they can also be viewed as representing a type of invariance in memory, different from the invariance by permanence represented in the meanings of object nouns. This invariance is due to repetition, and learning verb meanings can be viewed as storing the contents of repeatable events.

The other world properties involved in verbs are also related to change/permanence, in so far they are focused on: time, aspects, mode of progress in time of the process or change, nature of the states of no-change during the process, etc. They make various kinds of combination in the lexicon, the ones the various languages have historically assigned a meaning unit, more or less at chance for the secondary ones. They therefore provide speakers with a wide variety of tools for cutting out slices from the world. Their common feature is the general temporal framework, with the basic opposition between 'before', 'during' and 'after' (Desclés, 1990). Circumstances are a constitutive part of any situation, irrespective of whether they are expressed in a sentence by specific adjuncts as in 17., or remain implicit, with a variable contribution from the non-linguistic enunciative context. That they are often peripheral or poorly focussed, does not detract from their basic role in situation representations.

Conclusion

If cognition is considered to be a single object to be studied by multiple disciplines, that together form cognitive science, it is necessary to take advantage of the progress made in all of these disciplines to exchange new results and make them participate to common knowledge. This cross-cognitive approach must incorporate results obtained by empirical methods, experiment as in cognitive psychology and neurobiology, observation, as in linguistics, and construction of programs, as in artificial intelligence research, and other more conceptually based results, obtained by analytical or deductive methods, such as in logic or philosophy of the mind. Some notions are common to all these disciplines, they concern language and thought, and are at the top of the lattice of notions that encompass descriptions of cognition,. These include 'proposition', 'predicate' and 'concept'.

Recent research in the fields of logic, linguistics and cognitive psychology, on notions located lower, but still high, in the lattice, such as 'event', 'action', 'situation', 'constituent of situations', and 'constituent focusing', have shown such convergence that they seem to justify fundamental modification of the highest notions. Unification of the notion of 'predicate', and its super-unification with 'concept' might be of paramount importance for cognitive science. The content of the other notions analyzed in this paper depends more on empirical research, particularly experimental studies, which are still in progress.

References

Albrecht, J. E. & O'Brien, E. J. (1993). Updating a mental model: Maintaining both global and local coherence. *Journal of Experimental Psychology: Learning, Memory and Cognition,* 19, 1061-1070.

Baillargeon, R. (1987). Object permanence in 3.5 and 4.5-month-old infants. *Developmental Psychology*, 23, 655-664.

Baillargeon, R., Spelke, E. S., & Wasserman, S. (1985). Object permanence in five-month-old infants. *Cognition*, 20, 191-208.

Baillargeon, R., & DeVos, J. (1991). Object permanence in young infants: Further evidence. *Child Development*, 62, 1227-1246.

Baudet, S., Cordier, F., & François, J. (1997). L'identification des classes de procés en contexte. II: Etude expérimentale. In Jacques François et Guy Denhiére: *Sémantique linguistique et psychologie cognitive*. Grenoble: Presses Universitaires de Grenoble. pp. 249-282.

Berlyne, D. (1960) *Conflict, arousal and curiosity*. New York: McGraw Hill.

Britton, B. K. & Graesser, A. C. (1995). *Models of text understanding*. Hillsdale, NJ: Erlbaum.

Campion, N., & Rossi, J. P. (1999). *Inférences et compréhension de textes*. L'Année Psychologique, sous presse.

Carnap, R. (1947). *Meaning and necessity*. Chicago: University of Chicago Press.

Carreiras, M., Carriedo, N., Alonso, M. A., & Fernandez, A. (1997). The role of verb tense and verb aspect in the foregrounding of information during reading. *Memory and Cognition*, 25, 438-446.

Cordier, F. (1987). Analyse de la représentation typique des verbes de procés. *Revue canadienne de Psychologie*, 41, 490-500.

Cordier, F. (1993). *Les représentations cognitives privilégiées.* Lille: Presses Universitaires de Lille.

Cordier, F. & Le Ny, J. F. (1998). *La contribution de la signification à la reconnaissance de mots: nouvelles données.* Actes du Colloque de l'Association pour la recherche Cognitive, Saint-Denis.

Cordier, F. & Le Ny, J. F. (1999). *Contribution of meaning to word recognition in an isolated lexical decision task.* Manuscript. Orsay: LIMSI-CNRS.

Davidson, D. (1980). *Essays on Actions and Events.* Oxford, Clarendon Press.

Denis, M., & Cocude, M. (1992). Structural properties of visual images constructed from poorly or well-structured verbal descriptions. *Memory and Cognition*, 20, 497-506.

Denis, M. & Le Ny, J. F. (1986) Centering on figurative features during the comprehension of sentences describing scenes. *Psychological Research*, 48, 145-152.

Desclé[FC1]s, J. P. (1990). *Langages applicatifs, langues naturelles et cognition.* Paris: Hermés.

Desclés, J. P., Flageul, V., Kekenbosch, C., Meunier, J. M., & Richard, J. F. (1998). *Sémantique cognitive de l'action: une approche théorique et expérimentale.* Langages.

Dretske, F. (1995). *Naturalizing the mind.* Cambridge, Ma: MIT Press.

Ehrlich, M. F., Tardieu, H., & Cavazza, M. (eds) (1993) *Les modéles mentaux: Approche cognitive des représentations.* Paris: Masson.

Ericsson, K. A. & Kintsch, W. (1995). Memory in comprehension and problem solving: A long-term working memory. *Psychological Review,* 102, 211-245.

Fiez, J. A., et Tranel, D. (1997). Standardized stimuli and procedures for investigating the retrieval of lexical and conceptual knowledge for actions. *Memory and Cognition*, 543-569.

Fillmore, C. J. (1968). The case for case. In E. Bach & R. T. Harms (Ed.) *Universal of linguistic theory.* New-York: Holt, Rinehart & Winston, p.1-90.

Fincher-Kiefer, R. (1993). The role of predictive inference in situation model construction,. *Discourse Processes*, 16, 99-124.

François, J. (1991). La pertinence linguistique des représentations propositionnelles de la sémantique cognitive. *Sémiotiques* 1, 69-80.

François, J. (1997a). La place de l'aspect et de la participation dans les classements conceptuels des prédications verbales. In Jacques François et Guy Denhiére: *Sémantique linguistique et psychologie cognitive.* Grenoble: Presses Universitaires de Grenoble. pp. 119-156.

François, J. (1997b). La combinatoire des propriétés aspectuelles et participatives des prédications verbales. *Cahiers de Recherche Linguistique*, 9, 5-38.

François, J., & Tapiero, I. (1997). L'identification des classes de procés en contexte. In Jacques François et Guy Denhiére: *Sémantique linguistique et psychologie cognitive.* Grenoble: Presses Universitaires de Grenoble. pp. 225-248.

Frege, G. (1892). Sinn und Bedeutung. *Zeitschrift f r Philosophie und philosophische Kritik*, 100.

Gentner, D (1981). Verb semantic structures in memory for sentences: Evidence for componential representation. *Cognitive Psychology*, 13, 56-83.

Gernsbacher, M. A. (1990). *Language comprehension as structure building*. Hillsdale, NJ:Erlbaum.

Gernsbacher, M. A. (1997). Two decades of structure building. *Discourse Processes*, 23, 265-304.

Glenberg, A. M., Meyer, M., et Lindem, K. (1987). Mental models contribute to foregrounding during text comprehension. *Journal of Memory and Language*, 26, 69-83.

Graesser, A. C., & Bower, G. H. (Eds) (1990). *The psychology of learning and motivation: Inferences and text comprehension*. New York, San Diego: Academic Press.

Graesser, A. C., Singer, M. & Trabasso, T. (1994). Constructing inferences during narrative text comprehension. *Psychological Review,* 101, 371-395.

Hopper, P. J. (1979). Aspect and foregrounding in discourse. In T. Givon (Ed.), *Syntax and semantics. Vol. 12: Discourse and syntax* (pp. 213-241). New-York: Academic Press.

Hinton, G. E., & Shallice, T. (1991). Lesioning an attractor network: Investigations of acquired dyslexia. *Psychological Review*, 98, 74-95.

Jackendoff, R. (1983). *Semantics and Cognition*. Cambridge, Mass: MIT Press.

Johnson-Laird, P. N. (1983). *Mental models: Towards a cognitive science of language, inference, and consciousness*. Cambridge: Cambridge University Press.

Just, M. A., & Carpenter, P. A. (1992). A capacity theory of comprehension: individual differences in working memory. *Psychological Review*, 99, 122-149.

Kintsch, W. (1974) *The representation of meaning in memory*. Hillsdale, NJ:Lawrence Erlbaum.

Kintsch, W. (1988). The use of knowledge in discourse processing: A construction-integration model. *Psychological Review*, 95, 163-182.

Kintsch, W. (1998). *Comprehension: A paradigm for cognition*. Cambridge, UK: Cambridge University Press.

Kintsch, W. & van Dijk, T. (1978). Toward a model of discourse comprehension and production. *Psychological Review*, 85, 363-394.

Langacker, R. (1987, 1992). *Foundations of cognitive grammar*. Stanford: Stanford University Press. Vol. 1 et 2.

Langacker, R. (1991). Nouns and verbs. In R. Langacker: *Concept, Image and Symbol*. Berlin: Mouton/de Gruyter, pp. 59-100.

Lécuyer, R., Pécheux, M. G., et Streri, A. (1994). *Le développement cognitif du nourrisson*. Paris: Nathan.

Le Ny, J. F. (1976). Sémes ou mémes? *Bulletin de Psychologie: La mémoire sémantique*. 46-54.

Le Ny, J. F. (1979). *La sémantique psychologique*. Paris, Presses Universitaires de France.

Le Ny, J. F. (1989). *Science cognitive et compréhension du langage*. Paris: Presses Universitaires de France.

Le Ny, J. F. (1995). Mental lexicon and machine lexicon: which properties are shared by machine and mental word representations? Which are not? In Patrick Saint-Dizier and Evelyne Viegas (eds.), *Computational Lexical Semantics*, New-York: Cambridge University Press, 1995, pp 50-67.

Le Ny, J. F. (1997). Cross-cognitive aspects of discourse comprehension: semantic post-representation. *International Journal of Psychology*, 33, 397-410.

Le Ny, J. F. (1998). La sémantique des verbes et la représentation des situations. In F. Cordier, J. François & J. E. Tyvaert (Eds.) *Sémantiques du verbe; un état de la question.* Nancy: Landisco. Pp.87-100.

MacKoon, G. & Ratcliff, R. (1992). Inference during reading, *Psychological Review*, 99, 440-466.

Martins, D., & Le Bouédec, B. (1998). La production d'inférences lors de la compréhension de textes chez des adultes: une analyse de la littérature. *L'Année Psychologique*, 98, 511-543.

Miller, G. A. et Johnson-Laird, P. N. (1976). *Language and Perception.* Cambridge, Ma: Harvard Univ. Press.

Morrow, D. G. (1985). Prominent characters and events organize narrative understanding. *Journal of Memory and Language*, 24, 390-404.

Morrow, D. G., Bower, G. H., & Greenspan, S. L. (1989). Updating situation models during narrative comprehension. *Journal of Memory and Language,* 28, 292-312.

Morrow, D. G., Greenspan, S. L. et Bower, G. H. (1987). Accessibility and situation models in narrative comprehension. *Journal of Memory and Language*, 26, 165-187.

Newell, A. & Simon, H. A. (1972). *Human problem solving.* Englewood Cliffs, NJ:Prentice Hall.

Norman D. & Rumelhart, D. & the LNR group (1975) *Explorations in cognition.* San Francisco: Freeman.

Piaget, J. (1936). *La naissance de l'intelligence chez l'enfant.* Neuch'tel: Delachaux et Niestlé.

Piaget, J. (1937). *La construction du réel chez l'enfant.* Neuchâtel: Delachaux et Niestlé.

Quine, W. V. O. (1953). *From a logical point of view.* Cambridge, Mass: Cambridge University Press.

Quine, W. V. O. (1960). *Word and object.* Cambridge, Mass: Cambridge University Press.

Quine, W. V. O. (1970). *Philosophy of Logic.* Englewood Cliffs, NJ: Prentice Hall.

Recanati, C., & Recanati, F. (1997). *La classification de Vendler revue et corrigee.* Paris: CREA, rapport no 9718.

Rinck, M. & Bower, G. H. (1995). Anaphora resolution and the focus of attention in situation models. *Journal of Memory and Language*, 34, 110-131.

Rinck, M., H''hnel, A., Bower, G. H., & Glowalla, U. (1997). The metrics of spatial situation models. *Journal of Experimental Psychology: Learning, Memory and Cognition*, 23, 622-637.

Rinck, M., Williams, P., Bower, G. H., & Becker, E. S. (1996). Spatial situation models and narrative understanding: Some generalizations and extensions. *Discourse Processes*, 21, 23-55.

Rumelhart, D. E., & McClelland, J. L. (1986). *Parallel Distributed Processing: Explorations in the microstructure of cognition.* Vol. 1. Cambridge, Ma.: Bradford Books.

Schank, R. C. (1975). *Conceptual information processing.* Amsterdam: North Holland, New York: Elsevier.

Sharkey, N. E. (1990). A connectionist model of text comprehension. In D. A. Balota, G. B. Flores d'Arcais and K. Rayner (eds.). *Comprehension process in reading.* Hillsdale, NJ: Erlbaum.

Spelke, E. S. (1986). Preferential looking methods as tools for the study of cognition in infancy. In G. Gottlieb & N. Krasnegor (Eds). *Measurement of audition and vision in infancy*, pp. 323-363. New York: Ablex.

Spelke, E. S., Breilinger, K., Macomber, J., et Jacobson, K. (1992). Origins of knowledge. *Psychological Review*, 99, 605-632.

Tesniere, L. (1959). *El éments de syntaxe structurale*. Paris: Klincksieck.

Tijus, C. A., Legros, D., Hamilton, E., et Richard, J. F. (1997). *The hierarchical organization of verb meanings*. St-Denis: Université de Paris 8.

van Dijk, T. A. & Kintsch, W. (1983). *Strategies of discourse comprehension*. New-York: Academic Press.

Vendler, Z. (1967). Verbs and times. In: *Linguistics and Philosophy*. Ithaca: Cornell University Press, pp. 97-121.

Walker, C. H. & Yekovich, F. R. (1984). Script-based inferences: Effects of text and knowledge variables on recognition memory. *Journal of Verbal Learning and Verbal Behavior*, 23, 357-370.

Walker, C. H. & Yekovich, F. R. (1987). Activation and use of script based antecedents in anaphoric reference. *Journal of Memory and Language*, 26, 673-691.

Whitney, P., Ritchie, B. G., & Crane, B. S. (1992). The effect of foregrounding on reader's use of predictive inferences. *Memory and Cognition*, 20, 424-432.

Wilson, S. G., Rinck, M., McNamara, T. P., Bower, G. H., & Morrow, D. G. (1993). Mental models and narrative comprehension: Some qualifications. *Journal of Memory and Language*, 32, 141-154.

Zwaan, R. A., Langston, M. C., & Graesser, A. C. (1995). The construction of situation models in narrative comprehension: An event-indexing model. *Psychological Science*, 6, 292-297.

Zwaan, R. A., Magliano, J. P., & Graesser, A. C. (1995). Dimensions of situation model construction in narrative comprehension. *Journal of Experimental Psychology/ Learning, Memory and Cognition*, 21, 386-397.

Zwaan, R. A., & Radvansky, G. A. (1998). Situation models in language comprehension and memory. *Annual Review of Psychology*, 162-185.

Chapter 10

Coaches and Teachers at the Crossroads of Emerging Patterns and Direct Perception

Martinus J. Buekers[*]
Department of Kinesiology,
Faculty of Physical Education

Abstract

The last few decades the field of motor learning and control underwent an interesting theoretical shift. The classical cognitive based approach was challenged by the view that 'direct perception' or the 'dynamic emergence of patterns' could explain how complex movements are controlled. In spite of their evident merits, however, the translation of these concepts into practical principles or rules for teaching sport skills seems to be far from obvious. The present review is an attempt to tackle this problem as it formulates a number of training rules that are based on the assumptions of perception-action coupling and the dynamic system approach.

Key words: Sport practice, Perception-action coupling, Training rules, Ecological psychology.

Introduction

"The masters of martial arts will all tell you something that is very ironic. After years and years of practice, all movements will become intuitive; whilst you hardly 'know' what you are doing. You completely pass your mind, and you know as much of technique as the day you started. As such experts and beginners have something in common. Both are absolutely unaware of technique."

(Hoover, 1989, p. 407)

[*] Address: Katholieke Universiteit Leuven, Belgium, Katholieke Universiteit Leuven, Belgium, Tervuursevest 101, 3000 Leuven, Belgium; Phone number +32 16 329067; Fax number +32 16 329197; (e-mail: *martinus.buekers@flok.kuleuven.ac.be*)

Many coaches and teachers are familiar with this paradox, in particular as they are inclined to find optimal procedures and methodologies to guide their athletes and players from this initial level of unawareness to expert performance level. To do so they often had to fight themselves a way through an overwhelming amount of information piled up in training programs, articles, books, and even web-sites. There is nothing mysterious to the fact that a sound fighting spirit is required to successfully cope with this (scientific) information, as it is evenly true that the same esprit is needed to translate the theoretical concepts that recently emerged in the field of motor control and learning into efficient guidelines for practice. Apparently the flow from research findings to application principles, and vice versa, is not always as fluid as the cycles that are supposed to govern the perception-action chain. In contrast it appears that the distinct ways of thinking adopted by coaches and researchers form the embankments and dams that obstruct the smooth passage of information from one side to the other. This should not be too surprising as both actors seem to be driven by goals that never ceased to provoke a schism between theory and practice, between lab and field, between explanation and application. Whereas researchers in the field of motor control and learning try to understand the different aspects of motor behaviour, coaches are 'in essence' inclined to improve the performance level of their players and athletes. This difference in vocation seems to have an influence that stretches out beyond the borderlines of professional behaviour, provoking a fundamentally divergent perspective. Perhaps this difference in 'Weltanschauung' is best expressed by a phrase used elsewhere: "A coach has to be certain about things he is not certain about, and a scientist has to be uncertain about things he is certain about" (Buekers, 2000). Most likely this contrasting philosophy was an important factor in generating the so-called gap between researchers and practitioners. Although it can be anticipated from this latter argument that it will be troublesome to translate research findings into 'useful' methodologies, this difficulty should not prevent us from trying to tackle the problem.

The goal of the present review is to take up this challenge, as we want to examine how recent research findings can be interpreted from the viewpoint of practice. At the moment this attempt is very appealing given the important theoretical developments of the last few decades. These developments appear to be characterised by a reduction of the impact of cognition on a number of control processes that interfere during the production and execution of movements. This shift from a cognition-based methodology to rules that are based on direct perception or emerging patterns seems to engender a reflex of neglect or self-defence by the coaches, as they are prone to control as much as possible the interaction between athletes and environment. Obviously this urge to control the different parameters of the situation goes along well with a theoretical account that is founded on cognition. Therefore, motor programs and schema's (e.g., Schmidt, 1975; for a recent perspective on these concepts see Willingham, 1998) were embraced by coaches as ideal concepts, not in the least because of their clear, straightforward, and predictable functioning. In contrast to the comprehensible nature of the cognitive approach, the more recent theories appear to be less accessible, even though they proclaim to formulate 'simple laws' of control to explain complex behaviour.

The purpose of the following paragraphs is to make these new approaches more accessible. To do so, we will first plunge into some considerations on coaches and coaching. The second and third paragraph will briefly describe the basic assumptions of the dynamic systems and the ecological psychology approach. In the fourth, and most important paragraph we will try to derive from these approaches some principles or rules that could be implemented by coaches to organise and eventually enhance the efficiency of their practice sessions. In the last part of this review an attempt is made to consider the issue from a broader perspective.

Before entering into the debate I would like to set some boundaries of relevance for the present discussion, a procedure that is comparable to one I have used many times to define the 'team codes' for my players whenever a new season started. As these team codes were necessary to determine how the individual players had to operate within the bandwidths of the team rules, it is also essential to clearly set the orientation and limits of the present discussion to avoid creating false expectations by the readers. Let me therefore be explicit on this point and emphasise that the present review is written from the perspective of the coach who engages theoretical knowledge as an erudite counsellor in an attempt to apply the recommended insights to make practice sessions as efficient as possible. One could say that since reality, i.e., help the players to win and/or attain their highest possible performance level, is the principal goal of the coach, she/he will be inclined to incorporate all information valuable for achieving this goal. Note here that although 'winning' is nurtured by the soul of 'strong' competition it cannot be valued unless it is achieved within the limits of fair play and pedagogical principles.

Given their apparent importance for sport, it is evident that relevant research findings in the field of motor control and learning are instrumental to accomplish this goal, and should therefore find their way to the coaches. However, the transformation of these research findings should be accomplished using a language that is accessible by the coaches. It is my goal to attain this clarity in the present review, focussing on the consequences of the scientific findings for the problems a coach is confronted with.

SOME CONSIDERATIONS ON COACHES AND COACHING

> "An impression first strikes upon the senses, … . Of this impression there is a copy taken by the mind, which remains after the impression ceases; …"
>
> (Hume, 1739/1985, p. 55)

As stated in the introduction, a coach has to act such that his/her team or athlete attains the highest possible performance level. This formalisation is derived from definitions of the training process by authors who focussed on the training process (e.g., Carl, 1983; Weineck, 1982) and considered it as a process for which progress in a specific domain is the essential characteristic. Hagedorn (1981) extended this viewpoint to the theory of training, hereby emphasising the relevance of scientific data when stating that the theory of training is the systematic summary of relevant scientific findings described in rules and methodologies. Thus, it appears that scientific findings can provide the basic tools for the enhancement of performance. It is obvious that relevance is not the privilege of motor control and learning, but stretches out to different scientific disciplines, e.g., physiology, sociology, psychology, etc., confronting the coach with a multidisciplinary challenge. Given the focus of the present review, however, motor control and learning are crucial since these disciplines certainly provide pertinent information for 'technique training'. Nevertheless, we will return to the important multidisciplinary problem in the last section.

As for all aspects of training, technique training is affected by the search for perfection. Players need to totally master the different technical elements of the game (e.g., pass, set, attack, defence) in order to solve the game-problems they are constantly confronted with (Roth, 1990). Consequently, the coach will need to act such that the players will acquire these perfect movement executions. The acquisition process is often lengthy and can be extremely demanding for both the coach and the players, as the number of repetitions needed to attain this expert performance level can be immense. The years and years of practice mentioned by Hoover (1989) and the statement that 'expert performers have completed literally millions of trials of specific practice' (Abernethy, 1994), nicely demonstrates that reaching an expert performance level is (closely) related to the quantitative aspect of practice. Illustrative for this quantitative aspect is also the observation of Helsen et al. (1998) revealing that soccer players completed on average 10,000 practice hours to attain a professional level.

Although the need for repetition has been widely acknowledged by coaches and players, it would be simplistic and completely improper to reduce the search for perfect movement execution to a trivial matter of rehearsal and consider the coach as nothing more than a movement reproduction catalyst. In contrast, the very nature of the activities deployed in sport skills, especially in ball games, requires the potential to cope with constantly changing situations. Thus, instead of a rigorous implementation of well-learned movement sequences, plasticity and variability are needed to conform to the demands of the environment. The task of the coach then becomes increasingly difficult, as he/she needs to find the correct offset between the formation of prescribed movement patterns on the one hand, and the creation of adaptive behaviour on the other hand. In a broader sense this represents an interaction between rigidity and flexibility, and on an even larger scale one could imagine this interplay to symbolise the interaction between order and chaos. Although this latter consideration might require some flexibility from the reader, it is of particular interest given the important role these phenomena play in the dynamic system approach that will be discussed in the following paragraph.

When we observe the amazing skillfulness demonstrated by players in top-level competition games, it is hard to overlook the success of the training process. Coaches tend to link this performance quality to the quality of the training program and even to their competence in developing such a program. I do not intend to open a can of worms whilst concurring at least partially with this opinion. Rather, I want to assert that the applied methodologies and procedures were efficient; an argument that seems to hold unless one confines oneself to the conviction that quality emerges in spite of possible external interference. Given the former assumption, the question arises as to what makes the training program successful, or more specific, as to which (scientific) principles the successful coaches use to organise their practice sessions? A first explanation could be that these principles are a result of trial and error behaviour, the consequence of a continuous search for efficient drills and game situations, leading to a repertoire that seems to work well. It is interesting to note that the repertoire of top coaches might even become the training bible for coaches with a strong motivation to be part of the inner circle. Because of my personal experience with the advantages of these emerging procedures, I certainly do not want to devalue their merits. However, it appears to me that this experience-based knowledge should go hand in hand with scientifically based explanations that eventually evoke useful training rules or principles. As a matter of fact, the study by Mohr and Barrett (1962) on the advantage of knowledge of mechanical principles for learning swimming skills clearly illustrates the long history of this search for scientific explanation.

A typical example of a training principle that emerged from research can be found in the schema theory developed by Schmidt (1975). This theory was based on the assumption that movements are controlled by so-called generalised motor programs, i.e., schema's that specify the invariant characteristics of movements. Walmsley and Williams (1994) examined one of these invariants as they investigated the relative temporal structure of rapid pistol firing. They showed that, regardless of the overall speed of the movement, the relative timing of the successive shots did not change. In addition, plane trajectories showed marked consistency across timing conditions. In contrast to these invariants, movement parameters (e.g., force, direction, and duration) do vary according to the circumstances. A basic assumption of schema-theory was that the execution of movement patterns under a variety of circumstances would produce better-developed schema's. Since these stronger schema's were believed to lead to a more appropriate movement execution, it became mandatory to create training circumstances that confronted the players with an entire range of movement patterns. Although this 'variability of practice' hypothesis was not always confirmed (for a review, see Van Rossum, 1990), a number of studies (e.g., Kelso and Norman, 1978; Moxley, 1979, Boutmans et al., 1985) did approve the viability of this 'learning through variation' rule. Actually, the notion of variability has occupied an important niche in the coaching world. This is particularly applicable for young players whose 'movement education' follows the path of discovery along a diversity of movement patterns routes. It is important to note that variability is still required at the more advanced stages of learning, as it prepares the players to cope with the high variability of the actual game situations. Thus, the initial global variability becomes more and more confined by the bandwidths of specific sport disciplines. Stated differently, technique training is characterised by a shift from 'variation in variable situations to variation in specific situations' (Buekers, in press).

The above mentioned training principle has been conceived within the framework of cognitive psychology, as it is founded on the existence of motor programs and schema's being the memory structures that are thought to control movements. Regardless of its viability this motor program account inspired many researchers to examine how motor learning could be enhanced. Consequently, an enormous amount of studies focussed on issues as, for example, knowledge of results (e.g., Buekers and Magill, 1995; Wulf et al., 1993), modelling (e.g., Caroll and Bandura, 1987; Pollack and Lee, 1992; Reiser and Gagné, 1982, Weir and Leavitt, 1990), and contextual interference (for a recent review see Brady, 1998). It is not the purpose of the present review to elaborate these topics here. I just want to note however that since the findings of some of these studies could be easily translated to training practice (e.g., Wuyts and Buekers, 1994) the success of the approach should not be too surprising.

To summarise this paragraph, I want to rephrase an interpretation already presented in the introduction. As to my feeling, the comprehensibility and manageability of the cognitive view on motor skill learning was the reason for its positive reception by the coaches, who were seized by the notion of a computer program that can be written, refined and even rewritten in the player's memory structure. This latter aspect perfectly illustrates the desire of coaches to get hold of the circumstances and act according to principles that produce predictable training effects. It is feasible to suggest that this search for certainty, this need to control as much as possible the essential variables of the training process, provokes a resistance against theoretical frameworks that are based on self-organisation or characterised by a direct link between perception and action.

Linking Perception and Action

"This means that moving stimuli, ..., get preference over stationary stimuli: they're literally "perceived" more intensely "

(Calvin, 1986, p. 217)

In contrast to the cognitive view, the ecological psychology approach is opposed to the idea of memory representations, motor programs, and/or schema's being 'the' concepts to explain movement control. As formulated by Katz (1983), representations are not in the head, but in the external world. A basic premise of the approach is that human behaviour should be studied in direct relation to the environment in which it takes place (Gibson, 1979). Actually the system, e.g., a player that is trying to hit a ball, is thought to establish a direct interaction with the environment, without needing to progress through intermediate stages as for instance a visual representation of the environment. In his elucidating review, Warren (1998, p.183) expresses the essence of the ecological perspective well as he states that "Rather than being governed by an internal model and a precomputed plan, behavior is controlled 'online', if you will, by occurent information about the relation between the observer and the environment". Following this line of reasoning the individual would not need to learn motor programs. In contrast, it would suffice to recognise specific regularities (laws of control) whose application would automatically lead to the adequate response. In the perspective of training sport techniques the goal would thus be to teach the athletes or players to recognise or pick-up the relevant information that solicits the appropriate action. Or as Still and Good (1998, p. 40) stated: "..., learning is an education of attention as well as of articulation rather than the building of an internal model of the world". A possible venue to accomplish this learning or training goal might be the 'rehearsal under facilitating environmental conditions'. I will return to this latter solution in the fifth paragraph whilst discussing the practical implications of these theoretical insights. First, I want to elucidate the phenomenon of perception-action coupling using an example that has attractive power for coaches, especially for those involved in ball games, namely ball catching.

Traditionally, ball catching has been perceived as an 'activity of prophecy'. That is to say, players were supposed to 'read' the characteristics of the initial part of ball trajectory and predict from this information the future location of the ball. Accordingly, players had to move to this location as fast as possible in order to be in a comfortable position to capture or return the ball. In this view, the relations between perception and action are characterised by the concept of linearity, i.e., perception proceeds and determines action; the trajectory of the ball is scrutinised, followed by the execution of the required motor response. The perception-action approach relinquishes this linearity and 'restores' the importance of 'actions' as it postulates that perception and action are mutually dependent. Circularity, inter-dependency, coupling, and the status of the actor-environment system (Bootsma, 1998), are the key words that represent the functioning of the perception-action cycle. In other words, an intrinsic property of the process is its cyclic nature, i.e.; the actions of the player produce alterations in the perceptual flow. On its turn, this perceptual flow provides the information by which the required forces can be produced to achieve a designated status of the actor-environment system. For our example of ball catching this implies that, when a player is running towards a ball trying to capture it, his/her movement (as well as that of the ball) will have an affect on the status of the actor-environment system. Through the constant updating of this latter status, by means of picking up relevant, mainly visual information, movements can be continuously adjusted according to the current situation. This was exactly what Montagne et al. (1999) observed in a ball-catching experiment. In their study subjects had to move their hand along a horizontal axis in order to capture a ball that was projected form different angles of approach. The data revealed that, when the initial position of the hand was equal to the future interception point, typical curvilinear trajectories were traced by the hand. It was as if the hand was attracted by the ball and continuously pulled towards its current position. This finding is not congruent to the expectations of a predictive strategy for which the hand should have remained in its initial position.

THE EMERGENCE OF ORGANIZATION

"But, I can hear you say, ..., how could that chaos order itself? Be patient, I will explain"
(Cyrano de Bergerac, 1993)

It is not unusual for coaches to witness that, after trying to execute a particular complex skill for hundreds of times, all of a sudden a player is able to assemble the required movement pattern components correctly. It appears feasible to consider this feeling of recognition or identification as the cognitive awareness of the eventual successful performance, whereas the actual mechanism of the successful aggregation should be looked for elsewhere. One can say without any hesitation that, at present, the work of Bernstein (1967) provides the most attractive framework to consider this 'phenomenon of emerging co-ordination'. Embedded within the general concepts of 'chaos' theory (e.g., Prigogine and Stengers, 1984), this 'dynamic systems approach' considers co-ordination to be regulated by self-organisation, and not by a hierarchically prescribed mode of control. Since the publication of Kelso's study (1984) on the emergence of specific co-ordination pattern in bimanual co-ordination, numerous experiments have been conducted to better understand the fundamentals of co-ordination as well as how these patterns are learned (Beek and van Santvoord, 1992; Swinnen and Walter, 1988; Zanone and Kelso, 1992).

An illustrative practice-related example of the functioning of these emerging principles is provided in an experiment by Temprado et al. (1997). In their study the authors examine the spatio-temporal organisation of the movement patterns (i.e., the co-ordination) produced by players who execute a volleyball serve. The goal of their study was twofold. Apart from the identification of the essential co-ordination variables of this volleyball skill, the authors also wanted to determine whether learning to serve conformed to the idea of a direct transformation from disorder to a particular ordered co-ordination state. It should be noted that the argumentation on this latter issue was based on indirect evidence given its inference from the comparison of movement patterns of novice and expert players. Players were videotaped whilst executing a number of overhand serves to a target. Markers were placed on the arms of the subjects in order to analyse the trajectories of the shoulder, elbow, and wrist of the serving arm. The results of the study revealed the existence of a strong coupling between wrist and shoulder for novice players. It should be noted here that the occurrence of this rigid coupling seems to be almost inevitable when learning new movement patterns. Also, it gives rise to the well-know staggering movements that make it easy to differentiate the clumsy novice behaviour from the smooth, effortless movement elegance demonstrated by the experts. The rigid coupling is no longer present in the expert players, because they are able to dissociate the different components of the movement. The data shows that in contrast to novices, the maximal backward displacements of shoulder, elbow, and wrist of experts were clearly separated in time. A more revealing finding of the study is perhaps the presence of this typical expert co-ordination pattern in novices, even though it occurred only a limited number of times. The authors interpret this observation as support for the view that learning the volleyball serve is characterised by a transition from one state of co-ordination to another, rather than the development of a novel movement pattern.

FROM THEORETICAL CONCEPTS TO PRACTICAL TRAINING RULES

"Let's practice theory"

(Old Chinese proverb, 321 BC)

It is rather common that some time passes by before theoretical notions find their way to the 'field of practice'. This might partly be due to the initial scepticism that lengthens the inauguration of innovative concepts. Moreover, scientists generally do not share the same curiosity as the practitioner to explore potential application benefits of their concepts. Mutatis mutandis, the theoretical appetite of coaches usually suffers from the same lack of interest. Obviously we need scientists 'stigmatised' by a personal experience in coaching, or coaches propelled by a desire to delve into the scientific literature. Or, in Calvin's (1986, p. 469) lyric expression: "I can hardly wait for composers to start mining the neurophysiological literature for leads to catchy tunes". Being in the situation of a stigmatised researcher I will pick up the challenge in this paragraph and try to smoothly transform the theoretical serves into practical passes.

Let me first attempt to rephrase the basic presumptions of direct perception and the dynamic systems approach in terms of a training context, and in a second order try to translate these into training principles. The training-related description for direct perception could be formulated as: "THE TRAINING OF SPORT SKILLS CONSTITUTES THE ESTABLISHMENT OF AN INTERACTION BETWEEN AN ENVIRONMENT (E.G., FIELD, BALL, OPPONENTS, NET, TEAMMATES, ...) AND A PLAYER WHO IS EXPECTED TO ACT IN ACCORDANCE WITH THIS ENVIRONMENT". Based on this assumption I want to propose the following training rules. The first rule can be expressed as: "*Every action undertaken by a player should be executed within appropriate environmental settings*". Intrinsic to the word appropriate, the crucial element of this rule, is not the notion what the player does but how his/her actions are performed in relation to the conditions imposed by the circumstances of the game. In the domain of ball games, drills without balls (e.g., tennis service, volleyball defence, getting into specific tactical positions, ...) should therefore be banned, as they lack a basic component of this player-game interaction. A number of coaches would oppose to this idea, referring to the extreme difficulty learners have when trying to instantly execute complex movement patterns at 'game pace' level, i.e., within the spatio-temporal constraints dictated by the game. Although this argument has certain validity, this does not prevent us from using simplified drills that gradually lead to the acquisition of complex skills. Note that this first rule also discredits the advantage of training devices that are at present widely promoted as miraculous enhancers of the perceptual system of sportsmen. As Savelsberghs and van der Kamp (in press) already pointed out, the lack of an ecological relation of these perceptual skills with the perception-action cycles in the actual sport skill should be blamed for the uselessness of this method. As insinuated by these authors, a training program, i.e., the successive series of drills or games performed over a longer period of time, needs to be conceived within the framework of final game requirements. Thus even drills that serve to improve specific game components should not be isolated from the essence of the activity. In contrast, they should wear the signature of the 'actual drill - final game correspondence'. This relation is of considerable importance, not only for reasons of motor control, but also because it invites the coaches to create training programs that follow the route of intrinsic logic allowing the players to comprehend *what* they are doing, and, even more important *why* they are doing it. This latter argument is certainly valid for expert players as they have an extensive knowledge of the activities they are involved in.

The above mentioned aspect of simplification provides the basis for a second, learning related training rule. This rule can be formulated as follows: "*The learning of sport skills should be based on enhancing the crucial links between perception and action by means of environmental facilitators*". This rule is based on the premise that the inadequacy of novices to tune their actions to the pertinent properties of the environment can be overcome by adding features or changing circumstances such that learners are 'escorted' through the required perception-action cycle. In other words, learning tools can be circumscribed as manipulations that intensify the interaction between perception and action. An interesting example of such a manipulation is presented in a study by Ulrich et al. (1998), where infants with Downs syndrome were tested when walking on a treadmill. The results showed an increased walking proficiency when the infants wore socks with Velcro strips sewn on the bottom. According to the authors, this context manipulation caused the infants' feet to be in contact with the belt for a longer period of time during stance, hence enhancing the flow of information between the belt and the legs. Apparently, intensifying the interaction between environment and actor produces performance benefits. A similar performance benefit was observed by Clarckson et al. (1986) for foot placement during the execution of a ballet figure. When auditory signals were provided to indicate incorrect foot pronation, a prompt decrease of pronation times was observed. As both examples illustrate, the applied manipulations help the learner become more sensitive to the key elements of the perceptual flow, leading to a better performance. Note that learning benefits are not limited to these environmental facilitators. It has been shown repeatedly that cognitive interventions (instructions, KR), also possess the power to enhance performance. However, emphasising the functioning of environmental facilitators is an obvious consequence of the issue that I wanted to survey in this review. To summarise, this second training rule dictates the coaches should design and employ drills that provoke access to the crucial features of the perceptual flow. Of course, the application of this rule necessitates an in depth knowledge of the sport discipline. It is hard to imagine someone discussing crucial elements of the perception-action cycles for a given sport discipline when he/she lacks the experience of whirling around in these cycles.

The third and final perception-action based training rule pertains to prediction redundancy as it states: *"Request players to establish a 'close relation' with the ball, instead of asking him/her to move to a 'future' interception position."* The expression 'close relation with the ball' is somewhat poetic wording to emphasise the continuous character of the processes that interfere during ball capturing or hitting. Despite the apparent capacity of players to predict future positions (e.g., landing point, vortex) of ball trajectories, the prospective control process provides a parsimonious and less complex substitute for this distance based computation process. A player should therefore not be confronted with drills that require movements to designated positions where a ball must be captured or manipulated. In contrast, movements should be executed in such a way that a particular relationship between player and ball is established. A possible candidate for establishing this relationship is the LOT strategy proposed by McBeath et al. (1995). According to these authors, a successful catch is performed when the player moves such that the ball follows a linear optical trajectory. This and other strategies (for a recent review see Montagne et al., 1998) are illustrative of the redundant need to calculate in advance where the ball will be caught. Catching or hitting is based on the intrinsic capacities of the perceptual system for an online pick-up of the relevant information taking the player to the appropriate final location without previous knowledge of this position. This latter observation also conforms to the different velocity requirements at ball contact. Although certain game situations might benefit from a stationary player position, numerous occasions require actions to be performed whilst accelerating through the phase of ball contact. Some more thinking will be needed to explore the functioning of these smart perceptual mechanisms (Montagne, et al., 1998). However, the assimilation of 'prediction redundancy' by the coaches could already mark an interesting progress.

As for direct perception I will first provide a training-related version of the elementary hypothesis of the dynamic system approach to the reader. In considering self-organisation as the salient concept of this approach a possible description would be: "SPORT TRAINING NEEDS TO ENHANCE THE LEARNING OF TECHNICAL AND TACTICAL SKILLS IN DEVELOPING DRILLS THAT SOLICIT THE INTRINSIC CAPABILITIES OF THE PLAYER(S)". Starting from this definition, two additional training rules will be elaborated, one related to technique training, the other one to the tactical component of the game. Before doing so, the findings of an experiment by Davids et al. (in press) will be considered to introduce the fourth training rule. Expert volleyball players were asked to serve the ball overhand to a designated position on the court. Movement characteristics of hip, shoulder, hitting arm and ball were registered and analysed. A strong temporal coupling was observed between ball zenith and the initiation of the striking action. This observation led the authors to argue that: "practice of the serve should emphasise the development of an invariant peak height of ball toss" (Davids et al., in press). Moreover, the considerable variability observed for the left-right and forward-backward dimensions of ball trajectory indicated that consistency in these directions should not be accentuated during the learning process. As such, this finding is congruent with the functional aspect of variability. Movements and sport skills should not be considered as replica's of ideal, unique and prescribed movement patterns, but rather be seen as dynamic solutions to a specific game problem. Given this viewpoint, the best way to express the fourth rule is: "*Do not reproduce movements; just let movements produce themselves*". One should of course be cautious and not interpret this rule as a permission to adhere an attitude of "laissez-faire, laissez-aller". In contrast, a very thoughtful reflection is needed to find the correct balance between elements that should be free to vary and those that should be confined to invariance. As the volleyball example illustrates, players should be pressed to concentrate on specific features of the action to be performed, leaving room for solutions that go along well with the personal capacities of the player. Thus, instead of giving a detailed technical description of the skill the teacher or coach should reduce the informational load and make the players focus on a limited number of 'critical' criteria.

The fifth and last training rule is somewhat different from the previous ones as it implies movement patterns at a different level of organisation, i.e., the level of interaction between two or more players. Moreover, some caution will be needed here since the fifth training rule is also different because of the speculative character that is attached to it. To be successful at the level of tactics, players need to execute their movements within the spatio-temporal constraints of the overall 'inter-player co-ordination'. This requires a careful and often extremely well timed execution of the different parts of the play. It should therefore be no surprise that coaches tend to meticulously plan the actions of individual players and decorate these movements with the status of obligation. Of course, coaches will need to provide certain guidelines to get some order into the complex tactical patterns. As mentioned above, this order should not be obtained by means of an overdose of information, but rather looked for in small quantities of critical constraints. Thus, "*create circumstances that stimulate players to find their own solutions to the tactical situations they will be confronted with*" is a final training rule that emphasises the need to make use of the intrinsic capacities of players and teams.

CONCLUSION

> "Modern science led us to the point, both in physics and social sciences, ..., where we can no longer accept only one series of principles on how the world functions. We simply have to accept a number of alternative realities."
>
> (LeShan and Margenau, 1987, 33)

It is important to note that the rules mentioned in the previous paragraphs only cover a limited spectrum of the coaching principles. The orientation of this review to the practical consequences of a specific theoretical approach will certainly add to this restriction. In spite of this limitation however, these training rules do not solely represent a few additional handy tools. In contrast, they are the representatives of a 'non-cognitive' training philosophy that advocates the importance of the environment and the intrinsic dynamics of the moving system for the training process. In the light of the extensive influence of the cognitive approach during the last few decades this shift in orientation seems logic and even welcome. Especially since the excessive effect of cognitive psychology on training methods and procedures resulted in a marked indifference or even neglect towards the other actors in the game of motor control. On the other hand, biasing this review towards the intrinsic dynamics of the system and the functioning of this system in 'close proximity' with the environment should not engender a complete disregard of the good old training procedures. In contrast, the coaches should add these concepts and rules to their existing body of knowledge and bring up from this well of information the resources the situation calls for.

It is not difficult to imagine how laborious the task of coaches and teachers actually are, when one pleads in favour of the multidisciplinarity; i.e., the integration of knowledge from several related research fields (e.g., biomechanics, physiology, psychology, ...). The need to be open to different, even opposing, theoretical views within the – at least for sport coaching - 'royal' field of motor control and learning, can only add to the enormous challenge a coach or teacher is confronted with. This challenge should, however, not be an exclusive mission of teachers or coaches, as they would not have the time nor the necessary background to access the different disciplines that are covered with the inevitable mantle of scientific jargon. The challenge then flows back to the scientists who should translate the in depth insight of their preferred field of study, and make it accessible to workers in the field. It is as if both teams bring their experience to the midfield where the ball of knowledge can be played back and forth. After this exchange the actors can return to their proper occupation, often along with ideas, strategies and techniques that open some new routes for further exploration. I admit that the idea of information exchange is not an original one, but, similar to sport practice one can expect simple ideas to be empowered by repetition.

Bringing this review to a conclusion, I do want to recognise the boundaries of this contribution as it covers only a small sample of variables that constitute the complex puzzle of successful performance. This limitation even stretches out to the merits of scientific information, a statement all experienced teachers and coaches would probably agree upon. Thus, I am aware that the training rules proposed in the previous paragraphs cannot the considered as unique keys to open the doors of success. It would be naive or even a manifestation of good old hubris to comply with this reasoning. In contrast, the dictum: "Theory of training is an exact science nurtured by reason, coaching is an art cherished by the heart. A good coach therefore has to be a reasonable artist" perfectly expresses my point of view. As this metaphor suggests, the scientific knowledge and its practice-related derivatives should be escorted by the energy of emotions. Only when a coach or teacher succeeds in appeasing these two antagonists, exceptional performance will emerge.

REFERENCES

Abernethy, B. (1994). Introduction. *International Journal of Sport Psychology*, **25**, 241-248.

Beek, P.J. and van Santvoord, A.A.M. (1992). Learning the cascade juggling: A dynamics system analysis. *Journal of Motor Behavior*, **24**, 85-94.

Bernstein, N. (1967). *The co-ordination and regulation of movements*. Pergamon: Oxford.

Bootsma, R.J. (1998). Ecological movement principles and how much information matters. In *Models in Human Movement Science* (edited by A. Post, J.R. Pijpers, P. Bosch and M.S.J. Boschker), pp. 51-63. Enschede: Print Partners Ipskamp.

Boutmans, J., Buekers, M. and Cockaerts K. (1985). De schema-theorie: De 'variability of practice' bij een doelworp in basketbal. *Hermes*, **18**, 37-44.

Brady, F. (1998). A theoretical and empirical review of the contextual interference effect and the learning of motor skills. *Quest*, **50**, 266-293.

Buekers, M. (2000). Can we be so specific to claim that specificity is the solution for learning sport skills. *International Journal of Sport Psychology*. **31**, 485-489.

Buekers M. and Magill, R.A. (1995). The role of task experience and prior knowledge for detecting invalid augmented feedback while learning a motor skill. *Quarterly Journal of Experimental Psychology*, **48A**, 84-97.

Calvin, W.H. (1986). *The river that flows uphill: A journey from the big bang to the big brain*. New York: Macmillan.

Carl, K. (1983). *Training und Trainingslehre in Deutschland*. Schorndorf: Verlag Karl Hofmann.

Caroll, W.R. and Bandura, A. (1987). Translating cognition into action: The role of visual guidance in observational learning. *Journal of Motor Behavior*, **19**, 385-398.

Clarckson, P.M., James, R., Watkins, A. and Foley, P. (1986). The effect of augmented feedback on foot pronation during barre exercise in dance. *Research Quarterly for Exercise and Sport*, **57**, 33-40.

Davids, K., Bennett, S. and Handford, C. (in press). Acquiring Coordination in Self-Paced Extrinsic Timing Tasks: A Constraints-Led Perspective. *International Journal of Sport Psychology*.

De Bergerac Cyrano (1993). *De andere wereld*. Antwerpen, Houtekiet.

Gibson, J.J. (1979). *An ecological approach to visual perception*. Boston: Hougton-Mifflin.

Hagedorn, G. (1981). Modell des Manschaftstrainings. In *Training im Mannschaftsspiel.*: (edited by G. Hagedorn), pp 13-57. Berlin: Bartels and Wernitz.

Helsen, W.F., Starkes, J.L. and Hodges, N.J. (1998). Team sports and the Theory of Deliberate Practice. *Journal of Sport and ExercisePsychology*, **20**, 13-35.

Hoover, T. (1989). *Samurai*. Utrecht: Het Spectrum.

Hume, D. (1739/1985). *A treatise of human nature*. London: Penguin Books.

Katz S. (1983). R L Gregory and others: the wrong picture of the picture theory of perception. *Perception*, **12**, 269-279.

Kelso, J.A.S. (1984). Phase transitions and critical behavior in human bimanual coordination. *American Journal of Physiology, Regulatory, Integrative, and Comparative Physiology*, **15**, R1000-R1004.

Kelso, J.A.S. and Norman, P.E. (1978). Motor schema formation in children. *Developmental Psychology*, **14**, 153-156.

LeShan, L. and Margenau, H. (1987). *De ruimte van Einstein en de wereld van Van Gogh (Einstein's Space and van Gogh's Sky)*. Den Haag: Mirananda.

McBeath, M.K., Shaffer, D.M. and Kaiser, M.K. (1995). How baseball outfielders determine where to run to catch fly balls. *Science*, **268**, 569-572.

Mohr, D.R. and Barrett, M.E. (1962). Effect of knowledge of mechanical principles in learning to perform intermediate swimming skills. *Research Quarterly*, **33**, 574-580.

Montagne, G., Laurent, M. and Durey, A. (1998). Visual guidance of goal-oriented locomotor displacements: The example of ball interception tasks. *Ecological Psychology*, **10**, 25-37.

Montagne, G., Laurent, M., Durey, A. and Bootsma, R. J. (1999). Movement reversals in ball catching. *Experimental Brain Research*, **129**, 87-92.

Moxley, S.E. (1979). Schema: The variability of practice hypothesis. *Journal of Motor Behavior*, **11**, 65-70.

Pollack, B.J. and Lee, T.D. (1992). Effects of the model's skill level on observational motor learning. *Research Quarterly for Exercise and Sport*, **63**, 25-29.

Prigogine, I. and Stengers I. (1984). *Order out of chaos*. Toronto: Bantam Books.

Reiser, R.A. and Gagné, R.M. (1982). Characteristics of media selection models. *Review of educational research*, **2**, 499-512.

Roth, K. (1990). Ein neues "ABC" für das Techniktraining im Sport, *Sportwissenschaft*, **20**, 9-26.

Savelsbergh, G. and van der Kamp, J. (in press). Information in learning to co-ordinate and control movements: Is there a need for specificity of practice? *International Journal of Sport Psychology*.

Schmidt, R.A. (1975). A schema theory of discrete motor skill learning. *Psychological review*, **82**, 225-260.

Still, A. and Good, J. (1998). The ontology of mutualism. *Ecological Psychology*, **10**, 39-63.

Swinnen, S.P. and Walter, C.B. (1988). Constraints in coordinating limb movements. In *Cognition and action in skilled behaviour* (edited by A.M. Colley and J.R. Beech), pp. 127-143. Amsterdam: Elsevier.

Temprado, J., Della-Grasta, M., Farrell, M. and Laurent, M. (1997). A novice-expert comparison of (intra-limb) coordination subserving the volleyball serve. *Human Movement Science*, **16**, 653-676.

Ulrich, B.D., Ulrich, D.A. and Angulo-Kinzler, (1998). The impact of context manipulations on movement patterns during a transition period. *Human Movement Science*, **17**, 327-346.

Van Rossum, J.H.A. (1990). Schmidt's schema theory: The empirical base of the variability of practice hypothesis. *Human Movement Science*, **9**, 387-435.

Walmsley, A. and Williams, L.R.T. (1994) Rapid-fire pistol shooting as a dynamic problem, *Perceptual and Motor Skills*, **78**, 1019-1024.

Warren, W.H. (1998). Visually controlled locomotion: 40 years later. *Ecological Psychology*, **10**, 177-219.

Weineck, J. (1982). *Optimale training.* De Vrieseborch: Haarlem.

Weir, P.L. and Leavitt, J. (1990). Effects of model's skill level and model's knowledge of results on the performance of a dart throwing task. *Human Movement Science*, **9**, 369-383.

Willingham, D.B. (1998). A neuropsychological theory of motor skill learning. *Psychological Review*, **105**, 558-584.

Wulf G., Schmidt R.A. and Deubel H. (1993). Reduced feedback frequency enhances generalized motor program learning but not parameterization learning. *Journal of Experimental Psychology Learning Memory and Cognition*, **19**, 1134-1150.

Wuyts, I. and Buekers, M. (1995). The effect of visual and auditory models on the learning of a rhythmical synchronization skill. *Research Quarterly for Exercise and Sport*, **66**, 105-115.

Zanone, P.G. and Kelso, J.A.S. (1992). Evolution of behavioral attractors with learning: Nonequilibrium phase transitions. *Journal of Experimental Psychology Learning Memory and Cognition*, **18**, 403-421.

Chapter 11

DIVING AND COGNITIVE PERFORMANCE: THE SELECTIVE IMPAIRMENT OF CONTROL PROCESSES

A. Charles, F. Allimann
Université de Bordeaux 2, France
P. Ragot
Clinique des Cèdres, Mérignac, France

Numerous psychological and physiological problems arise in depth diving (Spira, 1999), and more specifically when depth is more than 550 feet. However, disturbed behavior or significant perturbation in psychological performance can nonetheless appear at 100 ft. (or even since 50 ft.) when compressed air is used (Fructus, Naquet, Gosset, Fructus & Brauer; 1969). A feeling of happiness or well-being is described by the divers in these conditions. These experiences can be described by three components: A loss of feeling of reality, an increase in useless movements, and a perturbation of cognitive capacities (for example, reasoning and immediate memory). These behavior problems increase with increasing depth and any form of work becomes impossible at a depth of 350 ft, although these sensations and experiences are reversible when air pressure is diminished.

The majority of the studies concerning these difficulties concern physiological issues, but some studies have addressed cognitive aspects of these problems. In a report to The Medical Research Council, Miles and MacKay (1959, in Bennett, 1982) showed no performance decrease in mental arithmetic and memory tests at 200 ft. However, using a pressure chamber, Kiessling and Maag (1962) showed a decline of 7.9% in cognitive capacities at a depth of 100 ft. They concluded that this mild impairment was of little practical importance. The results of several more recent studies, however, are straightforward concerning the magnitude of such deficits. Baddeley (1966) tested divers in the water and in pressure chamber. They showed that a dexterity test took 49% longer at 100 ft. when an impairment of only 6% was showed in a pressure chamber. Furthermore, Baddeley and Fleming (1967) showed an impairment of 10 % in an addition task, and a subsequent study by Baddeley (1967) and Baddeley, de

Figueredo, Hawskwell-Curtis & Williams (1968) suggested that the marked discrepancy between these two experimental conditions could be due to anxiety. Similarly, Abraini, Ansseau, Bisson, de Mendoza and Therme (1998) also suggested that personality factors play a crucial role in the occurrence of diving anxiety. Some recent experiments that have investigated the neuropsychological effects of diving showed a marked impairment in span and mental arithmetic performances in relation to depth of diving (Bugat, 1997), and Bast-Pettersen (1998) obtained a slight increase in a reaction time experiment in the context of repeated diving. However, no hypothesis has been proposed concerning the precise cognitive processes implicated in this decrement. Our interest in the following experiment concerns the hypothesis of a selective impairment of psychological performance in the artificial condition of depth pressure. Furthermore, the experiment will be performed in an air pressure chamber to a simulated depth of 165 feet. In this way, the possibility that diving-related stress (and subsequently, anxiety) may influence performance will be greatly diminished, leaving only the influence of air pressure.

A recent conceptualization of control processes of activity has been proposed by Norman and Shallice (1986).`` This model separates two processes, automatic and control processes. Automatic processing is performed by a contention-scheduling component: Actions are directly linked to the triggers who fired them out. The control processes are performed by the Supervisory Attentional System (SAS) which is used by the subject in the case of novel or dangerous activities, or in the case of activities which do not permit the use of an automatic process, i.e. in the case of complex activities. This latter component is integrated by Baddeley (1996) into the Central Executive component of his model of Working Memory. We hypothesize that while the automatic activities are not significantly slowed down or disturbed by depth of diving (i.e. performance is identical in the surface and in the depth conditions), controlled activities are impaired. That is, the more complex the activity is to be performed, the greater the observed decrease in performance will be. A complex activity necessitates the maintaining in memory of a greater mass of information and for a longer time. Any slowing down of information processing would be detrimental when several sources of information must be retrieved in Working Memory to perform a given response (although little or no consequences are seen for small or individual quantities of information).

In the present study, three separate tasks are administered. The first task is a simple reaction time study, and is rapidly performed in an automatic manner. We hypothesize that this first task will not be impaired by depth of performing. The second task is a digit span task which involves the Articulatory Loop of the Working Memory model proposed by Baddeley (1992). It has long been shown that this task is highly sensitive to speed of processing, due to the fact that rehearsal uses a limited duration loop. The most rapidly performed task is typically the repetition of the presented digits, and the greatest number of digits can be maintained in the articulatory loop (Ellis & Hennely; 1980). For this reason, it is hypothesized that a slowing down due to increased pressure would attenuate performance in the digit span task. The third task is a mental arithmetic task with two conditions differing in their level of difficulty. The involvement of the articulatory loop has already been quoted by Hitch (1978) in such tasks, and the SAS (or the CE) is likewise involved. Such involvement becomes greater as the task increases in difficulty. Therefore, we hypothesize that pressure will impair performance more in the complex condition than in the easy condition.

METHOD

Subjects

Twelve divers (26 to 39 years old) associated with different divers' clubs volunteered to participate in this experiment. Although no subject was a professional diver, all subjects possessed a certificate of diving competence. Subjects who reported having dived recently were excluded from this experiment.

Materials

The experiment is performed in an air pressure chamber divided in two rooms. The smaller of these rooms served as the experiment chamber. Two scuttles allow a visual communication with the experimenters situated outside of the experiment chamber. The tasks are presented on the screen of a PC microcomputer which is placed outside of the experimental room and in front of one scuttle. A hand-held box with 4 switches is placed on a shelf under the experimental scuttle, and the responses of the subject are given using these switches.

Description of the Tasks and Procedure

Reaction Time Task

The subject is seated in front of the scuttle, holding the box with the 4 response switches. In the center of the screen are presented 4 rectangles corresponding spatially to the each of the switches. Each time a star appears in one of the rectangles, the subject must push on the corresponding switch as rapidly as possible. Whenever (and only when) a correct response is given, a new star appears at random in one of the for rectangles. The total duration of the task is 45 seconds. The mean reaction time is automatically calculated.

Digit Span Task

Different series of digits of increasing size are presented to the subject. The first series contains 4 digits and the last contains 9. Each digit is presented for 3 sec. and is separated from the next digit by an interval of 1 sec. Two series of digits of equal length are successively presented. After seeing the series on the screen, the subject is instructed to copy the digits in the order they were presented on a separate sheet. The entire set of all presented series is then recorded so that the experimenter can correct the responses of the subject.

Mental Arithmetic Task

This tasks presents arithmetic operations on a computer screen in an incomplete manner. In the place of a given number appears a star. Two types of operations are proposed to the subject: Six simple operations of single-digit numbers (3 additions and 3 multiplications) and 6 difficult operations of two-digit numbers (which also include 3 additions and 3 multiplications). The entire set of 12 operations is presented in a random order, and two sets of 12 operations are used. Four possible solutions to replace the star in order to render the operation correct (example: * + 5 = 8 / *2, 3, 4, 5)* are presented under each operation. The four solutions are presented so as to correspond spatially to the four switches of the response box, as before, the subject chooses his or her solution by pushing the corresponding switch. The operation rests on the screen until the correct solution is chosen by the subject, and then the next operation is immediately presented. The time elapsed in choosing the correct solution is automatically recorded by the program.

All 12 subjects performed the three tasks at under air pressures corresponding to 1) surface pressure, and 2) to a depth of 165 feet. The 12 subjects are separated into 6 groups of 2 subjects. Three groups begin the experiment at the surface pressure and the other three groups begin at a 'depth' pressure of 165 feet. The order of the three tasks is counterbalanced across the different groups. The total time of 'diving' (pressure exposure) is approximately 70 mn including the 40 mn of decompression. For surface tasks, approximately 20 mn are needed to perform the tasks.

RESULTS

Reaction Time Task

The reaction times are very similar at the two diving depths: 563 msec at surface air pressure (sd = 92) and 589 msec. at 165 ft. (sd = 82). This difference is not significant, $F(1, 11) = .55$, $p>.10$. As hypothesized, therefore, the relatively automatic processes implied in this reaction time task are not disturbed by increasing air pressure. As noted above, half of the subjects began at depth of 165 ft. whereas the other half begin at a surface air pressure. Table 1 shows the results corresponding to these two different orders of administration for the reaction time task. There is neither effect of depth ($p < 1$), nor of order ($p < 1$), but there is a slight decrease in reaction time when the task is performed at surface air pressure if the subject had already completed the task at a depth of 165 ft (in comparison to performing it for the first time at the surface).

Table 1. Mean Reaction Times in Function of Order of Administration of Task

Depth of diving	First Administration of Task	Second Administration of Task
Surface	596*	531**
165 feet	554**	624*

*Task performed at surface, then at 165 feet
**Task performed at 165 feet, then at surface

These differences are not significant, $F(1, 10) = 1.51$ and $F(1, 10) = 2.38$; $p<.10$ (respectively for 0 and 165 ft. depth). However, the interaction between these two variables is significant, $F(1,10) = 5.38$, $p<.05$. This latter result must be replicated before being interpreted.

Digit Span Task

We also hypothesized that the span task would be affected by depth because span tasks are highly dependant on the speed of articulatory processing involved in rehearsal and because control processes are involved in the rehearsal of the order of presentation. The mean span achieved by subjects is 7.58 digits (sd = .79) at 0 ft. and of 6.75 digits (sd=.86) at 165 ft. This decrease of span size is highly significant $F(1, 10) = 16.18$, $p<.005$. In Table 2 is presented the span's size in function of order of administration. The interaction between depth and order is not significant $F(1\text{-}10) = 3.08$, $p>.10$.

Table 2. Mean Span Size in Function of Order of Administration of Task

	First Administration of Task	Second Administration of Task
Depth of diving	1	2
Surface	8*	7.16**
165 feet	6.83**	6.67*

*Task performed at surface, then at 165 feet
**Task performed at 165 feet, then at surface

Mental Arithmetic Task

The mean reaction times (in msec.) are presented in Table 3. Mental arithmetic is performed more slowly when simulated depth of diving is increased $F(1, 11) = 9.08$; $p<.025$. Similarly, it can be observed that simple operations are processed more rapidly than the complex operations. The increase in reaction time between these two conditions is highly significant $F(1, 11) = 23.87$, $p<.0005$. Furthermore, this increase is significant whatever the simulated depth to which the task is performed $F(1,11)=18.69$, $p<.005$ and $F(1,11) = 18.60$, $p<.005$ (respectively at 0 ft. and 165 ft.). However, the increase in reaction time between the two levels of difficulty of mental arithmetic is about the same at the two simulated depths, and the interaction between depth and difficulty of the task is not significant, $p > .1$). These results confirm only partially our hypothesis: Increasing depth slows down processing, but the slowing down is equivalent for the complex calculations and the simple calculations.

Table 3. Mean Reaction Times (and Standard Deviation) for the Two Types of Mental Arithmetic

Depth of Diving	Simple Operations	Complex Operations
Surface	3733 (1226)	6276 (2385)
165 ft.	4663 (2174)	7118 (2585)

The number of errors committed by the subjects are shown in table 4. The simple operations are more error-free than the complex operations, $F(1, 10) = 6.14$, $p<.05$. However, this effect is not the same at surface air pressure and at deep air pressure. At a surface air pressure, there is no significant difference between the two types of operations, $F(1, 10) = 0.34$, $p>.10$ whereas there is an important and significant increase when an air pressure of 165 ft. is simulated, $F(1, 10) = 7.49$, $p<.025$. The interaction between level of air pressure and complexity of the problem to be solved is also significant, $F(1, 10) = 6.09$, $p<.05$. When complex problems are globally more time consuming than simple arithmetical problems (whatever the depth of processing), the increase of air pressure impairs only the processing of complex problems.

Table 4. Number (and Standard Deviation) of Errors for the Two Types of Mental Arithmetic

Depth of Diving	Simple Operations	Complex Operations
Surface	1.33 (1.67)	1.58 (1.90)
165 ft	1.41 (1.7)	3.83 (3.91)

DISCUSSION

This experiment addressed the question of how hyperbaric conditions influence the efficiency of cognitive processing of information. Similarly to the majority of prior work on this topic, we hypothesized that the more complex the task is to perform, the more detrimental would be the effect of increasing air pressure. Our goal was to replicate these findings in an air pressure chamber and to identify the cognitive processes more specifically implicated in such detrimental influences. Therefore, consistent with the models proposed by Norman and Shallice (1986) and by Baddeley (1996), we hypothesized that a decrease in processing speed would imply a decrement in performance in complex tasks leaving simple and more automatic tasks unimpaired. The results can be summarized as follows: Digit span performance is better at an air pressure corresponding to "surface" conditions than at a depth of 165 ft. Moreover, the speed of mental calculation is higher at the "surface" than at a greater depth. However, although complex problems take longer to solve than simple problems, the increase in processing time due to increasing depth is only additive. That is, there is no interaction between depth and difficulty of the problem being solved. The results are quite different, however, if the quality of the response is taken into account. Simple operations are accomplished in the same manner whatever the air pressure, but the processing of complex operations is significantly impaired by increasing air pressure. An important finding is that reaction time task is a relatively automatic task and that it is *not* impaired by depth. This result demonstrates simultaneously that there is no slowing down of the response and, also, no increase in the number of errors (which should have increase reaction times).

For this reason, when the task does not involve attentional control or resources, there is no impairment linked to the increasing air pressure. The increase in air pressure impairs performance only in complex tasks, i.e if an attentional resource is involved, as is the case for mental arithmetic tasks or span tasks. These types of tasks, which involve attentional control, can be defined as complex tasks. These are tasks for which no response is available or for which the available response does not correspond do the requirements of the task. To the contrary, the reaction time task can be easily automated because there is a clear and constant spatial correspondence between the localization of the star on the screen and the localization of the appropriate switch on the response box. No such automatization is possible for the two other tasks. The span task is a mixture of automatic processing of the digits and of controlled processing (implied in the rehearsal process). The unusual presentation format of mental arithmetic problems (a missing digit replaced by a star) leads us to suspect that controlled processing is implied in solving these problems, especially when complex problems must be performed. Complexity is then at the core of understanding the results of this experiment: When a mental representation of a continuously changing situation is required to solve a cognitive problem, depth impairs performance. There are clear implications of these results for professional divers who often perform underwater cognitive tasks (for example, when constructing bridges or off-shore platforms), as well as for potentially all divers when conducting more general calculations (such as determining the speed of returning to the surface and decompression levels).

REFERENCES

Abraini, J.H., Ansseau, M., Bisson, T., de Mendoza, J.L. & Therme, P. (1998) Personality patterns of anxiety during occupational deep dives with long-term confinement in hyperbaric chamber. *Journal of Clinical Psychology, 54, 825-830.*

Baddeley, A. D. (1966) Influence of depth on the manual dexterity of free divers: A comparison between open sea and pressure chamber testing. *Journal of Applied Psychology, 50,* 81-85.

Baddeley, A. D. (1967) Diver performance and the interaction of stresses. *Underwater Association Report, 2,* 35-38

Baddeley, A. D. (1996) Exploring the Central Executive. *The Quarterly Journal of Experimental psychology, 49,* 5-28.

Baddeley, A. D. & Flemming, N.C. (1967) The efficiency of divers breathing oxy-helium. *Ergonomics, 10,* 311-319.

Baddeley, A. D., de Figueredo, J.W., Hawskwell-Curtis J.W. & Williams A.N. (1968) Nitrogen narcosis and performance underwater. *Ergonomics, 11,* 157-164.

Bast-Pettersen, R. (1998) Long-term neuropsychological effects in non-saturation construction divers. *Aviation Space & Environmental Medicine, 70,* 51-57.

Bennett, P.B. (1982) Inert gas narcosis. In P.B. Bennett & D.H. Elliott (Eds.), *The physiology and medicine of diving* Baillière Tindall: London.

Bugat, R. (1997) La plongée humaine à grande profondeur *Bulletin de Psychologie, 50, 627-637.*

Ellis, N.C. & Hennely, R.A. (1980) A bilingual word-length effect: Implications for intelligence testing and the relative ease of mental calculation in Welsh and English. *British Journal of Psychology, 71,* 43-52.

Fructus, X., Naquet, R., Gosset, A., Fructus P. & Brauer, R.W. (1969) Le syndrome nerveux des hautes pressions. *Marseille Médical, 6,* 509-512.

Hitch (1978) The role of short-term working memory in mental arithmetic. *Cognitive Psychology, 10,* 302-323.

Kiessling R.J. & Maag, C.H. (1962) Performance impairment as a function of nitrogen narcosis. *Journal of Applied Psychology, 46,* 91-95.

Norman & Shallice (1986) Attention to action: Willed and automatic control of behavior. In R.J. Davidson, G.E. Schwartz & D. Shapiro (Eds.), *Consciousness and Self-regulation (Vol. 4)* New-York: Plenum Press.

Spira, A. (1999) Diving and marine medicine review part II: Diving diseases. *Journal of Travel Medicine, 6, 180-198.*

INDEX

A

B

C

D

E

N

O

P

T

U

V

W